Halides of the Transition Elements

Halides of the Lanthanides and Actinides

Halides of the Transition Elements

is a series of three volumes consisting of
the present title together with the following

Halides of the Second and Third Row Transition Metals
by J. H. Canterford and R. Colton
and
Halides of the First Row Transition Metals
by J. H. Canterford and R. Colton

Halides of the Lanthanides and Actinides

D. Brown

Chemistry Division,
Atomic Energy Research Establishment,
Harwell, Berks.

A WILEY-INTERSCIENCE PUBLICATION
John Wiley & Sons Ltd.
London New York Sydney

Library of Congress catalog card No. 68-57464

SBN 470 10840 1

Printed in Great Britain at the Pitman Press, Bath

Preface

Apart from certain specialist reviews the chemistry of the halides of the *d*- and *f*-transition elements have not previously been collectively discussed. This book, which deals with the halides of the lanthanide and actinide elements, is one in a series of publications designed to provide information useful to both research scientists and university lecturers.

The lanthanide elements, except promethium, and the actinide elements up to and including uranium have been available for many years but it is relatively recently that the 'man-made' transuranium elements have become available, firstly neptunium in 1939 and finally lawrencium in 1961. The inorganic, physical and structural chemistry of the halides of both series of elements has been extensively studied during the past two decades and despite the fact that only submilligram amounts of certain of the transuranium elements are currently available many aspects of their chemistry have already been explored.

I have discussed the halides, oxyhalides and halogenocomplexes of the lanthanide and actinide elements in parallel in order to emphasize the similarities and differences in behaviour of the two series of elements. For convenience the halides of lanthanum and of actinium, which elements are more correctly considered as the precursors of the respective series, and those of scandium and yttrium are also reviewed. Following a general introduction the four classes of halogen compounds are dealt with in separate Chapters. Preparative methods are assessed, the chemical and physical properties of the compounds discussed and structural information provided in these Chapters. Thermochemical data and infrared vibrational frequencies are collected together in Appendices A and B respectively, and the mixed halides are briefly discussed in Appendix C.

Every attempt has been made to provide adequate literature coverage to the end of 1967 although in order to restrict the bibliography to a reasonable size selected references have been quoted to provide coverage of much of the earlier work. In such an active field of chemistry it is

v

virtually impossible to publish a completely up-to-date review as witnessed, for example, by the recently reported trivalent thorium oxyfluoride and berkelium halides, which compounds, although listed in the appropriate Tables in Chapter I, are not discussed in the succeeding Chapters and by the newly discovered divalent states of several transuranium elements.

I am greatly indebted to Professor K. W. Bagnall for critically reading the drafts of each Chapter and for the many helpful suggestions which he made for their improvement. I wish to thank my wife for her assistance in preparing and reading the manuscript and correcting the proofs, and I am grateful to a number of my colleagues for reading the manuscript and proofs, and for their suggestions and corrections.

I wish to thank the many authors who have provided me with copies of original publications and with advance details of their current studies and also the Chemical Society, The American Chemical Society and the publishers of *Acta Crystallographica, Journal of the American Ceramic Society, Journal of Inorganic and Nuclear Chemistry, The Chemistry of Uranium, NNES,* Div. VIII, Vol. 5, *Crystal Structures* by R. W. G. Wyckoff, *Man-Made Transuranium Elements* by G. T. Seaborg, *Structural Inorganic Chemistry* by A. F. Wells, and *U.S. Report* TID-5290, for permission to reproduce illustrations from their journals or books.

Chilton, July 1968   D. B.

Contents

vii

Chapter 1

Introduction

The lanthanide* and actinide† elements, in which the 4f and 5f electron shells respectively are being filled, form halides and complex halides which in many cases can be classified in isostructural series, particularly in the trivalent state, and since many of the preparative methods are common to both groups of elements it is convenient to discuss them together. Since scandium and yttrium behave in many respects like lanthanum, the precursor of the 4f series, the halides of these elements are also included in this volume. The most noticeable difference in the chemistry of the two series of elements is the greater range of valence states encountered amongst the earlier actinides and in this opening chapter an attempt will be made to present a general picture of the current state of the halide chemistry of these two series of elements, together with a brief discussion of the special problems inherent in the investigation of the preparative chemistry of the more intensely radioactive members of the actinide series.

VALENCE STATES

The electronic configurations[1] of the gaseous lanthanide and actinide elements and the characterized valence states are listed in Tables 1.1 and 1.2 respectively. The most stable aqueous state is italicized and parentheses indicate that the only known halides are 'metallic' rather than salt-like in behaviour (p. 9). From a comparison of their electronic configurations it is clear that the close chemical similarity between the lanthanides, all of which are predominantly trivalent in solution, is not due to a common electronic configuration, $5d6s^2$, since this configuration is the exception rather than the rule. The stability of the trivalent state in solution does in fact[2] depend mainly on the hydration and ionization energies. The special stabilities of the $4f^0$, $4f^7$ and $4f^{14}$ electronic configurations and the tendency to approach such configurations have often

* Elements 58–71 inclusive.
† Elements 90–103 inclusive.

1

TABLE 1.1

Electronic Configurations for Gaseous Atoms of Actinide and Lanthanide Elements (Predicted configurations are in parentheses)

Atomic number	Element	Electronic configuration[a]	Atomic number	Element	Electronic configuration[b]
89	Actinium	$6d7s^2$	57	Lanthanum	$5d6s^2$
90	Thorium	$6d^27s^2$	58	Cerium	$4f5d6s^2$
91	Protactinium	$5f^26d7s^2$	59	Praseodymium	$4f^36s^2$
92	Uranium	$5f^36d7s^2$	60	Neodymium	$4f^46s^2$
93	Neptunium	$5f^46d7s^2$	61	Promethium	$4f^56s^2$
94	Plutonium	$5f^67s^2$	62	Samarium	$4f^66s^2$
95	Americium	$5f^77s^2$	63	Europium	$4f^76s^2$
96	Curium	$5f^76d7s^2$	64	Gadolinium	$4f^75d6s^2$
97	Berkelium	$(5f^86d7s^2$ or $5f^97s^2)$	65	Terbium	$4f^96s^2$
98	Californium	$(5f^{10}7s^2)$	66	Dysprosium	$4f^{10}6s^2$
99	Einsteinium	$(5f^{11}7s^2)$	67	Holmium	$4f^{11}6s^2$
100	Fermium	$(5f^{12}7s^2)$	68	Erbium	$4f^{12}6s^2$
101	Mendelevium	$(5f^{13}7s^2)$	69	Thulium	$4f^{13}6s^2$
102[e]	—	$(5f^{14}7s^2)$	70	Ytterbium	$4f^{14}6s^2$
103	Lawrencium	$(5f^{14}6d7s^2)$	71	Lutetium	$4f^{14}5d6s^2$

[a] In addition to the electronic structure of radon (element number 86), whose electronic configuration is: $1s^22s^22p^63s^23p^63d^{10}4s^24p^64d^{10}4f^{14}5s^25p^65d^{10}6s^26p^6$.

[b] In addition to the electronic structure of xenon (element number 54), whose electronic configuration is: $1s^22s^22p^63s^23p^63d^{10}4s^24p^64d^{10}5s^25p^6$.

[e] The name of element 102 has not yet been definitely fixed.

been cited as important in explaining the existence of divalent and tetra-valent lanthanide compounds. However, the recent studies reported by Corbett and his colleagues, e.g. references 3–7, clearly indicate that electronic structure can no longer be considered the sole factor in determining the existence of reduced or oxidized lanthanide ions. Thus, it is now known that neodymium[3,4] and dysprosium[6] form stable divalent chlorides ($4f^4$ and $4f^{10}$ respectively) which are salt-like in character and are in fact isostructural with samarium and ytterbium dichloride respectively. Corbett and co-workers[7] have suggested that the changes in the heats of sublimation of the metals are primarily responsible for the irregular trends observed in the degree of reaction of the lanthanide metals with their respective trichlorides. This aspect is discussed in more detail later (p. 167).

The existence of stable higher oxidation states of the earlier actinides, as compared with the extreme difficulty of oxidizing the corresponding

TABLE 1.2

Characterized Lanthanide and Actinide Oxidation States[a]

	La	Ce	Pr	Nd	Pm	Sm	Eu	Gd	Tb	Dy	Ho	Er	Tm	Yb	Lu
Atomic number	57	58	59	60	61	62	63	64	65	66	67	68	69	70	71
Oxidation states[c]	(2)[b] *3*	(2) *3* 4	(2) *3* 4	(2) *3*	*3*	2 *3*	2 *3*	(2) *3*	*3* 4	2 *3*	*3*	*3*	2 *3*	2 *3*	*3*

	Ac	Th	Pa	U	Np	Pu	Am	Cm	Bk	Cf	Es	Fm	Md	(?)	Lw
Atomic number	89	90	91	92	93	94	95	96	97	98	99	100	101	102	103
Oxidation states[c]	*3*	(2) 3 *4*	3 4 *5*	3 *4* 5 6	3 4 *5* 6	3 *4* 5 6	*3* 4 5 6	*3* 4	*3* 4	*3*	*3*	*3*	2 *3*	2 *3*	*3*

[a] Most stable aqueous state is italicized.
[b] Parentheses indicate that the known halides are 'metallic' rather than salt-like in behaviour.
[c] Not including those divalent states observed only in a fluoride matrix (page 101).

lanthanides, is usually ascribed to the poorer shielding of the $5f$ electrons from external fields by the outerlying electrons than is the case with the lanthanides. Since the atomic radii of corresponding pairs of lanthanide and actinide elements are not appreciably different, the $5f$ orbitals must be somewhat further out from the nucleus than the $4f$ orbitals, and so must extend spatially into the $6d$ and $7s$ orbital regions. This accords with the view that the $4f$ orbitals, although energetically favourable for covalent bonding, are not involved because they are too small spatially to overlap with bonding orbitals from another atom. However, as pointed out by Bagnall[45] these factors are not sufficient to explain the relative ease of oxidation of the earlier actinides and it is probable that the main factor is the smaller effective nuclear charge experienced by the actinide $5f$ electrons, as compared with that experienced by the $4f$ electrons of the corresponding lanthanide homologue, which results from the screening of the $5f$ electrons from the nucleus by the additional underlying $4f$ and $5d$ shells. Although few ionization potential data are available for the actinides, the situation is clearly comparable to the differences in ease of oxidation of first, second and third row d-transition elements (e.g. Fe, Ru, Os) but without the influence of the increased atomic radii which occurs in passing from the $3d$ to the $4d$ series, so that the comparison is more exactly with Ru and Os rather than the Fe group as a whole.

The shielding of one f electron by another in a given shell is very poor, owing to the shapes of the orbitals, so that as the f shell fills up, the effective nuclear charge experienced by each f electron increases at each increase of nuclear charge, so reducing the size of the whole f^n shell (actinide contraction) and increasing the difficulty of achieving higher oxidation states. As the atomic number increases, it is not surprising therefore to find that after berkelium there is no evidence at all for higher oxidation states.

For the earlier members of the actinide series the question of assignment of electrons to $5f$ or $6d$ orbitals is difficult, since here the energy separations apparently lie within the range of chemical binding energies. It is possible therefore that the electronic configuration of a given valence state of an element will vary from compound to compound or even with the physical state of a given compound. Magnetic susceptibility studies, which are of considerable value in elucidating electronic configurations in the d-transition and $4f$- transition elements, are of much less value for the actinides since the exact behaviour to be expected for a given $5f$ or $6d$ configuration has not yet been successfully calculated. Experimentally determined values frequently lie between those expected on the basis of Russell–Saunders coupling and the spin only value for a $6d^n$ configuration.

TABLE 1.3

The Lanthanide Halides

	La	Ce	Pr	Nd	Pm	Sm	Eu	Gd	Tb	Dy	Ho	Er	Tm	Yb	Lu
Fluorides	—	—	—	—	—	SmF_2	EuF_2	—	—	—	—	—	—	YbF_2	—
	LaF_3	CeF_3	PrF_3	NdF_3	PmF_3	SmF_3	EuF_3	GdF_3	TbF_3	DyF_3	HoF_3	ErF_3	TmF_3	YbF_3	LuF_3
		CeF_4	PrF_4						TbF_4						
Chlorides				$NdCl_2$		$SmCl_2$	$EuCl_2$			$DyCl_2$			$TmCl_2$	$YbCl_2$	
	$LaCl_3$	$CeCl_3$	$PrCl_3$	$NdCl_3$	$PmCl_3$	$SmCl_3$	$EuCl_3$	$GdCl_3$	$TbCl_3$	$DyCl_3$	$HoCl_3$	$ErCl_3$	$TmCl_3$	$YbCl_3$	$LuCl_3$
Bromides						$SmBr_2$	$EuBr_2$						$TmBr_2$	$YbBr_2$	
	$LaBr_3$	$CeBr_3$	$PrBr_3$	$NdBr_3$	$PmBr_3$	$SmBr_3$	$EuBr_3$	$GdBr_3$	$TbBr_3$	$DyBr_3$	$HoBr_3$	$ErBr_3$	$TmBr_3$	$YbBr_3$	$LuBr_3$
Iodides	(LaI_2)	(CeI_2)	(PrI_2)	NdI_2		SmI_2	EuI_2	(GdI_2)					TmI_2	YbI_2	
	LaI_3	CeI_3	PrI_3	NdI_3	—	SmI_3	EuI_3	GdI_3	TbI_3	DyI_3	HoI_3	ErI_3	TmI_3	YbI_3	LuI_3

TABLE 1.4

The Actinide Halides

	Ac	Th	Pa	U	Np	Pu	Am	Cm	Bk	Cf	Es	Fm	Md	(?)	Lw
Fluorides	AcF_3	ThF_4	— PaF_4 Pa_2F_9 PaF_5	UF_3 UF_4 U_4F_{17} U_2F_9 UF_5 UF_6	NpF_3 NpF_4 — — NpF_6	PuF_3 PuF_4 Pu_4F_{17} — PuF_6	AmF_3 AmF_4	CmF_3 CmF_4	BkF_3	—	—	—	—	—	—
Chlorides	$AcCl_3$	$ThCl_4$	$PaCl_4$ $PaCl_5$	UCl_3 UCl_4 UCl_5 UCl_6	$NpCl_3$ $NpCl_4$	$PuCl_3$	$AmCl_3$	$CmCl_3$	$BkCl_3$	$CfCl_3$	—	—	—	—	—
Bromides	$AcBr_3$	$ThBr_4$	$PaBr_4$ $PaBr_5$	UBr_3 UBr_4 UBr_5	$NpBr_3$ $NpBr_4$	$PuBr_3$	$AmBr_3$	$CmBr_3$	$BkBr_3$	—	—	—	—	—	—
Iodides	AcI_3	(ThI_2) ThI_3 ThI_4	PaI_3 PaI_4 PaI_5	UI_3 UI_4	NpI_3	PuI_3	AmI_3	CmI_3	BkI_3	—	—	—	—	—	—

TABLE 1.5

The Lanthanide Oxyhalides

	La	Ce	Pr	Nd	Pm	Sm	Eu	Gd	Tb	Dy	Ho	Er	Tm	Yb	Lu
Fluorides	LaOF	CeOF	PrOF	NdOF	PmOF	SmOF	EuOF	GdOF	TbOF	DyOF	HoOF	ErOF	TmOF	YbOF	LuOF
Chlorides	LaOCl	CeOCl	PrOCl	NdOCl	PmOCl	SmOCl	EuOCl	GdOCl	TbOCl	DyOCl	HoOCl	ErOCl	TmOCl	YbOCl	LuOCl
Bromides	LaOBr	CeOBr	PrOBr	NdOBr	PmOBr	SmOBr	EuOBr	GdOBr	TbOBr	DyOBr	HoOBr	ErOBr	TmOBr	YbOBr	LuOBr
Iodides	LaOI	—	—	—	PmOI	SmOI	EuOI	—	—	—	—	ErOI	TmOI	YbOI	—

TABLE 1.6
The Actinide Oxyhalides

	Ac	Th	Pa	U	Np	Pu	Am	Cm	Bk	Cf
Fluorides	AcOF	ThOF ThOF$_2$	— $\{$ Pa$_2$OF$_8$ —	— U$_2$OF$_8$ — UO$_2$F$_2$	— — NpOF$_3$ NpO$_2$F$_2$	PuOF PuO$_2$F$_2$	— AmO$_2$F$_2$	—	—	—
Chlorides	AcOCl	ThOCl$_2$	PaOCl$_2$ Pa$_2$OCl$_8$ $\{$ Pa$_2$O$_3$Cl$_4$ PaO$_2$Cl	UOCl UOCl$_2$ UOCl$_3$ UO$_2$Cl$_2$	NpOCl$_2$	PuOCl	AmO$_2$F$_2$ AmOCl	—	—	CfOCl
Bromides	AcOBr	ThOBr$_2$	PaOBr$_2$ $\{$ PaOBr$_3$ PaO$_2$Br	UOBr$_2$ UOBr$_3$ UO$_2$Br UO$_2$Br$_2$	NpOBr$_2$	PuOBr	—	—	BkOBr	—
Iodides	[AcOI]	ThOI$_2$	PaOI$_2$ $\{$ PaOI$_3$ PaO$_2$I	[UO$_2$I$_2$]	—	PuOI	—	—	BkOI	—

[ ] : existence not proven.

Consequently, a detailed evaluation of all the factors which may contribute to the quenching of the orbital contribution of the f- as well as the d-electrons is necessary for each individual compound. However, in certain instances, for example[8,9] UCl_3 and $PuCl_3$, electronic configurations have been predicted from a comparison of the observed magnetic behaviour with that of the compounds of corresponding lanthanide elements, in this case Nd (III) and Sm (III) respectively. Paramagnetic resonance, an alternative method of obtaining information about electronic structures, has been employed to a small extent with the actinide halides. For example results for[10] U^{3+} and[11] Cm^{3+} incorporated in $LaCl_3$ single crystals have been found to be consistent with the electronic configurations $5f^3$ and $5f^7$ respectively, the ground states being $^4K_{9/2}$ and $^8S_{7/2}$. Other data are discussed in the appropriate chapters.

Tables 1.3 to 1.6 summarize the presently known lanthanide and actinide binary halides and oxyhalides; many mixed halides, particularly of uranium (IV), are also known (Appendix C). All four trivalent binary halides are known for both scandium and yttrium and of their trivalent oxyhalides only ScOl remains uncharacterized. Tetravalent and divalent halides of these two elements are unknown. It is obvious from Tables 1.2, 1.4 and 1.6 that the increasing stability of the lower oxidation states of the actinides with increasing atomic number is reflected by their halides. As one would expect, the multiplicity in oxidation states of uranium, neptunium and plutonium results in the largest array of halides for these elements and the fluorides exhibit the greatest stability in the higher oxidation states.

Although certain of the lanthanide dihalides undoubtedly contain the M^{2+} ion (e.g. $NdCl_2$, $SmBr_2$ and $DyCl_2$) the properties of the recently reported[12,13] diiodides of lanthanum, cerium and praseodymium (p. 228) and[14,15] of the actinide diiodide ThI_2 (p. 229) are best explained on the basis of the presence of 'free electrons' in such compounds and they are formulated as $M^{3+}(e^-)(I^-)_2$ or $M^{4+}(e^-)_2(I^-)_2$. Compounds of this type are enclosed in parentheses in Tables 1.3 and 1.4.

The fact that certain trivalent oxyhalides and promethium triiodide (Tables 1.3 and 1.5) are unknown can be attributed to lack of investigation rather than any difficulties associated with their preparation. Similarly all trivalent halides and oxyhalides of the elements uranium to lawrencium inclusive are undoubtedly capable of existence, as probably are other uranium (V) and neptunium (V) oxyfluorides and oxychlorides. It is also possible that future work will lead to the characterization of the unknown trivalent protactinium halides. The absence of compounds of elements beyond californium is, of course, due largely to the very limited amounts of these elements which are available (p. 14). However, the fact that neptunium and plutonium pentafluoride and neptunium pentachloride (Table 1.4) have not yet been characterized is more surprising. Thermodynamic calculations[16,17] suggest that certainly the neptunium compounds and possibly plutonium pentafluoride should be capable of existence, but despite several attempts to prepare these halides they still remain unknown. Similarly, unsuccessful attempts have been made to prepare solid plutonium tetrachloride but this is not so surprising since it is calculated[18] that the dissociation pressure of chlorine over solid plutonium tetrachloride at room temperature is about 10^7 atmospheres. Plutonium tetrabromide and tetraiodide are calculated to be increasingly less stable and in view of the instability of uranium tetraiodide (p. 214) it is unlikely that a pentaiodide is capable of existence.

Plutonium tetrachloride has, however, been stabilized as amide complexes such as[19] $PuCl_4 \cdot 2.5$ DMA (DMA = N,N dimethylacetamide) and $PuCl_4 \cdot 6A$ (A = acetamide) and[19-21] hexachloroplutonates (IV), $M_2^I PuCl_6$ (M^I = Cs, Rb) are known. Similarly fluoro complexes of neptunium (V) and plutonium (V) have recently been reported[22,23], e.g. $CsMF_6$ and Rb_2MF_7 (M = Np and Pu) but hexachloroneptunates (V) are still unknown.

TABLE 1.7

Ionic Radii of Actinide and Lanthanide Elements[24,29,44]

Lanthanide series				Actinide series			
Element	Radius (Å)	Element	Radius (Å)	Element	Radius (Å)	Element	Radius (Å)
La³⁺	1.061			Ac³⁺	1.11	—	—
Ce³⁺	1.034	Ce⁴⁺	0.92	Th³⁺	(1.08)	Th⁴⁺	0.99
Pr³⁺	1.013	Pr⁴⁺	0.90	Pa³⁺	(1.05)	Pa⁴⁺	0.96
Nd³⁺	0.995			U³⁺	1.03	U⁴⁺	0.93
Pm³⁺	(0.979)			Np³⁺	1.01	Np⁴⁺	0.92
Sm³⁺	0.964			Pu³⁺	1.00	Pu⁴⁺	0.90
Eu³⁺	0.950			Am³⁺	0.99	Am⁴⁺	0.89
Gd³⁺	0.938			Cm³⁺	0.986	Cm⁴⁺	—
Tb³⁺	0.923	Tb⁴⁺	0.84	Bk³⁺	(0.981)		
Dy³⁺	0.908			Cf³⁺	0.976		
Ho³⁺	0.894						
Er³⁺	0.881						
Tm³⁺	0.869						
Yb³⁺	0.858						
Lu³⁺	0.848						

X-ray diffraction studies, which have played an important role in the characterization of many actinide halides, show that analogous compounds of the elements actinium to americium are often isostructural and that there is a monotonic decrease in lattice dimensions with increasing atomic number. The derived[24,29,44] ionic radii (Table 1.7) show that there is an actinide contraction for both M^{3+} and M^{4+} ions, analogous to the well-known lanthanide contraction, as the positive charge on the nucleus increases. This, of course, is a consequence of the addition of successive electrons to an inner ($4f$ or $5f$) electron shell and the crystal data on the actinide element halides have been instrumental in classifying them as a series of elements in which the $5f$ electron shell is being filled. The lanthanide and actinide contractions are illustrated in Figure 1.1.

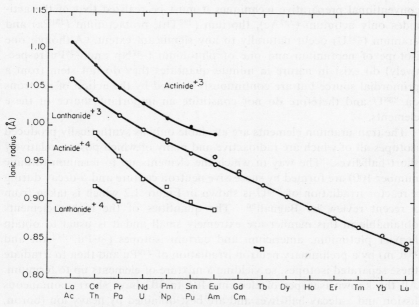

Figure 1.1   The ionic radii of actinide and lanthanide elements

In certain instances there is a change in structure-type in proceeding along a series of analogous compounds. For example the tribromides of the elements La—Pr and Ac—Np inclusive possess the uranium trichloride-type (or yttrium hydroxide) structure whereas those of the elements Nd—Eu and Pu—Cm inclusive possess the plutonium tribromide-type structure (p. 197). Such changes in structure-type are consistent with the lanthanide and actinide contractions which result in an increased ratio of anion:cation radii. Zachariasen has proposed[25], on the basis of crystal-chemical properties, that the term 'thoride' rather than 'actinide' be used for the 5f series of elements in their tetravalent state and has suggested that the term actinide might be confined to those elements in their trivalent state. However, this terminology is not widely accepted.

SPECIAL PROBLEMS ASSOCIATED WITH THE RADIOACTIVE ACTINIDE ELEMENTS

Whereas the lanthanide elements, except promethium, which is available in gram amounts as the fission product isotope ^{147}Pm, are available in kilo quantities and the investigation of their chemistry is possible by conventional preparative techniques it must be realized that of the actinides only actinium (^{227}Ac), thorium (^{232}Th), protactinium (^{231}Pa) and uranium (^{238}U) occur naturally to any significant extent. Although one isotope of neptunium and one of plutonium (^{237}Np and ^{239}Pu respectively) do exist in nature in minute quantities they do not stem from a primordial source but are continuously formed by the action of neutrons on ^{238}U and therefore do not constitute an important source of these elements.

The transuranium elements are available only as synthetically produced isotopes all of which are radioactive and many of which possess relatively short half-lives. The way in which the elements up to fermium (atomic number 100) are formed by successive neutron capture and β-decay during a reactor irradiation of ^{235}U is shown in Figure 1.2 which is taken from a recent review by Bagnall[26]. The quantities of the higher elements obtainable in this manner are extremely small and it is usual to obtain heavier plutonium, americium and curium isotopes (^{242}Pu, ^{243}Am and ^{244}Cm) by a preliminary neutron irradiation of ^{239}Pu and then to irradiate these separated isotopes, so yielding a mixture of elements up to fermium. All the known isotopes of elements 101 to 103 have short spontaneous fission and α-decay half-lives and are best obtained by heavy ion (boron, carbon, nitrogen, oxygen or neon) bombardment of curium or transcurium isotopes using a cyclotron or linear accelerator. The yields are extremely small, often of the order of a few atoms only, and it is hardly surprising that no halides of these elements have yet been prepared. An interesting and relatively rapid method, e.g. references 27 and 28, of obtaining the heavier elements is to explode a thermonuclear device in an enclosed space such as an underground cave. This technique yields very high neutron fluxes ($\sim 3 \times 10^{24}$ neutrons/cm^2) and is equivalent to many years

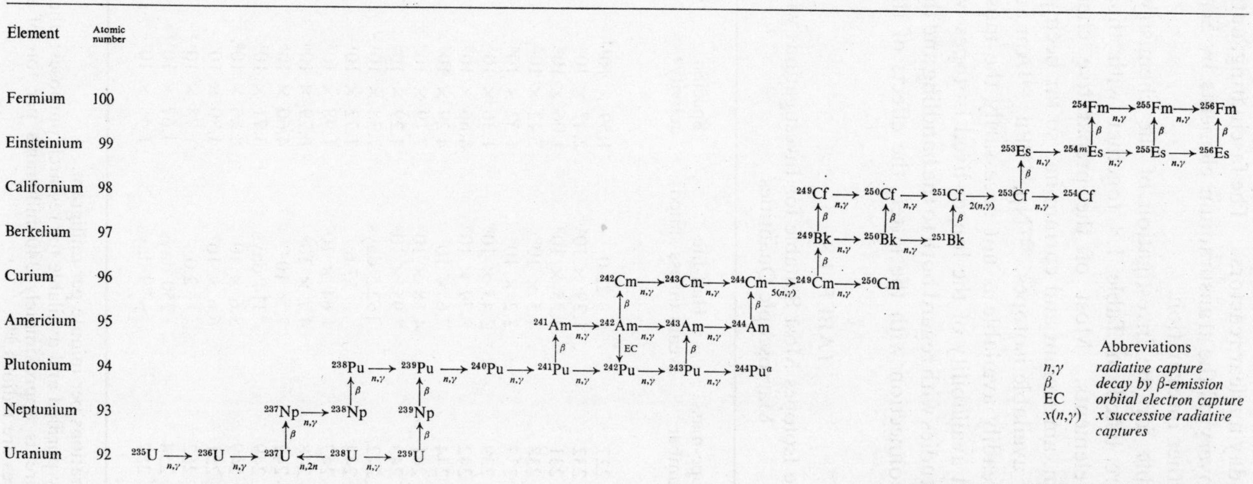

Figure 1.2 The formation of transuranium isotopes by neutron irradiation

Element	Atomic number
Fermium	100
Einsteinium	99
Californium	98
Berkelium	97
Curium	96
Americium	95
Plutonium	94
Neptunium	93
Uranium	92

Abbreviations

n,γ radiative capture
β decay by β-emission
EC   orbital electron capture
$x(n,\gamma)$ x successive radiative captures

[a] α-emitter of ~7·6 × 10⁷ year half-life; neutron capture gives rise to ^{245}Pu and ^{246}Pu, both of which decay by β-emission to ^{245}Am and ^{246}Am, and then to ^{245}Cm and ^{246}Cm

irradiation in present-day nuclear reactors.   The fascinating, authoritative summary of the discovery of the transuranium elements by Seaborg[29] is recommended for further information.

The isotopes suitable for the investigation of the chemistry of these synthetic elements are listed in Table 1.8 together with those of the naturally occurring elements.   Most of the preparative chemistry of neptunium, plutonium, americium and curium has so far been performed with the most readily available isotopes, ^{237}Np, ^{239}Pu, ^{241}Am and ^{242}Cm. However, the most readily available is not necessarily the most suitable isotope and increased availability of the longer-lived isotopes will greatly simplify preparative studies with regard both to the handling and the stability of the product.   In connection with the latter, the effects of the emitted

TABLE 1.8

Some Actinide Isotopes Most Suitable for Investigations with Macroscopic Quantities

Element	Isotope-mass number	Half-life (years unless stated)	Specific activity^a	Quantities available
Actinium	227	22.0	1.59×10^{11}	>mg
Thorium	232	1.39×10^{10}	2.47×10^{2}	>kg
Protactinium	231	3.25×10^{3}	1.06×10^{8}	>g^b
Uranium	238	4.5×10^{9}	7.42×10^{2}	>kg
Neptunium	237	2.2×10^{6}	1.52×10^{6}	>kg
Plutonium	239	2.43×10^{3}	1.36×10^{8}	>kg
	242	3.79×10^{5}	8.66×10^{6}	>g
	244	7.6×10^{7}	4.29×10^{4}	>μg
Americium	241	4.58×10^{2}	7.19×10^{9}	>g
	243	7.95×10^{3}	4.29×10^{8}	>g
Curium	242	162.5 days	7.38×10^{12}	>mg
	244	17.6	1.72×10^{11}	>g
	247	1.64×10^{7}	1.98×10^{5}	<mg
	248	4.7×10^{5}	6.79×10^{6}	<mg
Berkelium	247	7×10^{3}	4.60×10^{8}	>μg
	249	314 days	3.71×10^{12}	>μg
Californium	249	3.6×10^{2}	8.86×10^{9}	>μg
	251	6.6×10^{2}	4.80×10^{9}	>μg
	252	2.0	1.75×10^{12}	>g
Einsteinium	254	250 days	4.67×10^{12}	<μg
Fermium	257	~79 days	1.48×10^{13}	<μg

^{a} Number of α-disintegrations per minute per milligram.
^{b} The quantity currently purified and available for research purposes; the extent of its occurrence in nature is approximately 340 milligrams per ton of uranium but extraction procedures are tedious.[30]

radiation complicate the observed chemistry since in the solid state radiation decomposition may lead to changes in valence state and the elimination of simple anions. For example, although uranium hexafluoride (^{238}U) is stable, plutonium hexafluoride (^{239}Pu) undergoes radiation decomposition ($\sim 1.5\%$ per day) to the tetrafluoride. Similarly, curium tetrafluoride, CmF_4, has only been prepared using the longer-lived isotope ^{244}Cm because the intense radiation associated with ^{242}Cm results in a greater rate of radiation decomposition. Obviously, therefore, the possibility of attaining valence states greater than $+3$ for the transplutonium elements depends to some extent on the isotope used.

In addition to the chemical difficulties posed by the α-emitting transuranium isotopes there are the problems associated with health hazards to be considered. The main danger in handling the α-emitters arises from the possibility of accidental ingestion of the radioactive material with the resulting radiation damage to those organs in which it may be selectively retained. For example, plutonium, americium and curium accumulate principally in the bone, the kidneys and bone, and in the gastrointestinal tract, respectively. The maximum permissible body burden for ^{239}Pu, ^{241}Am or ^{242}Cm is approximately 0.04 microcuries (1 curie $= 2.2 \times 10^{12}$ α dis. per min), equivalent to 6.2×10^{-7}, 1.5×10^{-8} and 1.3×10^{-11} grams respectively, and, as pointed out by Bagnall[26], these isotopes are therefore many orders of magnitude more toxic, weight for weight, than hydrocyanic acid. Various accounts are available, see, for example, references 1, 31 and 32, of the glove-box procedures necessary for the safe handling of these isotopes and of those which are also γ-emitters, or undergo spontaneous fission, and this aspect will not be discussed here. It will be obvious, however, that working on the milligram scale or less, may be from a choice influenced by health considerations rather than from the limitations on supply.

The above considerations, quantity available and health hazards, have led to the development of elegant microtechniques which have frequently been applied to the preparation of the actinide halides. The 'capillary technique', described in detail by Fried and his associates, e.g. references 33 and 34, can be successfully applied to preparations involving microgram to milligram amounts of material. Essentially this technique employs an x-ray capillary as the reaction vessel and all manipulations and transfer of the product are minimized. The apparatus used for the first successful preparation of plutonium trichloride is illustrated in Figure 1.3. Although kilogram amounts of neptunium and plutonium can now be obtained it must be remembered that much of the halide chemistry of these elements was elucidated when only milligram quantities were available and that

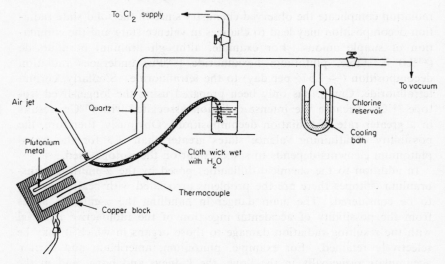

Figure 1.3   Apparatus used in the first preparation of PuCl₃, February 1944. The compound was prepared by treating a 50 microgram piece of plutonium metal with chlorine gas. After placing the plutonium in the capillary tube, the system was evacuated. Chlorine gas was added and a small amount condensed in the reservoir as shown. The system was closed and remained filled with chlorine at a pressure of about 60 mm Hg. The copper block was heated to 450°C and the reaction product was formed in the top of the capillary tube. The section of the capillary containing the product was sealed off and the compound formed was identified by x-ray diffraction. (*After* Glenn T. Seaborg, *Man-Made Transuranium Elements*, © 1963. Reprinted by permission of Prentice-Hall, Inc., Englewood Cliffs, N.J., U.S.A.)

such techniques still find application in connection with the higher actinides.

More recently, Cunningham[35] and his associates have developed elegant methods for studying the preparative chemistry, crystallography and magnetic properties of berkelium and californium compounds using only nanogram (10^{-9} g) quantities of the isotopes. The essential point in this technique is that the isotope is first adsorbed on a single bead of ion exchange resin to facilitate transfer operations. Thus, for the preparation of californium trichloride, $CfCl_3$, and the oxychloride, CfOCl, the ion exchange bead, volume approximately 10^{-5} μl, was first ignited in air at 1300° to convert the californium to the oxide, Cf_2O_3, which was subsequently reacted with hydrogen chloride gas, or a mixture of hydrogen chloride and water vapour, at 450° (Figure 1.4). Using this technique the chlorides were prepared and characterized[35] by x-ray powder diffraction analysis with 0.1 to 0.2 μg of ^{249}Cf. Cunningham[35] has also described balances for submicrogram weighing, an apparatus suitable for reduction of nanogram quantities of actinide fluorides to their metals and a method for determining the melting point of the resulting metal.

X-ray powder photography has played, and will continue to play, an extremely important role in the characterization of the halides of the actinide elements and in this connection mention must be made of the

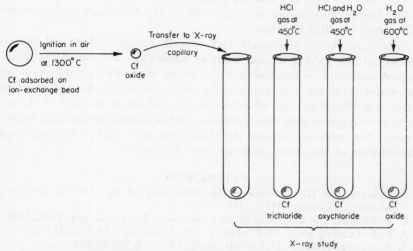

Figure 1.4 Experimental set-up for preparation of heavy element compounds as, for example, those of californium, on the submicrogram scale. (*After* Glenn T. Seaborg, *Man-Made Transuranium Elements*, ©, 1963. Reprinted by permission of Prentice-Hall, Inc., Englewood Cliffs, N.J., U.S.A.)

invaluable work of Zachariasen,* who made the first identification of numerous compounds when only microgram amounts were available. With the intensely γ-active isotopes, such as americium-241, and those which rapidly generate γ-emitting daughters, such as actinium-227, it is necessary to use relatively small x-ray specimens to combat fogging of the films. Where only microgram amounts of the higher actinides are available an x-ray powder diffraction technique similar to that described by Cunningham[35] is essential if satisfactory patterns are to be obtained.

It is not pertinent to discuss further the problems specific to the investigation of the preparative chemistry of the actinide elements but it is hoped that this brief account of the special techniques required will have been sufficient to stimulate the interest of the reader.

* References to this work are given where appropriate in the succeeding chapters.

HALOGENATING AGENTS AND DRY ATMOSPHERE BOXES

The preparation and safe handling of the more commonly used halogenating agents, particularly those employed in fluoride preparations, are well documented and it is unnecessary to duplicate the discussion of these aspects. Detailed accounts are available, e.g. references 36–40, for fluorine, hydrogen fluoride and other common fluorinating agents such as the halogen fluorides and the preparation and properties of these and other halogenating agents are dealt with in *The Handbook of Preparative Chemistry*[41], edited by Brauer.

The majority of lanthanide and actinide binary halides and indeed many actinide oxyhalides and complex halides are unstable in the atmosphere and must be handled in either dry or inert atmospheres. General procedures for the safe handling of such compounds are discussed in several publications, e.g. references 41–43, and the availability of Hersch cells for oxygen content determinations and of suitable commercial hygrometers for moisture content determinations now makes the continual monitoring of box atmospheres a simple matter.

REFERENCES

1. J. Katz and G. T. Seaborg, *The Chemistry of the Actinide Elements*, Methuen, London, 1957, p. 464.
2. T. Moeller, *The Chemistry of the Lanthanides*, Reinhold, New York, 1963.
3. L. F. Druding and J. D. Corbett, *J. Am. Chem. Soc.*, **83**, 2462 (1961).
4. R. A. Sallach and J. D. Corbett, *Inorg. Chem.*, **3**, 993 (1964).
5. L. F. Druding, J. D. Corbett and B. N. Ramsey, *Inorg. Chem.*, **2**, 869 (1963).
6. J. D. Corbett and B. C. McCollum, *Inorg. Chem.*, **5**, 938 (1966).
7. J. D. Corbett, D. L. Pollard and J. E. Mee, *Inorg. Chem.*, **5**, 761 (1966).
8. P. Handler and C. A. Hutchinson, *J. Chem. Phys.*, **25**, 1210 (1956).
9. J. K. Dawson, C. J. Mandelberg and D. Davies, *J. Chem. Soc.*, **1951**, 2047.
10. C. A. Hutchinson, P. M. Llewellyn, E. Wong and P. Dorain, *Phys. Rev.*, **102**, 292 (1956).
11. P. Fields, A. Friedman, B. Smaller and W. Low, *Phys. Rev.*, **105**, 757 (1957).
12. J. D. Corbett, L. F. Druding, W. J. Burkhard and C. B. Lindahl, *Discussions Faraday Soc.*, **32**, 79 (1962).
13. A. S. Dworkin, R. A. Sallach, H. R. Bronstein, M. A. Bredig and J. D. Corbett, *J. Phys. Chem.*, **67**, 1145 (1963).
14. R. J. Clark and J. D. Corbett, *Inorg. Chem.*, **2**, 460 (1963).
15. D. E. Scaife and A. W. Wylie, *J. Chem. Soc.*, **1964**, 5450.
16. L. Brewer, *U.S. Report UCRL*-633, 1950.
17. L. Brewer, L. Bromley, P. W. Giles and N. Logfren in 'The Transuranium Elements' (G. T. Seaborg, J. J. Katz and W. M. Manning, Eds.) *Nat. Nucl. Energy Ser.*, *Div. IV*, McGraw-Hill, New York, **14B**, 1111 (1949).
18. L. Brewer, L. Bromley, P. W. Giles and N. Logfren in 'The Transuranium Elements' (G. T. Seaborg, J. J. Katz and W. M. Manning, Eds.) *Nat. Nucl. Energy Ser.*, *Div. IV*, McGraw-Hill, New York, **14B**, 873 (1949).

19. K. W. Bagnall, A. M. Deane, T. L. Markin, P. S. Robinson and M. A. A. Stewart, *J. Chem. Soc.*, **1961**, 1611.
20. V. V. Fomin, N. A. Dimitrieva and V. E. Reznikova, *Zh. Neorgan. Khim.*, **3**, 1999 (1958).
21. R. Benz and R. M. Douglas, *J. Inorg. Nucl. Chem.*, **23**, 134 (1961).
22. L. B. Asprey, T. Keenan and R. A. Penneman, *Inorg. Nucl. Chem. Letters*, **2**, 19 (1966).
23. R. A. Penneman, G. D. Sturgeon, L. B. Asprey and F. H. Kruse, *J. Am. Chem. Soc.*, **87**, 5803 (1965).
24. J. J. Katz and G. T. Seaborg, *The Chemistry of the Actinide Elements*, Methuen, London, 1957, p. 437.
25. W. H. Zachariasen in 'The Actinide Elements' (G. T. Seaborg and J. J. Katz, Eds.) *Nat. Nucl. Energy Ser.*, *Div. IV*, McGraw-Hill, New York, **14A**, 769 (1954).
26. K. W. Bagnall, *Science Progress*, **52**, 66 (1964).
27. R. W. Hoff and D. W. Dorn, *U.S. Report UCRL*-7347 (1963).
28. *U.S. Report UCRL*-14500 (1966).
29. G. T. Seaborg, *Man-made Transuranium Elements*, Prentice Hall, New Jersey, 1964.
30. D. Brown and A. G. Maddock, *Quart. Rev. (London)*, **17**, 289 (1963).
31. G. N. Walton (Ed.), *Glove Boxes and Shielded Cells*, Butterworths, London, 1958.
32. K. W. Bagnall, *Chem. Brit.*, **1965**, 143.
33. S. Fried and H. Davidson, ref. 17, p. 1072; *J. Am. Chem. Soc.*, **72**, 771 (1950).
34. S. Fried and W. H. Zachariasen, *Proc. Intern. Conf. Peaceful Uses At. Energy*, 1*st Geneva*, **7**, 235 (1956).
35. B. B. Cunningham, *Microchem. J. Symp. Ser.*, **1**, 69 (1961); *Proc. R. A. Welch Foundation Conf. Chem. Res.*, **VI**, 237 (1962).
36. R. D. Peacock in *Advances in Fluorine Chemistry* (Eds. M. Stacey, J. C. Tatlow and A. G. Sharpe), Butterworths, London, 1965, Vol. 4, p. 31.
37. E. L. Muetterties and C. W. Tullock in *Preparative Inorganic Reactions* (Ed. W. L. Jolly), Vol. 2, Interscience, 1965, p. 237.
38. R. Landau and R. Rosen, *Ind. Eng. Chem.*, **39**, 281 (1947).
39. G. H. Cady in *Fluorine Chemistry* (Ed. J. H. Simons), Vol. 1, Academic Press, New York, 1950, p. 311.
40. B. Weinstock, *Record Chem. Progr.* (*Kresge-Hooker Sci. Lib.*), **23**, 23 (1962).
41. G. Brauer (Ed.), *Handbook of Preparative Chemistry*, Vol. 1, 2nd ed., Academic Press, New York, 1963.
42. J. C. Stewert in *Technique of Inorganic Chemistry* (Eds. H. B. Johassen and A. Weissberger), Vol. III, Interscience, New York, p. 167.
43. C. J. Barton in *Technique of Inorganic Chemistry* (Eds. H. B. Johassen and A. Weissberger), Vol. III, Interscience, New York, 1963, p. 257.
44. J. L. Green, U.S. Report, UCRL-16516 (1965).
45. K. W. Bagnall, personal communication (1968).

Chapter 2

Fluorides and Oxyfluorides

The presently known fluorides and oxyfluorides of the 4f- and 5f-transition elements have been listed earlier (pp. 6 and 8). They have been studied more extensively than the other halides and oxyhalides, particularly the actinide element compounds, and the literature is so extensive that it is impossible in this treatment to provide a complete coverage. The reader is therefore referred to the earlier review articles[1-14] which contain many valuable references to the original literature. These include the pertinent volumes of the *National Nuclear Energy Series*[1-3], which deal with the work of The Manhattan Project, the appropriate volumes[4-7] of *Nouveau Traité de Chimie Minérale*, general reviews by Sharpe[8], Hodge[9], Bagnall[10] and Simons[11] and more specific articles by Steindler[12] (PuF₆), DeWitt[13] (UF₆) and Tananaev and colleagues[14] (uranium fluorides).

HEXAVALENT

The hexafluorides of uranium, neptunium and plutonium are known but americium hexafluoride has not been prepared. The short-lived isotope ^{241}Am ($t_{\frac{1}{2}} = 458$ years) was used for the attempted[15] preparation of the last and the failure to isolate the compound may have been due to radiation decomposition caused by the intense α-emission of this isotope and perhaps the use of the longer-lived^{243}Am ($t_{\frac{1}{2}} = 7,600$ years) might prove more successful. Complex fluorides of uranium (VI), $M^I UF_7$ and $M^I_2 UF_8$ ($M^I =$ univalent cation), but not of neptunium (VI) or plutonium (VI), have been reported.

Oxyfluorides of the type MO_2F_2 (M = U, Np and Pu) are well characterized and recently americyl fluoride, AmO_2F_2, has been prepared. Numerous hexavalent oxyfluoro complexes of uranium are known but little work has been reported on the analogous neptunium (VI), plutonium (VI) and americium (VI) compounds.

20

Hexafluorides

The preparation of uranium hexafluoride has been extensively investigated since it was first reported by Ruff and Heinzelmann[16] who reacted uranium metal or uranium carbide with fluorine. It has been demonstrated that fluorine will convert almost all simple uranium compounds to the hexafluoride (see Table 2.1 and references 1, 5, 9, 10 and 14) but as

TABLE 2.1

Conditions for the Conversion of Uranium Compounds to Uranium Hexafluoride by Fluorine[a]

Reaction	Temperature (°C)
(1) $U + 3F_2 \rightarrow UF_6$	20
(2) $UF_4 + F_2 \rightarrow UF_6$	220–400
(3) $UO_2F_2 + 2F_2 \rightarrow UF_6 + O_2$	270
(4) $UC_2 + 7F_2 \rightarrow UF_6 + 2CF_4$	350
(5) [b]$U_3O_8 + 4C + 9F_2 \rightarrow 3UF_6 + 4CO_2$	300
(6) [b]$U_3O_8 + 17F_2 \rightarrow 3UF_6 + 8OF_2$	370
(7) [b]$UO_2 + 3F_2 \rightarrow UF_6 + O_2$	>500
(8) [b]$UO_3 + 6F_2 \rightarrow UF_6 + 3OF_2$	400
(9) $2UCl_5 + 5F_2 \rightarrow UF_4 + UF_6 + 5Cl_2$ (?)	−40

[a] References to the original literature are collected in reference 1, p. 396.
[b] Hydrofluorination of UO_2, U_3O_8 and UO_3 to yield respectively UF_4, $UF_4 + UO_2F_2$ and UO_2F_2 may be used as a preliminary step and the products then fluorinated by either reaction (2) or (3) shown above.

would be expected certain of the reactions are uneconomical in their utilization of fluorine. The only satisfactory preparations of neptunium and plutonium hexafluorides involve the use of elementary fluorine.

One general preparative method for the three hexafluorides involves oxidation of a lower fluoride by fluorine at elevated temperatures. Uranium tetrafluoride is oxidized[17], by way of the pentafluoride, at about 220° but the preparation of neptunium and plutonium hexafluoride requires[18-21] increasingly higher temperatures. In order to prevent thermal decomposition of the resulting hexafluorides special reaction vessels are employed[18-21] to permit rapid condensation of the volatile products close to their point of preparation (e.g. Figure 2.1). In contrast to these reports other workers[22,23] have found that it is possible to fluorinate plutonium tetrafluoride at temperatures as low as 200°. The reasons for this discrepancy are not clear but may be associated with the variable reactivity of different tetrafluoride samples and the dependence

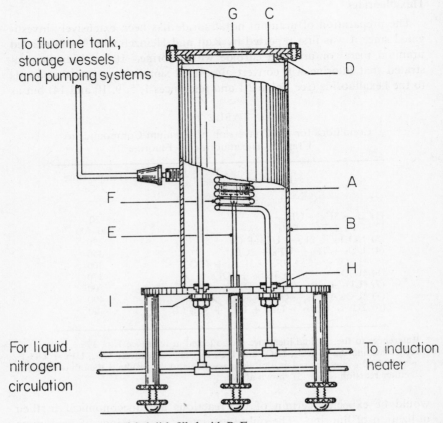

To fluorine tank,
storage vessels
and pumping systems

G   C

D

A

B

H

I

For liquid
nitrogen
circulation

To induction
heater

A Nickel dish filled with PuF₄
B Brass reactor can
C Removable cover for loading reactor
D Tongue and groove; Teflon gasket for closure
E Nickel supporting rod for nickel dish
F Coil of 3/16-in. o.d. copper tubing
G Fluorothene window
H Teflon seal and insulator
I Micalex insulator

Figure 2.1   Reactor for the preparation of plutonium hexafluoride.[18]   (*After* B. Weinstock and J. G. Malm, *J. Inorg. Nucl. Chem.*, **2**, 380 (1956))

of the reaction rate on the fluorine pressure. Neptunium and plutonium hexafluoride have also been prepared by reacting the appropriate metal or lower fluoride with platinum hexafluoride[24] but the reported yields were low.

A suitable laboratory-scale conversion of U_3O_8 to the hexafluoride was described recently[25]. Plutonium dioxide reacts readily with fluorine[26] to yield plutonium hexafluoride and, although the reactions have not been reported, it is probable that the various neptunium oxides could likewise be converted to the hexafluoride.

Bromine trifluoride[27-29], chlorine trifluoride[30,31,443,444] and bromine pentafluoride[445] are also capable of converting uranium metal, uranium tetrafluoride, uranium oxides and uranyl fluoride into the hexafluoride. Some early work has indicated that chlorine trifluoride may be useful[32] for the preparation of plutonium hexafluoride but bromine trifluoride is rapidly oxidized by the last with the formation of plutonium tetrafluoride. Sulphur tetrafluoride converts[33,34] uranium oxides and uranyl fluoride to the hexafluoride at 300°, the former reaction being analogous to the behaviour of molybdenum and tungsten trioxides. Above 500°, however, uranium hexafluoride is reduced to the tetrafluoride. The relative instability of plutonium hexafluoride, or its greater oxidizing power, is further demonstrated by the fact that at 30° it is reduced[34] by sulphur tetrafluoride. Thus it is possible to use this reagent to facilitate the separation[35] of uranium and plutonium.

A further preparation of uranium hexafluoride not involving elemental fluorine is the reaction between dry oxygen[36,37] and the tetrafluoride at 800°,

$$2UF_4 + O_2 \rightarrow UF_6 + UO_2F_2$$

It has been suggested[20] that a similar reaction could be employed for the preparation of plutonium hexafluoride but other work[38,39] indicates that it is not formed at 800°. Plutonium trifluoride, in fact, reacts[39,40] with oxygen at 600° to yield the tetrafluoride and the dioxide in the reversible reaction,

$$4PuF_3 + O_2 \rightleftharpoons 3PuF_4 + PuO_2$$

Plutonium dioxide has been shown[20] to react with a mixture of hydrogen fluoride and oxygen to yield a volatile substance but the product was not identified.

Less useful preparations of uranium hexafluoride include those involving conversion of the hexachloride[41] and pentachloride[16] to the corresponding fluorides by reaction with hydrogen fluoride; the pentafluoride obtained in the latter is being decomposed above 175°,

$$3UF_5 \xrightarrow{175°} U_2F_9 + UF_6$$

$$U_2F_9 \xrightarrow{>200°} UF_4 + UF_6$$

The oxidation of the tetrafluoride by cobaltic fluoride[42] at 250° also yields the hexafluoride.

Provided water and hydrogen fluoride are absent the three hexafluorides can be safely handled in Pyrex or quartz; in this connection sodium and potassium fluoride can be employed as 'getters'. Neptunium and plutonium hexafluoride, but not the uranium compound, are susceptible to photodecomposition. Radiation decomposition of solid plutonium hexafluoride, due to the α-emission from ^{239}Pu, is quite extensive ($\sim 1.5\%$ per day), and consequently it is advisable to store plutonium hexafluoride in the vapour state. It is also readily decomposed by γ-radiation whereas uranium hexafluoride is relatively stable.

The compounds are usually purified by distillation[43,44] but provided fluorine is present the formation and subsequent thermal decomposition[45] of alkali metal fluoro complexes can be utilized.

Structures. The crystalline actinide hexafluorides possess orthorhombic symmetry[19,46,47,49], space group D_{2h}^{16}–*Pnma* with four molecules per unit cell (Table 2.2). The molecules in the crystals do not appear to be perfect

TABLE 2.2

Crystallographic Data for the Actinide Hexafluorides[a]

	Colour	Unit cell dimensions (Å)			Density (calc.) gcm^{-3}	Reference
		a_0	b_0	c_0		
UF$_6$	White	9.900	8.962	5.207	5.06	46
NpF$_6$	Orange	9.910	8.970	5.210	5.00	47, 49
PuF$_6$	Reddish-brown	9.950	9.020	5.260	4.86	19

[a] These are all orthorhombic, space group D_{2h}^{16}–*Pnma*, $n = 4$.

octahedra but the results of vibrational spectra[12,13,47] and electron diffraction[47,48] studies clearly indicate that in the liquid (Raman spectra[50], UF$_6$ only) and vapour states the hexafluorides possess regular octahedral symmetry. The various spectral studies on uranium hexafluoride have been summarized by DeWitt[13]. An apparent[51] zero dipole moment for uranium hexafluoride lends support to the point group symmetry O_h. The metal–fluorine bond distances of the actinide and certain d-transition element hexafluorides, derived by electron diffraction studies[47] on the vapours, are shown in Table 2.3.

TABLE 2.3

Metal–Fluorine Distances in some Hexafluoride Vapours[47]

Compound	M–F (Å)	Compound	M–F (Å)
WF_6	1.826	UF_6	1.994
OsF_6	1.830	NpF_6	1.981
IrF_6	1.833	PuF_6	1.969

Weinstock and Goodman[52] have recently reviewed and discussed in detail the available information on the vibrational spectra of the actinide and other hexafluoride molecules. Unlike certain of the *d*-transition element hexafluorides none of the hexafluorides of the actinide elements exhibit Jahn–Teller distortion. The fundamental vibrational frequencies of these hexafluorides are compared with those of the *d*-transition element hexafluorides in Table 2.4 and force constant calculations are to be found in the above mentioned review. Both fundamentals have been observed[53,54,451] in the infrared spectra of the actinide hexafluorides. Raman data are available only for uranium hexafluoride since the extensive photodecomposition observed with the plutonium compound has

TABLE 2.4

Fundamental Vibrational Frequencies (Infrared and Raman) of Hexafluoride Molecules (cm^{-1})[52,451]

Hexa-fluoride	ν_1 (R)	ν_2 (R)	ν_3 (IR)	ν_4 (IR)	ν_5 (R)	ν_6 (inact.)
MoF_6	741	643	741	262	(312)	(122)
TcF_6	(712)	(639)	748	265	(297)	[174]
RuF_6	(675)	(624)	735	275	(283)	[186]
RhF_6	(634)	(592)	724	283	(269)	(189)
WF_6	(771)	(673)	711	258	(315)	(134)
ReF_6	755	(671)	715	257	(295)	(193)
OsF_6	(733)	(668)	720	272	(276)	[205]
IrF_6	(701)	(646)	719	276	(258)	(206)
PtF_6	(655)	(600)	705	273	(242)	(211)
UF_6	667	(535)	624	186	(201)	(140)
NpF_6	(648)	(528)	624	198	(205)	(165)
PuF_6	(628)	(523)	616	206	(211)	(173)
CrF_6	(720)	(650)	790	266	(309)	[110]

prevented similar measurements. Frequencies listed in parentheses (Table 2.4) are derived from combination bands and square brackets denote values estimated from systematics.

Properties. The actinide hexafluorides are all low-melting solids, the volatilities of the liquids increasing in the order $PuF_6 < NpF_6 < UF_6$ but neptunium hexafluoride is anomalous in possessing the highest vapour pressure of the three in the solid state. This last observation is not clearly understood. Some physical properties[47] of the hexafluorides are compared in Table 2.5 and the most reliable vapour pressure data[55,56] are shown in Table 2.6.

Because of the great technological importance of uranium hexafluoride in the separation of the fissile ^{235}U from natural uranium by gas phase diffusion its physical properties have been extensively investigated. On

TABLE 2.5

Some Physical Properties of the Actinide Hexafluorides[47]

	UF_6	NpF_6	PuF_6
Colour of solid	white	orange	dark brown
Colour of vapour	colourless	colourless	brown
Triple point (°C)	64.0	55.1	51.59
Boiling point (°C)	56.54	55.18	62.16
Vapour pressure at 0°C (mm Hg)	17.65	20.8	17.9
Vapour pressure at 25°C (mm Hg)	111.9	126.8	104.9
Vapour pressure at triple point (mm Hg)	1139.6	758.0	533.0

TABLE 2.6

Vapour Pressure Data for Actinide Hexafluorides[55,56]

Compound	Temperature range (°C)	State	Vapour pressure equation $\log_{10} p_{mm} =$
UF_6	0–64	Solid	$6.38363 + 0.0075377t - 942.76/(t + 183.416)$
	64–116	Liquid	$6.99464 - 1126.288/(t + 221.963)$
	116–230	Liquid	$7.69069 - 1683.165/(t + 302.148)$
NpF_6	0–55.10	Solid	$18.48130 - 2892.0/t - 2.6990 \log t$
	55.10–76.82	Liquid	$0.01023 - 1191.1/t + 2.5825 \log t$
PuF_6	0–51.59	Solid	$0.39024 - 2095.0/t + 3.4990 \log t$
	51.59–77.17	Liquid	$12.14545 - 1807.5/t - 1.5340 \log t$

TABLE 2.7
Selected Physical Constants for Uranium Hexafluoride

Property	Value	Temperature (°C)
Critical temperature (°C)	230.2 ± 0.2	—
Critical pressure (atm)	45.5 ± 0.5	—
Density, solid (g cm⁻³)	5.060 ± 0.005	25
Density, liquid (g cm⁻³)	3.595	70
Viscosity, liquid (centipoise)	0.91	70
Viscosity, gas (micropoise)	199.9	80
Surface tension (dyne cm⁻¹)	16.8 ± 0.3	70
Refractive index, liquid (4360 Å)	1.383	70
Molecular refraction (cc, 4360 Å)	22.59 ± 0.08	85
Dielectric constant, liquid	2.18	65
Dielectric constant, gas	1.00292 ± 0.000003	67.4
Thermal conductivity, gas (cal sec⁻¹ cm² °C cm⁻¹)	1.0 × 10⁻⁵	5

the other hand those of plutonium and particularly of neptunium hexa-fluoride have been comparatively neglected. Selected physical constants for uranium hexafluoride are listed in Table 2.7. It is beyond the scope of this book to discuss these properties and the numerous investigations in detail and the reader is referred to available compilations[1,5,6,12–14], particularly those of DeWitt[13] (UF₆) and Steindler[12] (PuF₆). Some thermodynamic values are listed in Tables A5–A9 (p. 241).

Uranium[57] and plutonium hexafluoride[58] exhibit weak, almost tem-perature-independent paramagnetism; the two non-bonding 5f-electrons in the latter are postulated as occupying the lowest (f_β) energy level with paired spins. The molar susceptibility of neptunium hexafluoride[59] is only 443 × 10⁻⁶ cgs units at 300°K (cf. the calculated spin only value is 1240 × 10⁻⁶ cgs units). Measurement of the paramagnetic resonance spectrum[60] of this compound has yielded a 'g' value of −0.604 and one interpretation[61] of the magnetic susceptibility results, on the basis of a $5f^1$ electronic configuration, gives a value, −0.621, in agreement with this experimental observation.

Although the physical properties of the hexafluorides, particularly those of uranium hexafluoride, have been investigated in some detail, their chemical behaviour has been relatively little studied. Again the majority of the reported data refer to uranium hexafluoride. They all react violently with water at room temperature: a reaction which, under controlled conditions, can be made to yield[19,20,62] the oxyfluorides, MO_2F_2. They decrease in stability from uranium to plutonium. For example, whereas uranium hexafluoride is stable towards oxygen, nitrogen, carbon dioxide, chlorine and bromine, plutonium hexafluoride is reduced to the tetrafluoride by the last[63] and by iodine[64], and there is some evidence[64] that it reacts with both carbon monoxide and nitrogen. Similarly uranium hexafluoride is stable towards bromine trifluoride, from which it is readily separated by fractional distillation[65,66], whereas plutonium hexafluoride is reduced[18,63] to the tetrafluoride with the formation of bromine pentafluoride (cf. the reactions with sulphur tetrafluoride, p. 23). Plutonium hexafluoride does in fact oxidize uranium tetrafluoride[18] to UF_6 above 200°. However, attempts to obtain plutonium pentafluoride by reacting together plutonium tetra- and hexafluoride have been unsuccessful.

Uranium hexafluoride is reduced to the tetrafluoride when heated in hydrogen[67] but the reaction has a high energy of activation, and even at 600° proceeds only slowly. It is more readily reduced[68] by hydrogen chloride (250°) and by hydrogen bromide (80°), the latter reaction providing[69] a useful preparation of the β-form of uranium pentafluoride. Carbon tetrachloride and carbon disulphide reduce both uranium[2,70,71] and plutonium[12] hexafluoride to their respective tetrafluorides. The latter reagent also reduces molybdenum hexafluoride to the pentafluoride but it does not react with tungsten hexafluoride. Hydrogen sulphide[70] similarly reduces uranium hexafluoride and should likewise reduce plutonium hexafluoride which is thermodynamically less stable than the uranium compound.

Uranium hexafluoride, like the hexafluorides of molybdenum and tungsten, does not react with antimony trifluoride but it is reduced to the tetrafluoride by phosphorus trifluoride (cf. $MoF_6 \rightarrow MoF_5$ but WF_6 is only reduced in the presence of hydrogen fluoride) and by arsenic trifluoride (cf. MoF_6 and WF_6 which do not react). It is also reduced by molybdenum pentafluoride and by tungsten tetrafluoride. By considering the various oxidation–reduction reactions discussed above, together with the fact that molybdenum hexafluoride is reduced to the tetrafluoride by tungsten tetrafluoride, the order of reactivity $WF_6 < MoF_6 < UF_6 < PuF_6$ is established. Neptunium hexafluoride, for which such information is not currently available, will presumably be intermediate between the uranium and plutonium compounds in this series. Thus, since chromium pentafluoride is more reactive than molybdenum hexafluoride, there is a marked decrease in reactivity of the higher fluorides with increase in atomic number of the transition element in passing from chromium to molybdenum. However uranium, which was originally considered to be a member of this subgroup, does not follow this pattern and this purely chemical evidence[25] supports the now accepted classification of uranium as an *f*-transition element.

The chlorides $SiCl_4$, $SbCl_3$ and $AsCl_3$ and the bromides PBr_5 and BBr_3 all reduce[25] uranium hexafluoride to the tetrafluoride with formation of the appropriate fluoride and free halogen, whereas $AlCl_3$ and BCl_3 convert it to the hexachloride. It is well established[72,73] that titanium tetrachloride reacts with tungsten hexafluoride to yield the hexachloride and titanium tetrafluoride but the results of independent studies of the corresponding reaction between uranium hexafluoride and titanium tetrachloride are in conflict. Thus it was originally reported[74] that even at liquid air temperature the reactants combine to form the unusual complex $UF_6 \cdot 2TiCl_4$ but more recently[25] it was found that in the presence of excess titanium tetrachloride the products were UF_6, UF_4, TiF_4 and Cl_2 whereas with excess uranium hexafluoride UF_4, TiF_4 and Cl_2 were formed.

The reaction between uranium hexafluoride and ammonia gas has been reported to yield either[75] NH_4UF_6 or a mixture[76] of the pentafluoride and NH_4UF_5 whereas that with anhydrous liquid ammonia[77] at $-70°$ yields the uranium (IV) salt, NH_4UF_5. Nitrosyl chloride[78] reduces uranium hexafluoride (cf. MoF_6 which behaves in the same way and WF_6 which does not react) as does nitric oxide, yielding the quinquevalent fluoro complex $NOUF_6$. Nitrogen dioxide also reacts to yield a hexafluorouranate (v), NO_2UF_6 (cf. there is no reaction between NO_2 and WF_6 or MoF_6[79]).

The phase relations of several binary uranium hexafluoride systems have been studied. The UF_6–BrF_3, UF_6–BrF_5 and UF_6–Br_2 systems are simple eutectic types which, like the UF_6–ClF_3 system, show some positive deviation from ideality. The solid phases in all the systems are the pure components. These, and the important UF_6–HF system, are discussed in detail by DeWitt[13] who provides a complete literature coverage.

Complex Fluorides

The formation of hexavalent fluoro complexes has been demonstrated only for uranium but the absence of neptunium (VI) and plutonium (VI) analogues is probably due only to the lack of investigation. It was originally shown[80,81] that uranium hexafluoride formed a variety of complexes with sodium, potassium, rubidium, silver and thallium fluorides. The sodium complex was reported then and subsequently[82-84] as Na_3UF_9 but ^{18}F exchange studies[85] and later[86] improved preparative techniques have shown that the complex is actually Na_2UF_8. Katz[86] first prepared the pure complex by reacting solid sodium fluoride with uranium hexafluoride vapour and it has since[87] been prepared, as has the heptafluoro salt, $NaUF_7$, by reacting the component halides in perfluoroheptane,

C_7F_{16}. Analogous molybdenum (VI) and tungsten (VI) complexes have been prepared by the former method. Although no ^{18}F exchange was observed[85] between uranium hexafluoride and potassium fluoride the potassium salts KUF_7 and K_2UF_8 have now been prepared[87] by reactions in C_7F_{16} and the caesium and ammonium heptafluoro complexes have been isolated[88] from reactions in chlorine trifluoride. NH_4UF_7 also forms[89] when the hexafluoride is condensed in a suspension of ammonium fluoride in tetrachloroethane and the corresponding hydrazinium salt, $N_2H_5UF_7$, separates as yellow crystals[90] when the component fluorides are allowed to react in anhydrous hydrofluoric acid. Nitrosonium and nitronium salts, $NOUF_7$ and NO_2UF_7 respectively, are obtained[78] by direct union of the component halides, a reaction which can be employed to prepare the molybdenum (VI) and tungsten (VI) analogues.

Single crystal studies have shown[87] that Na_2UF_8 possesses a body-centred tetragonal cell, space group D_{4h}^{17}–$I4/mmm$ with $a_0 = 5.27$ and $c_0 = 11.20$ Å. The preliminary results indicate that each uranium atom is surrounded by 8 equidistant fluorine atoms with a U–F distance of 2.29 Å. Ammonium heptafluorouranate (VI) possesses cubic symmetry with $a_0 = 13.11$ Å. It decomposes to a mixture of α- and β-uranium penta-fluoride at 170° in a vacuum and to the tetrafluoride at 450°. Whether the last results from the disproportionation of uranium pentafluoride is not clear. $NOUF_7$ possesses pseudo-cubic symmetry[78] with $a_0 = 5.29$ Å and NO_2UF_7 is tetragonal.

The thermal decomposition of sodium octafluorouranate (VI) has been studied[86] and the pressure of uranium hexafluoride over Na_2UF_8 can be expressed by the relationship,

$$\log p_{mm} = 9.25 \pm 0.02 - \frac{4.18 \times 10^3}{Tk}.$$

Nitronium and nitrosonium heptafluorouranate (VI) both undergo reversible thermal dissociation. Measurements between 0° and 70° have shown[78] that for the former

$$\log p_{mm} = 11.194 - 3018/T$$

and for the latter

$$\log p_{mm} = 11.795 - 3633/T$$

Others[446] have recently reported that Na_2UF_8 loses uranium hexa-fluoride above 170° and forms Na_3UF_9 which is stable to 310°.

Oxyfluorides

Compounds of the type MO_2F_2 are known for uranium (VI), neptunium (VI), plutonium (VI) and americium (VI). A uranium compound of composition $U_3O_5F_8$ has also been reported[447] as a product of the reaction between UF_6 and traces of water. Uranyl fluoride was reported by Berzelius[91] as early as the beginning of the last century; neptunyl fluoride was first identified[92] as a product of the reaction of sodium neptunyl acetate with hydrogen fluoride at 300–325° and the formation of the plutonyl compound was initially observed[19,20] on hydrolysis of plutonium hexafluoride. Americyl fluoride, AmO_2F_2, has only recently been prepared[93] by evaporation of an americium (VI) solution in hydrofluoric acid followed by treatment of the residue with anhydrous hydrofluoric acid. The neptunium and plutonium compounds were each identified crystallographically in the first instance.

Uranyl and neptunyl fluoride are both readily obtained by heating their respective trioxides in hydrogen fluoride[94–97] (300°) or in fluorine[96,97] (270–350°) and undoubtedly the action of fluorine on the recently discovered[98] plutonium trioxide hydrate, $PuO_3 \cdot 0.8H_2O$, would be an effective way of preparing PuO_2F_2. Obviously the formation of lower oxides during such reactions must be avoided or the product will contain the tetrafluoride. Neptunyl fluoride is also obtained by vacuum drying the precipitate[97] obtained by adding aqueous hydrofluoric acid to $NpO_3 \cdot H_2O$ but anhydrous uranyl fluoride is more difficult[99] to obtain from aqueous solutions. The hydrated plutonium compound which precipitates[100] on the addition of methanol and hydrofluoric acid to an aqueous plutonium (VI) solution has been converted to anhydrous plutonyl fluoride by washing with anhydrous hydrofluoric acid followed by drying over phosphorus pentoxide[101].

TABLE 2.8

Crystallographic Properties of the Oxyfluorides, $MO_2F_2{}^a$

	Colour	Lattice Parameters a_0 (Å)	α	Calculated density (g cm^{-3})	Reference
UO_2F_2	Yellow	5.764	42° 13′	6.37	104
NpO_2F_2	Pink	5.795	42° 16′	6.41	105
PuO_2F_2	White	5.797	42°	6.50	101

^{a} All are rhombohedral, space group $D_{3d}^5 - R\bar{3}m$, $n = 1$.

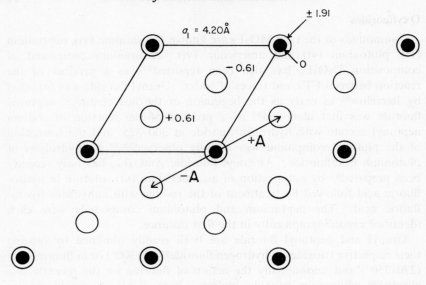

Uranyl groups, uranium atoms in plane of paper and oxygen atoms 1.91 Å above and below this plane.

Fluorine atoms, 0.61 Å above or below plane of uranium atoms.

+A, −A, vectors representing the two possible horizontal displacements between consecutive layers in a stack.

Figure 2.2   The structure of a layer in UO_2F_2[104]. (*After* W. H. Zachariasen, *Acta Cryst.*, **1**, 277 (1948))

Uranyl fluoride is one of the products of the reaction between dry oxygen and uranium tetrafluoride (p. 23) and it has also been prepared by the action of hydrogen fluoride on uranyl acetate[102] at 250° or uranyl phosphate hydrate[96] at 350–500° and by treating uranyl chloride with liquid anhydrous hydrogen fluoride at room temperature[103].

Structures. The hexavalent dioxydifluorides possess rhombohedral symmetry[101,104,105] with one molecule per unit cell (Table 2.8). Zachariasen[104] has shown that each uranium atom is bonded to six fluorine atoms, U–F distance = 2.50 Å, and two oxygen atoms each at a distance of 1.91 Å with the latter forming a linear UO_2^{2+} group. The structure is shown in Figure 2.2. The infrared stretching vibrations for the MO_2^{2+} group in these oxyfluorides have been recorded at around 980 cm^{-1}, the M–F bands occurring at longer wavelengths (Table B.2).

Properties. Little is known of the chemical or physical properties of neptunyl and plutonyl fluoride but those of uranyl fluoride have been investigated in more detail. The compounds are soluble in water and the UO_2F_2–H_2O[106], UO_2F_2–HF–H_2O[107] and PuO_2F_2–HF–H_2O[108] phase diagrams have been studied in detail. The hydrates[102] $UO_2F_2 \cdot 2H_2O$ and $UO_2F_2 \cdot 3H_2O$ are reported to be dehydrated at 110° without loss of hydrogen fluoride but above 300° uranyl fluoride decomposes[109] to U_3O_8. Uranyl fluoride is reduced to UO_2 in an excess of hydrogen at 450° but at higher temperatures some uranium tetrafluoride is formed[110], presumably as a result of the reaction of the hydrogen fluoride so formed with the dioxide. The reduction of neptunyl fluoride with the stoicheiometric amount of hydrogen for the reaction $2NpO_2F_2 + H_2 \rightarrow 2NpO_2F + 2HF$ at 300–325° has recently been shown[97] to yield NpO_2F contaminated with the dioxide. Sulphur reduces[111] uranyl fluoride to a mixture of UO_2 and UF_4 at 500–600° but only the tetrafluoride is formed when uranyl fluoride and sulphur are heated together in hydrogen fluoride at 300–400°.

Fluorine converts uranyl fluoride to the hexafluoride above 270° and will presumably react similarly at higher temperatures with neptunyl and plutonyl fluoride.

Adducts of uranyl fluoride with ammonia, $UO_2F_2 \cdot xNH_3$ ($x = 2$, 3 or 4), have been reported[112] but no other complexes are presently known; the complexes formed by neptunyl and plutonyl fluoride have similarly been neglected.

Hexavalent Oxyfluoro Complexes

It is possible to obtain a variety of hydrated and anhydrous uranyl fluoro complexes from aqueous solution. Thus $NaUO_2F_3 \cdot xH_2O$ ($x = 2$ and 4), $CsUO_2F_3$, $CsUO_2F_3 \cdot H_2O$, $K_3UO_2F_5$, $(NH_4)_3UO_2F_5$, $Cs_3UO_2F_5$, $Cs_2UO_2F_4 \cdot H_2O$, $K_3(UO_2)_2F_7$ and $K_5(UO_2)_2F_9$ have been isolated and studied and numerous complexes of types $M^IUO_2F_3 \cdot xH_2O$, $M^I(UO_2)_2F_5 \cdot xH_2O$ and $M^I(UO_2)_3F_7 \cdot xH_2O$ (where M^I = an organic base and x varies between 0 and 6) have been recorded. The earlier work on these complexes has been reviewed[113].

By varying the conditions it is possible to prepare the different classes of compound with a given alkali metal cation. Thus[113], the addition of potassium fluoride to a solution of uranyl nitrate yields a precipitate of the pentafluoro salt $K_3UO_2F_5$. Recrystallization of this product from an aqueous solution containing less than 13 % KHF_2 results in the formation of $K_5(UO_2)_2F_9$ and if either this product or $K_3UO_2F_5$ is recrystallized from uranyl nitrate solution $K_3(UO_2)_2F_7$ is obtained. Similarly[114], by varying the ratio of $CsF:UO_2F_2$ the complexes $CsUO_2F_3 \cdot H_2O$, $CsUO_2F_3$,

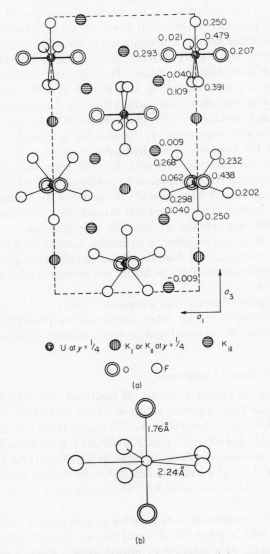

Figure 2.3 The structure of $K_3UO_2F_5$.[120] (a) View of the structure along a_2. Only half of the atoms in the unit cell are shown, the whereabouts of the omitted atoms being obvious since the lattice is body centred. (b) The $UO_2F_5{}^{2-}$ complex. The oxygen and fluorine atoms are at the vertices, and the uranium atom at the centre of a pentagonal bipyramid. The uranyl group forms the axis of the bipyramid. (*After* W. H. Zachariasen, *Acta Cryst.*, **7**, 783 (1954))

$Cs_2UO_2F_4 \cdot H_2O$ and $Cs_3UO_2F_5$ can all be prepared from dilute aqueous hydrofluoric acid. Recently[115] $(NH_4)_3UO_2F_5$ was shown to be the solid phase in equilibrium with ammonium fluoride and uranyl fluoride in water, and[116] the preparation of the hydrazinium complex $(N_2H_5)_3UO_2F_5 \cdot 1.5H_2O$ was described. Relatively few neptunium (VI) oxyfluoro complexes have been characterized. Thus, evaporation of a concentrated hydrofluoric acid solution containing equimolar amounts of CsF and $NpO_3 \cdot H_2O$ leads to the crystallization of $CsNpO_2F_3$ and by heating a $3:1$ mixture, similarly obtained, in dry nitrogen at 275° $Cs_3NpO_2F_5$ can be prepared[117].

Attempts to prepare certain plutonyl complexes from aqueous solution led to the characterization[100] of the pink quinolinium salt $C_9H_7NHPuO_2$-$F_3 \cdot H_2O$, which is analogous to the uranyl complex: precipitates obtained with other univalent cations were not identified.

In contrast to the known hexavalent oxychloro complexes, for which the type $M_2^I M^{VI} O_2 Cl_4$ (M^I = a univalent cation; M^{VI} = U, Np and Pu) is most commonly encountered, anhydrous complexes of the general formula $M_2^I M^{VI} O_2 F_4$ have not yet been reported.

The uranium (VI) fluoro complexes $NaUO_3F$ and KUO_3F are known[118,119] The latter, an orange-red solid, is made by heating uranium trioxide with excess potassium fluoride at 850° and then extracting unreacted potassium fluoride with water.

Structures and properties. X-ray powder diffraction studies[120] have shown that $K_3UO_2F_5$ possesses tetragonal symmetry (Table 2.9) and that the structure comprises pentagonal bipyramidal $UO_2F_5^{2-}$ units which are held together by potassium ions (Figure 2.3). $K_5(UO_2)_2F_9$ is monoclinic[121]; the available x-ray powder data for other complexes[114,117] have not been interpreted.

The hydrated hydrazinium[116] complex $(N_2H_5)_3UO_2F_5 \cdot 1.5H_2O$ decomposes to the tetravalent compound $N_2H_5UF_5 \cdot H_2O$ at 200° in a

TABLE 2.9

Crystallographic Data for Anhydrous Actinide (VI) Oxyfluoro Complexes[120,121]

Compound	Colour	Symmetry/space group	Lattice parameters (Å)		
			a_0	b_0	c_0
$K_3UO_2F_5$	Yellow	Tetragonal—C_{4h}^6–I4$_1$/a	9.160	—	18.167
$K_5(UO_2)_2F_9$	Yellow	Monoclinic—C_{2h}^6–C2/c	19.79	6.13	11.59
				$\beta = 101.2°$	

vacuum and at $400°$ uranium tetrafluoride is formed. In contrast to this behaviour the anhydrous ammonium dioxypentafluoro complex[122], $(NH_4)_3UO_2F_5$, yields initially $NH_4UO_2F_3(> 200°)$ which is converted to anhydrous uranyl fluoride above $300°$.

PENTAVALENT

Protactinium and uranium pentafluoride have been prepared but although thermodynamic calculations suggest that certainly neptunium pentafluoride[123] and possibly plutonium pentafluoride[124] should be capable of existence they remain uncharacterized. However, fully fluorinated complexes of the types $M^IM^VF_6$, $M_2^IM^VF_7$, and $M_3^IM^VF_8$ (M^I = univalent cation; M^V = Pa, U, Np and Pu) are known for all four elements.

Simple oxyfluorides are limited to those of protactinium uranium and neptunium, Pa_2OF_8, U_2OF_8, $NpOF_3$ and 'NpO_2F' respectively, although it has been suggested[125] that UOF_3 may also exist. The dihydrate of the neptunium (v) analogue, $NpOF_3 \cdot 2H_2O$ is also known. Alkali metal complexes of the dioxyhalides $M^IM^VO_2F_2$ (M^V = Np, Pu and Am) have been prepared from aqueous solution.

Pentafluorides

Protactinium pentafluoride is best prepared by fluorine oxidation of the tetrafluoride[126] at $700°$; it can also be obtained by the action of fluorine[127] or hydrogen fluoride on the pentachloride at $200°$ and, in an impure state[127,128], by thermal decomposition of the fluoro complexes NH_4PaF_6 and $(NH_4)_2PaF_7$. However, dehydration of the crystalline dihydrate $PaF_5 \cdot 2H_2O$, first reported[129] by von Grosse, leads to the formation of the oxyfluoride, Pa_2OF_8, above $160°$ even in the presence of excess hydrogen fluoride[126,127]

Uranium pentafluoride can be prepared by a variety of different reactions in either of two crystal modifications, α, the high-temperature form or β, the low-temperature form. Thus, the interaction[130,131] of uranium tetra- and hexafluoride (which can also be made to yield the intermediate fluorides U_2F_9 and U_4F_{17}, p. 49) gives β-UF_5 below $125°$ and α-UF_5 at $230–250°$. Fluorine oxidation[132] of the tetrafluoride at $150°$, one of the simplest preparative methods, results in the formation of α-UF_5 (at higher temperatures UF_6 is the product, p. 21) and β-UF_5 is obtained by the action of anhydrous liquid hydrogen fluoride on uranium penta- or hexachloride at room temperature[132]. Gaseous hydrogen fluoride converts[133] the pentachloride to α-UF_5 at $300°$. By suitable temperature control either α- or β-UF_5 can be prepared by reduction of the hexafluoride with hydrogen bromide vapour[69,134] and a mixture of the two forms is obtained[89] by the thermal decomposition of NH_4UF_7 at 170° in a vacuum. The β-modification precipitates[135] when boron trifluoride is added to a solution of the pentavalent fluoro complex $NOUF_6$ in anhydrous hydrogen fluoride.

Recently[148] it was shown that fluorine oxidation of a suspension of uranium tetrafluoride in anhydrous hydrogen fluoride proceeds rapidly at room temperature to yield β-UF_5. Very little oxidation to the hexafluoride occurs; under the same conditions neptunium tetrafluoride is not oxidized[271].

Structures. Protactinium pentafluoride has only been observed in one crystal form, which[126] is isostructural with the low temperature, β-form of uranium pentafluoride although the former was prepared at 700°. β-UF_5 possesses tetragonal symmetry[136] and a structure involving the bonding of each uranium atom to seven fluorine atoms has been suggested on the basis of x-ray powder diffraction data. The mean U–F distance is 2.23 Å. The high temperature form, α-UF_5, also possesses tetragonal symmetry and Zachariasen[136] has interpreted its x-ray powder pattern on the basis of an octahedral arrangement of fluorine atoms around each uranium atom with two at 2.23 Å and four at 2.18 Å with the ionic radius of U^{5+} as 0.87 Å. Obviously, these structures remain tentative in the absence of single crystal data; the unit cell dimensions for the pentafluorides are listed in Table 2.10. These compounds are not isostructural with the d-transition element pentafluorides which have been studied.

TABLE 2.10

Crystallographic Properties of Protactinium and
Uranium Pentafluoride[126,136]

Compound	Colour	Symmetry	Space group	Unit cell dimensions (Å)		Calculated density (g cm^{-3})
				a_0	c_0	
PaF$_5$	White	Tetragonal	D_{2d}^{12}–$I\bar{4}2d$	11.53	5.19	6.28
α-UF$_5$	Bluish-white	Tetragonal	C_{4h}^{5}–$I4/m$	6.525	4.472	5.81
β-UF$_5$	Bluish-white	Tetragonal	D_{2d}^{12}–$I\bar{4}2d$	11.473	5.209	6.45

Properties. Rather surprisingly protactinium pentafluoride is found to be[126] relatively involatile, subliming only above 500° in a vacuum (cf. VF_5, NbF_5 and TaF_5). Uranium pentafluoride disproportionates[137] slowly above 150° but a melting point of 348° has been reported[131] for α-UF$_5$ and vapour pressure data, obtained in the presence of uranium hexafluoride to limit disproportionation, are consistent with the equations,

$$\log p_{mm} \text{ (solid)} = \frac{-8001}{T} + 13.99$$

$$\log p_{mm} \text{ (liquid)} = \frac{-5388}{T} + 9.82$$

The magnetic susceptibility of β-UF$_5$ shows Curie–Weiss dependence over the temperature range 125–420°K with θ, the Weiss constant, $= -75.4°$ giving[138] an effective magnetic moment of 2.24 B.M.

Apart from complex fluoride formation (discussed below) the chemical properties of the pentafluorides have been little investigated. Both protactinium and uranium pentafluoride are moisture-sensitive, the former deliquescing rapidly and the latter undergoing disproportionation to uranium (IV) and uranium (VI). In aqueous solution uranium penta-fluoride gives a precipitate of the hydrated tetrafluoride and a solution of uranyl fluoride. Complexes with oxygen or nitrogen donor ligands are unknown apart from the dihydrate, PaF$_5$·2H$_2$O, which crystallizes from aqueous hydrofluoric acid[126,129], and the monohydrate PaF$_5$·H$_2$O which is formed[127] when protactinium (V) hydroxide is treated with hydrogen fluoride at 60°, a reaction which yields PaF$_5$·2H$_2$O at room temperature.

Pentavalent Fluoro Complexes

Three classes of pentavalent fluoro complexes, $M^IM^VF_6$, $M_2^IM^VF_7$ and $M_3^IM^VF_8$ (M^I = univalent cation) have been reported for the actinide elements protactinium, neptunium and uranium and examples of the first two classes are known for plutonium. Representatives of each class of complex have been prepared from aqueous solution in the case of protactinium (V), which is extremely stable to hydrolysis in aqueous hydrofluoric acid, but only *hexa*fluorouranates (V) can be prepared in this manner and then only by maintaining a high concentration of hydro-fluoric acid to prevent disproportionation. Other uranium (V) complexes have been prepared by a variety of solid-state reactions and the neptunium (V) and plutonium (V) salts only by fluorine oxidation of a lower valence state fluoride. Some metal–fluorine stretching vibrations for certain of the complexes are listed in Table B.3.

Alkali metal hexafluoro complexes of protactinium (V) can be prepared by crystallization from hydrofluoric acid[127,139–142] containing equimolar amounts of Pa (V) and the appropriate univalent fluoride. To obtain a pure sample the first crop of crystals, which always contains some hepta-fluoro salt, should be discarded[127,141,142]. They are more readily obtained[142,143] pure either by fluorine oxidation of an equimolar mixture of protactinium tetrafluoride and an alkali metal fluoride, a method which has also been used to prepare $CsNpF_6$[144] and $CsPuF_6$[145], or[142] by heating the solid obtained on evaporation of an aqueous solution containing a 1:1 mixture with fluorine.

The corresponding hexafluorouranates (v) are best prepared by crystallization from concentrated aqueous hydrofluoric acid[146] containing uranium pentafluoride and the appropriate alkali metal fluoride (CsF and RbF), by treatment of an equimolar mixture of uranium pentafluoride and MF (M = Li, Na, K, NH_4, Rb and Cs) with anhydrous hydrofluoric acid[147] or by fluorine oxidation[148] of uranium tetrafluoride–MF mixtures in anhydrous hydrofluoric acid. These and other hexafluorouranates (v) have also been prepared by a variety of different methods. The first report of a hexafluorouranate (v) complex appears to be that of Ogle and co-workers[149] who prepared the greenish-white nitrosonium salt, $NOUF_6$, by reacting uranium hexafluoride with nitric oxide (cf. MoF_6 which reacts in this way but WF_6 does not). Later publications described the preparation of $NOUF_6$ by this method and also by reacting the pentafluoride with nitrosyl fluoride[150], NOF, or the hexafluoride with nitrosyl chloride[78,135]. Similarly nitronium hexafluorouranate (v), NO_2UF_6, is obtained[151] when the hexafluoride is treated with nitrogen dioxide and also when the pentafluoride is allowed to react with nitryl fluoride, NO_2F.

Nitrosonium hexafluorouranate (v) can be converted to the corresponding lithium, sodium, potassium or silver salt and to $Ba(UF_6)_2$ by heating it with the appropriate metal nitrate[135] above 80°; dinitrogen tetroxide is evolved. Alternatively[135], treatment of $NOUF_6$ with 48%, or anhydrous hydrofluoric acid containing the appropriate univalent metal fluoride results in the precipitation of sodium, potassium or silver hexafluorouranate (v). Ammonium hexafluorouranate (v) reacts similarly with KF[75] or RbF[152] in 48% hydrofluoric acid to yield the pale green KUF_6 and $RbUF_6$ respectively. The white ammonium salt, NH_4UF_6, can be made[75] by reacting uranium hexafluoride with ammonia at 33°, but it is best prepared[152,153] by heating uranium pentafluoride with one mole of ammonium fluoride at 80–85° or[154] by heating the hexafluoride with ammonium fluoride at 120°. It decomposes at 150° in argon or in a vacuum with the liberation of fluorine. The silver, sodium, potassium, rubidium and caesium salts have also been prepared[153,155] by heating together stoicheiometric amounts of the appropriate fluorides at 300°. Hydrazinium bishexafluorouranate (v), $N_2H_6(UF_6)_2$ is obtained when

4

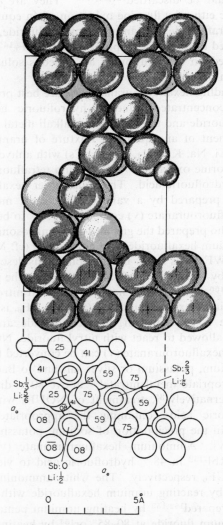

Figure 2.4 Two projections of the hexagonal LiSbF₆ arrangement. In the upper the antimony atoms do not distinctly show within the octahedra of enveloping fluorines. (*After* R. W. G. Wyckoff, *Crystal Structures*, Vol. 3, 2nd Ed., Wiley, New York, 1963)

excess uranium hexafluoride and hydrazinium fluoride are reacted in anhydrous hydrogen fluoride and in the presence of excess hydrazinium fluoride the product is $N_2H_6UF_7$[452,453]. The ultimate thermal decomposition product from $N_2H_6(UF_6)_2$ is uranium tetrafluoride.

Available crystallographic data for the hexafluoro complexes are listed in Table 2.11. $LiUF_6$[153] and one form of $NaUF_6$ (prepared by *slow* evaporation of hydrofluoric acid[147]) possess the rhombohedral $LiSbF_6$-type structure (Figure 2.4). A second crystal modification of $NaUF_6$, prepared by *rapid* removal of hydrofluoric acid and also by heating the component fluorides at 300° in a vacuum, is isostructural with $NaTaF_6$ possessing[147,153] face-centred cubic symmetry. Sodium hexafluoroprotactinate (v) is found to be tetragonal[142] whilst the potassium, ammonium and rubidium salts of protactinium (v) and of uranium (v) are isostructural[142], possessing orthorhombic symmetry. Preliminary lattice parameters[140,156] based on powder results alone indicated that the unit cell

TABLE 2.11

Crystallographic Properties of Some Pentavalent
Actinide Hexafluoro Complexes[142,147,153,155]

Compound	Colour	Symmetry	Structure-type or space group	Unit cell dimensions (Å)		
				a_0	b_0	c_0
$LiUF_6$	Pale blue	Rhombohedral[a]	$LiSbF_6$	5.262	—	14.295
$NaPaF_6$	White	Tetragonal		5.35	—	3.98
$NaUF_6$	Pale blue	Rhombohedral[a]	$LiSbF_6$	5.596	—	15.526
$NaUF_6$	Pale blue	f.c. cubic	$NaTaF_6$	8.608	—	—
$KPaF_6$	White	Orthorhombic	D_{2h}^{18}–$Cmca$	5.64	11.54	7.98
KUF_6	Yellow-green	Orthorhombic	D_{2h}^{18}–$Cmca$	5.61	11.46	7.96
NH_4PaF_6	White	Orthorhombic	D_{2h}^{18}–$Cmca$	5.84	11.90	8.03
NH_4UF_6	Yellow-green	Orthorhombic	D_{2h}^{18}–$Cmca$	5.83	11.89	8.03
$RbPaF_6$	White	Orthorhombic	D_{2h}^{18}–$Cmca$	5.86	11.97	8.04
$RbUF_6$	Yellow-green	Orthorhombic	D_{2h}^{18}–$Cmca$	5.82	11.89	8.03
$CsPaF_6$	White	Orthorhombic	D_{2h}^{18}–$Cmca$	6.14	12.56	8.06
$CsUF_6$	Pale blue	Rhombohedral[a]	$KOsF_6$	8.04	—	8.39
$CsNpF_6$	Pink-violet	Rhombohedral[a]	$KOsF_6$	8.017	—	8.386
$CsPuF_6$	Green	Rhombohedral[a]	$KOsF_6$	8.006	—	8.370
$AgUF_6$	—	Tetragonal		5.42	—	7.95

[a] Hexagonal cell dimensions are quoted for the rhombohedral complexes.

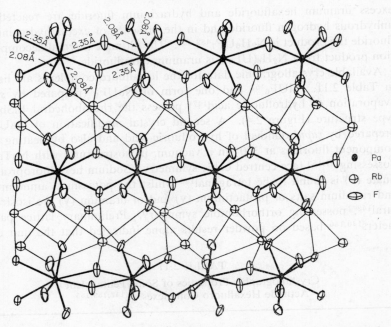

Figure 2.5   A view of the structure of RbPaF₆[157]. Atoms are represented by their ellipsoids of thermal motion; the smallest ellipsoids are Pa atoms, to which are bonded (heavy lines) eight F atoms. The Rb atoms appear nearly spherical and are connected to the F atoms by light lines. (*After* J. H. Burns, H. A. Levy and O. L. Keller, U.S. Report ORNL–4146 (1967); Full report to be published)

contained only one molecule of complex. However, the results of single crystal studies[142,148] show that the unit cell is actually tetramolecular with b_0 and c_0 being twice the size indicated by the powder results. The structure of rubidium hexafluoroprotactinate (v), $RbPaF_6$, is illustrated in Figure 2.5. Each protactinium (v) atom[157] is surrounded by eight fluorine atoms, four of which are shared in pairs with adjacent protactinium atoms. Pa–F bond distances are 2.35 Å for the bridging fluorines and 2.08 Å for the four non-bridging fluorines. $TlUF_6$ and $CsPaF_6$ also possess this structure but $CsUF_6$[153], $CsNpF_6$[144] and $CsPuF_6$[145] have the rhombohedral $KOsF_6$-type structure. The rhombohedral unit cell of $CsUF_6$ is illustrated[158] in Figure 2.6 together with the various bond distances. The optical absorption spectrum of this complex has been analysed[158] in terms of crystal field theory and shown to be consistent with a $5f^1$ configuration for the uranium (v) atom.

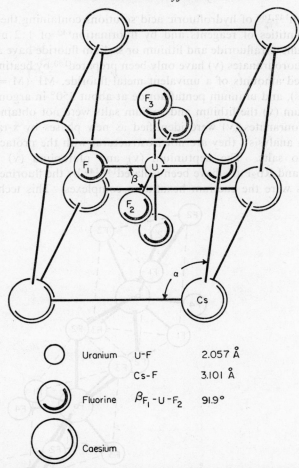

Uranium   U-F   2.057 Å

           Cs-F   3.101 Å

Fluorine $\beta_{F_1-U-F_2}$ 91.9°

Caesium

Figure 2.6   Rhombohedral unit cell of CsUF$_6$[158]. (Bond lengths and angles from reference 454). (*After* M. Reisfield and G. A. Crosby, *Inorg. Chem.*, **4**, 65 (1965))

Potassium heptafluoroprotactinate (v), K$_2$PaF$_7$, was first prepared[129] by reacting PaF$_5$·2H$_2$O with potassium fluoride in water but this and the ammonium, rubidium and caesium salts, which are increasingly soluble in water, can be more conveniently prepared[140] by precipitating them from 17M hydrofluoric acid solution with acetone. The lithium salt does not precipitate because of the smaller size of the lithium cation and with sodium fluoride only the octafluoroprotactinate (v), Na$_3$PaF$_8$, crystallizes[140,159]. Other attempts to prepare Li$_2$PaF$_7$ and Na$_2$PaF$_7$ by direct evaporation[142,160] of hydrofluoric acid solutions containing the stoicheiometric quantities of reagents and by fluorination[142] of 1:2 mixtures of protactinium tetrafluoride and lithium or sodium fluoride have also failed.

Heptafluorouranates (v) have only been prepared[153] by heating together the required amounts of a univalent metal fluoride, MF (M = NH₄, K, Rb and Cs), and uranium pentafluoride at about 350° in argon. As with protactinium (v) the lithium and sodium salts were not obtained. These heptafluorouranates (v) were identified as new phases by x-ray powder diffraction analysis; they are not isostructural with the protactinium (v) heptafluoro salts. The neptunium (v) and plutonium (v) complexes Rb_2NpF_7 and Rb_2PuF_7 have been isolated[144,145] by the fluorine oxidation method as were the caesium hexafluoro complexes. This technique has

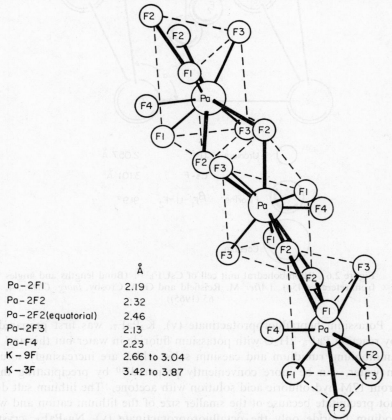

	Å
Pa – 2 F1	2.19
Pa – 2 F2	2.32
Pa – 2 F2 (equatorial)	2.46
Pa – 2 F3	2.13
Pa – F4	2.23
K – 9 F	2.66 to 3.04
K – 3 F	3.42 to 3.87

Figure 2.7 The PaF_9 chain and bond distances in K_2PaF_7[161]. (*After* D. Brown, S. F. A. Kettle and A. J. Smith, *J. Chem. Soc.* (*A*), **1967**, 1429)

also been applied to the preparation of certain[142,143] protactinium (v) heptafluoro complexes but the above wet preparations are much simpler.

Potassium heptafluoroprotactinate (v) possesses monoclinic symmetry, space group C_{2h}^6–$C2/c$, with 4 molecules per unit cell; the cell dimensions are $a_0 = 13.960$; $b_0 = 6.742$; $c_0 = 8.145$ Å; $\beta = 125.17°$. In the structure[161] (Figure 2.7) each protactinium atom is surrounded by nine fluorine atoms in an arrangement which may be idealized as a trigonal prism with three equatorial fluorines added. These PaF_9 units are linked in infinite chains parallel to [001] by two fluorine bridges. Interatomic distances are listed at the foot of Figure 2.7. Ammonium, rubidium and caesium heptafluoroprotactinate (v) possess the same structure, the unit cell of the last having[432] $a_0 = 14.937$, $b_0 = 7.270$, $c_0 = 8.266$ Å and $\beta = 125.32°$.

The rubidium heptafluoro complexes of uranium (v), neptunium (v) and plutonium (v) possess monoclinic symmetry[144,145] and are isostructural with K_2NbF_7 and K_2TaF_7 (Figure 2.8) being therefore seven coordinate

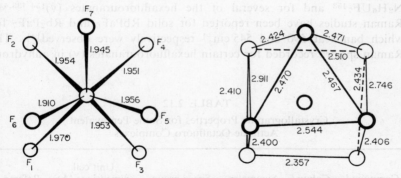

Figure 2.8   Interatomic distances in the NbF_7^{2-} ion. (*After* G. M. Brown and L. A. Walker, *Acta Cryst.*, **20**, 220 (1966))

unlike K_2PaF_7. Available unit cell dimensions with $\beta = 90°$ in each case are, $a_0 = 6.26$, $b_0 = 13.42$, $c_0 = 8.90°$ Å (for Rb_2NpF_7) and $a_0 = 6.27$, $b_0 = 13.41$, $c_0 = 8.88$ Å (Rb_2PuF_7).

Protactinium (v) octafluoro complexes, $M^I_3PaF_8$, can be prepared by precipitation from aqueous hydrofluoric acid[140,142,159] ($M^I = Li$, Na or Rb), by heating the heptafluoro complex with the appropriate alkali metal fluoride at 400° in argon[140] ($M^I = K$ and Cs) or[142] by heating a 3 : 1 mixture, evaporated from aqueous solution, in fluorine ($M^I = Li$ and Na). Analogous uranium (v) complexes result when[153] 3 : 1 mole ratios of a univalent metal fluoride, MF, (M = Na, K, Rb, Cs, Ag and Tl) and uranium pentafluoride are heated at 300°; Na_3UF_8 has also been prepared by fluorine oxidation[162] of Na_3UF_7 at 350° and Na_3NpF_8, Rb_3NpF_8 and Cs_3NpF_8 have been prepared in a similar fashion[163]. Lithium octafluoroprotactinate (v), when prepared by evaporating to dryness a hydrofluoric acid solution containing a 3:1 mixture followed by heating at 400° in argon, possesses tetragonal symmetry[140,159] but the product obtained on heating the mixture in fluorine possesses a different[142], unidentified structure. The sodium salts are all tetragonal whilst the potassium, rubidium and caesium octafluoroprotactinates (v) are face-centred cubic as, apparently[164], are their uranium (v) analogues. Available cell dimensions are shown in Table 2.12. It was recently pointed out[164] that the uranium (III), (IV) and (v) salts, $M_3^IUF_6$, $M_3^IUF_7$ and $M_3^IUF_8$ respectively, (M^I = K, Rb and Cs) are all of face-centred cubic symmetry and that within each series for a given alkali metal cation the unit cell dimensions vary little with the change in oxidation state of uranium and the increase in the number of fluorine atoms.

Magnetic susceptibility data have been recorded for Na_3UF_8[162], $N_2H_6UF_7$[453] and for several of the hexafluorouranates (v)[154,433,453]. Raman studies have been reported for solid $RbPaF_6$ and Rb_2PaF_7 for which bands at 590 and 545 cm^{-1} respectively were observed[141]. The Raman spectra recorded for certain hexafluorouranates (v) in anhydrous

TABLE 2.12

Crystallographic Properties for Some Pentavalent
Actinide Octafluoro Complexes

Compound	Colour	Symmetry	Space group	Unit cell dimensions (Å)		Reference
				a_0	c_0	
Li_3PaF_8	White	Tetragonal	$D_{4h}^6-P42_12$	10.386	10.89	140
Na_3PaF_8	White	Tetragonal	$D_{4h}^{17}-I4/mmm$	5.487	10.89	140
Na_3UF_8	Pale blue	Tetragonal	$D_{4h}^{17}-I4/mmm$	5.470	10.94	162
Na_3NpF_8	Lilac	Tetragonal	$D_{4h}^{17}-I4/mmm$	5.410	10.89	163
K_3PaF_8	White	f.c. cubic	O_h^5-Fm3m	9.235	—	140
K_3UF_8	Pale blue	f.c. cubic	O_h^5-Fm3m	9.20	—	164
Rb_3PaF_8	White	f.c. cubic	O_h^5-Fm3m	9.60	—	142
Cs_3PaF_8	White	f.c. cubic	O_h^5-Fm3m	9.937	—	140
Ag_3UF_8	—	Cubic		4.36	—	155
Tl_3UF_8	—	Cubic		4.75	—	155

hydrogen fluoride all showed[453] a single peak at 628 cm^{-1}. Infrared data are summarized in Table B.3 (p. 251).

Pentavalent Oxyfluorides

Diprotactinium (v) oxyoctafluoride, Pa_2OF_8, is a white, moisture-sensitive solid which is easily prepared by a variety of reactions including[126] the thermal decomposition of $PaF_5 \cdot 2H_2O$ at 160°, oxidation of the tetrafluoride by fluorine in the presence of oxygen, fluorination of the pentoxide by a mixture of hydrogen fluoride and oxygen and[127] by heating vacuum-dried protactinium (v) hydroxide in hydrogen fluoride at 160°. The analogous uranium (v) oxyfluoride, also a white solid, has been observed to form[165] when uranium tetrafluoride is heated in an intermittent oxygen flow. Plutonium (v) and neptunium (v) analogues are unknown but a hydrated oxytrifluoride, $NpOF_3 \cdot 2H_2O$, has been prepared[97] by reacting the pentoxide, Np_2O_5, with gaseous hydrogen fluoride at 40°. This compound is converted to the anhydrous oxytrifluoride, $NpOF_3$, at 150° in gaseous hydrogen fluoride. Apart from phases close to the composition NpO_2F, obtained by controlled hydrogen reduction[97] of NpO_2F_2, other oxyfluorides are presently unknown. Attempts to prepare protactinium (v) oxytrifluoride, $PaOF_3$, by reacting the oxytribromide with hydrogen fluoride at room temperature[127] have resulted in the formation of only Pa_2OF_8. This observation may be significant in view of the relative stabilities of Pa_2OCl_8 and $PaOCl_3$ (p. 131).

Properties. The oxyoctafluorides Pa_2OF_8 and U_2OF_8 are thermally unstable; the former[126] decomposes at 800° to give a sublimate of protactinium pentafluoride leaving an unidentified residue, whilst U_2OF_8[165] disproportionates at 300° in a vacuum to form a mixture of uranyl fluoride, uranium hexafluoride and uranium tetrafluoride. Diprotactinium (v) oxyoctafluoride is isostructural with Pa_2F_9 (p. 49), crystallizing with body-centred cubic symmetry, $a_0 = 8.406$ Å, and the presence of bridging oxygen atoms in the molecule is confirmed by its infrared spectrum (Table B.2). Neptunium oxytrifluoride dihydrate is a stable, green solid which contains discrete NpO^{3+} groups (Table B.2).

Pentavalent Oxyfluoro Complexes

Only complexes of the type $M^IM^{IV}O_2F_2$ (M^I = K, NH_4 and Rb, M^V = Np, Pu and Am) have been reported[166,167]. The grey-green neptunyl (v) and lavender plutonyl (v) rubidium salts precipitate[167] when a cooled solution of the pentavalent actinide element in dilute acid is added to a saturated aqueous solution of rubidium fluoride at ice temperature. The ammonium complex, $NH_4PuO_2F_2$, crystallizes[167] when solid ammonium fluoride is added to a plutonium (v) solution at pH = 6. The americium (v) salts[166,167], which vary in colour from white to tan are likewise precipitated from aqueous solution. All these complexes possess rhombohedral symmetry (Table 2.13).

TABLE 2.13

Crystallographic Properties of the Actinyl (v) Fluoro Complexes[166,167]

Compound	Colour	Lattice dimensions[a]	
		a_0 (Å)	$\alpha°$
$KAmO_2F_2$	Tan	6.760	36.25
$NH_4PuO_2F_2$	Lavender	6.817	36.16
$RbNpO_2F_2$	Grey-green	6.814	36.18
$RbPuO_2F_2$	Lavender	6.796	36.17
$RbAmO_2F_2$	Tan	6.789	36.15

[a] All are rhombohedral, space group $D_{3d}^5 - R\bar{3}m$, $n = 1$.

Obviously although much recent work has been concentrated on the fully fluorinated actinide (v) complexes the simple oxyfluorides and the oxyfluoro complexes have received much less attention and the present state of knowledge in this field leaves much to be desired.

TETRAVALENT

Three lanthanide tetrafluorides are known at present, the cerium, terbium and praseodymium compounds, the last only recently reported. Of the actinide elements, thorium to curium inclusive form tetrafluorides. Numerous complex fluorides of the types $M_2^I M^{IV} F_9$, $M^I M^{IV} F_5$, $M_2^I M^{IV} F_6$, $M_3^I M^{IV} F_7$, $M_4^I M^{IV} F_8$, $M_7^I M_6^{IV} F_{31}$ and $M^{II} M^{IV} F_6$ (M^I = univalent cation; M^{II} = divalent cation; M^{IV} = Ce, Pr, Tb, Th, Pa, U, Np, Pu, Am and Cm) are known, although examples of each class have not been reported for all the tetravalent elements listed.

The only known tetravalent oxyfluoride is $ThOF_2$ and this compound has hardly been studied; oxyfluoro salts have not been reported.

Some protactinium, uranium and plutonium fluorides intermediate in valence state between 4+ and 5+ have been characterized and these will be discussed before the tetrafluorides in this section.

Intermediate Fluorides

The first indication that intermediate uranium fluorides were capable of existence was found during an attempt to prepare uranium pentafluoride by the interaction of the tetra- and hexafluorides (p. 36) when a black product, referred to as 'black UF_4', was observed[168]. Further investigations[169-171] have shown that in addition to α- and β-UF_5, it is possible to prepare two intermediate uranium fluorides by the controlled interaction of uranium tetra- and hexafluoride; these are diuranium enneafluoride, U_2F_9 (the original 'black UF_4' phase) and a compound of composition U_4F_{17}. Conditions suitable for the preparation of U_2F_9, U_4F_{17} and the two crystal modifications of uranium pentafluoride are shown in Table 2.14. A more recent study[172] has confirmed these results and indicated that a third phase, of composition U_5F_{22}, can be prepared by reactions below 90°.

TABLE 2.14

Conditions for the Preparation of UF_5, U_2F_9 and U_4F_{17} by Reaction between UF_4 (s) and UF_6 (g)[171,173]

p_{mm} UF_6	Temperature (°C)		
	100°	200°	300°
17.7	β-UF_5	U_2F_9	U_4F_{17}
120–140	β-UF_5	α-UF_5	U_2F_9

Like uranium pentafluoride these intermediate compounds disproportionate when heated[173] (Figure 2.9). Diuranium enneafluoride possesses body-centred cubic symmetry[169] with $a_0 = 8.4545$ Å and powder results[174] have been interpreted on the basis of each uranium atom being surrounded by 9 fluorines with 3 at 2.26 Å, 3 at 2.31 Å and 3 at 2.34 Å. U_4F_{17} gives an x-ray powder diffraction pattern which is very similar to that of monoclinic uranium tetrafluoride.

A dark brown solid observed to form during hydrofluorination of Pa_2O_5 with H_2/HF mixtures[127,175] in the ratio 1:2 and[127] as a product of the thermal decomposition of ammonium heptafluoroprotactinate (v), $(NH_4)_2PaF_7$, has variously been referred to as Pa_2F_9 and Pa_4F_{17}. Analysis[175] indicates that the composition is close to Pa_4F_{17} but others[127] refer to the compound as Pa_2F_9 since it is isomorphous with U_2F_9. The cubic unit cell has $a_0 = 8.494$ Å. Further work, possibly involving interaction of protactinium tetra- and pentafluoride and a determination of the average valence state of the product, is necessary to clarify the present situation.

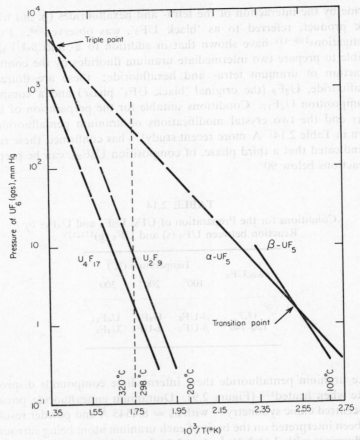

Figure 2.9   Disproportionation pressures of U_4F_{17}, U_2F_9, α-UF$_5$ and β-UF$_5$. (*After* J. J. Katz and E. Rabinowitch (Eds.), *Nat. Nucl. Energy Ser. VIII*, Vol. 5, McGraw-Hill, New York, 1951, p. 386)

One other intermediate fluoride, presumed to be Pu_4F_{17}, has been reported[40]; this brick-red solid was observed as a residue during the fluorination of plutonium tetrafluoride. Little is known concerning this material and further studies are obviously required.

Tetrafluorides

As discussed earlier (pp. 21 and 23) fluorine will oxidize various uranium, neptunium and plutonium compounds to their respective hexafluorides and protactinium tetrafluoride to the pentafluoride. It has been reported[176] that at 300° it will convert plutonium trifluoride to the tetrafluoride but it cannot be employed for the preparation of the tetrafluorides of the remaining three elements. However, the stability of thorium (IV) and the increasing stability of the lower oxidation states of the actinide elements beyond plutonium permit the use of fluorine for the preparation of thorium, americium and curium tetrafluoride. Indeed, it is essential for the preparation of the last two and also of terbium and cerium tetrafluoride.

The preparation of thorium and uranium tetrafluoride has been extensively studied and these compounds will be discussed together since many of the methods can be applied to both. Hydrated tetrafluorides are precipitated from aqueous hydrofluoric acid solutions of the tetravalent elements. It has been reported[177-179] that such compounds can be dehydrated by heating *in vacuo* (see references 1, 3–5, 7–10 and 14 for discussion) but it seems likely that some hydrolysis is inevitable, particularly with large amounts of material. The dehydration can, however, be successfully achieved by heating the solids above 500° in anhydrous hydrogen fluoride, for example see references 180 and 181, but obviously this procedure offers no advantage over direct hydrofluorination of the dioxides.

One of the simplest and most satisfactory preparations of thorium[182,183] and uranium tetrafluoride involves conversion of the dioxides by heating them in excess anhydrous hydrogen fluoride. Highly reactive forms of the dioxides can be obtained by thermally decomposing the oxalates at about 400°; this combination of a preliminary decomposition followed by hydrofluorination is superior to the direct conversion[182] of the oxalates. Methods suitable for either the industrial or laboratory scale conversion of uranium dioxide have been described[184-187].

A mixture of hydrogen fluoride and ammonia is reported[188] to convert uranium trioxide, UO_3, directly to the tetrafluoride at 500–750°. This reaction is superior to the UO_3– or U_3O_8–H_2/HF reactions. Alternative reagents for the direct conversion of uranium trioxide include[189] ammonium fluoride, ammonium bifluoride and[190] a mixture of ammonium bifluoride and hydrazine fluoride. The first has also been used successfully for the conversion[191] of uranyl nitrate or ammonium diuranate. Ammonium bifluoride reacts with thorium[192] and uranium[193,194] dioxide to yield initially the tetravalent fluoro complexes $NH_4M^{IV}F_5$ (M^{IV} = Th and U) which decompose to the tetrafluorides above 350°.

The Freons (fluorinated hydrocarbons) will also afford simultaneous reduction and fluorination of uranium trioxide[188,195-197] and of such compounds as[198] $(NH_4)_3U_4O_{16}F_3$ (a form of U (VI) peroxide). Freon 12 (CCl_2F_2), which reacts at about 400°, is particularly useful in this respect and it has also been used[196,197] for the preparation of thorium tetrafluoride from the dioxide. In view of the formation of gaseous by-products such as chlorine and phosgene it is essential to introduce excess reagent during the early stages of the reaction in order to minimize tetrachloride formation.

The above methods for converting thorium and uranium oxides to the tetrafluorides have found wide industrial application, particularly since the tetrafluorides can be efficiently reduced to the metals. This aspect is important for uranium in view of the special position it holds in connection with the production of nuclear power. The large-scale industrial processes are discussed in the collected papers of the Geneva Conferences[187,199-201] and the reader is referred to these publications for further details. The direct reduction of uranium hexafluoride to the tetrafluoride has been studied in connection with the production of isotopically enriched uranium fuels. Although this can be achieved by hydrogen at 600° the reaction has a high energy of activation and a more satisfactory reduction is obtained using either carbon tetrachloride[202] or trichloroethylene[203]. The main disadvantage of such reagents is that the product may be contaminated with chloride.

Where smaller quantities of the tetrafluoride are required, for example for research purposes, and the appropriate metal is available, alternative reactions may be more attractive. Thus thorium and uranium tetrafluoride can be prepared[204] by heating the metal with anhydrous, liquid hydrogen fluoride in a sealed tube at 225-250° or by treating the hydride[205,206] with hydrogen fluoride vapour at 250-350°. As an alternative to first converting uranium metal to the hydride (by heating[207,208] in hydrogen at 250°) it is recommended[206] that the metal is heated at 250° with an equimolar mixture of hydrogen and hydrogen fluoride. It is interesting to note that hydrogen chloride (p. 151) and hydrogen bromide (p. 196) both react with uranium hydride to yield trivalent halides.

Thorium tetrafluoride has been prepared by the action of fluorine[209] or hydrogen fluoride[210] on the tetrachloride or tetrabromide. Uranium tetrafluoride has also been prepared by a variety of reactions of lesser importance. These include[189] the reaction of uranium trichloride or $UO_2(HPO_4) \cdot H_2O$ with anhydrous hydrogen fluoride at 450° and 800° respectively, reaction of uranium tetrachloride with liquid hydrogen fluoride, the reduction of uranium hexafluoride by reagents such as hydrogen chloride, ammonia or thionyl chloride and the reaction[211] between uranium tetraacetate and ammonium bifluoride at 450° *in vacuo*.

Protactinium pentoxide or dioxide can be quantitatively converted to the tetrafluoride[212] by heating in a mixture of hydrogen and hydrogen fluoride at 500 to 600°. In the former instance a large excess of hydrogen is essential to prevent the formation of the intermediate fluoride Pa_2F_9 (p. 49) which, once it has formed, is only reduced at higher temperatures ($>650°$). It is also advisable to use low-fired ($<400°$) pentoxide to ensure complete conversion to the tetrafluoride. In contrast to the behaviour with protactinium dioxide, and in keeping with the decreasing stability of the higher oxidation states of the actinide elements with increasing atomic number, hydrogen fluoride reacts with neptunium and plutonium dioxide to yield the respective trifluorides. In the presence of oxygen, however[213,214], the tetrafluorides are obtained from this reaction at about 500°. Similarly hydrogen fluoride–oxygen mixtures oxidize the trifluorides above 500° and, as mentioned earlier, plutonium trifluoride is converted to the tetrafluoride by fluorine at 300°. However, in view of the reported oxidation of plutonium tetrafluoride to the hexafluoride by this reagent at temperatures as low as 200° (p. 21) it is probably not a satisfactory method for preparing the tetrafluoride.

Plutonium tetrafluoride can also be made by[215] treating the tetrafluoride hydrate with a mixture of hydrogen fluoride and oxygen at 350° and, like thorium and uranium tetrafluorides, by heating[216,217] the dioxide with ammonium bifluoride. In the latter reaction the intermediate fluoro complex NH_4PuF_5 decomposes at about 280°. Sulphur tetrafluoride, which can be employed for the preparation of uranium hexafluoride (p. 23) reduces plutonium hexafluoride at 30° and it has been used to convert the dioxide to plutonium tetrafluoride at 500°.

Fluorine oxidation of the appropriate trifluoride is the only method reported for the preparation of terbium[218] and curium[219,220] tetrafluoride. Radiation damage prevents the preparation of the latter when the short-lived isotope curium-242 is used, the only successful preparations having involved the longer-lived curium-244. Cerium and americium tetra-fluoride, also conveniently prepared in this manner[221,222], can be obtained[222,223] by fluorination of the appropriate dioxide at 500°. Cerium tetrafluoride was first obtained[224] by treating the trichloride with fluorine at room temperature. Attempts[223] to dehydrate cerium tetrafluoride hydrate, obtained from aqueous solution, invariably lead to decomposition and hydrofluorination of the dioxide yields a mixture of cerium tri- and tetrafluorides.

Fluorine is reported[218] to be incapable of oxidizing praseodymium, neodymium, dysprosium and samarium trifluoride at 300° but several tetravalent fluoro complexes of praseodymium have been prepared by fluorination of alkali metal fluoride–PrF_3 mixtures. It has been suggested[218] that praseodymium tetrafluoride cannot exist as a stable solid at room temperature but this was recently disproved. Thus, by treating sodium hexafluoropraseodymate (IV), Na_2PrF_6, with anhydrous liquid hydrogen fluoride at room temperature Soriano and co-workers[225] have isolated the anhydrous tetrafluoride. This work has been confirmed by others[226] who have obtained x-ray powder diffraction data for the tetrafluoride and find it to be isostructural with the other lanthanide and with the actinide tetrafluorides.

Crystal structures. Available crystallographic data for the tetra-fluorides are listed in Table 2.15. These compounds, which are all iso-structural with zirconium and hafnium tetrafluoride, possess monoclinic

TABLE 2.15

Crystallographic Properties of the Lanthanide and Actinide Tetrafluorides

Compound	Colour	Lattice parameters[a] (a_0, b_0 and c_0 in Å; β in °)				Density (g cm^{-3})	Reference
		a_0	b_0	c_0	β		
CeF_4	White	12.60	10.60	8.30	126	4.80	105
PrF_4	White	12.47	10.54	8.18	126.4	4.94	226
TbF_4	White	12.10	10.30	7.90	126	5.88	218
ThF_4	White	13.10	11.01	8.60	126	6.19	105
PaF_4	Brown	12.86	10.88	8.54	126.34	6.36	272
UF_4	Green	12.82	10.74	8.41	126.16	6.70	105
NpF_4	Green	12.67	10.62	8.31	126.16	6.80	105
PuF_4	Brown	12.59	10.55	8.26	126.16	7.00	105
AmF_4	Tan	12.49	10.47	8.19	126.16	7.34	222
CmF_4	Brownish-tan	12.45	10.45	8.16	126	7.49	219

[a] All the tetrafluorides possess monoclinic symmetry, space group C_{2h}^6–$C2/c$, with 12 molecules per unit cell.

symmetry, space group C_{2h}^6–$C2/c$ with twelve molecules per unit cell. A full structure analysis has been reported[227] for uranium tetrafluoride. Each uranium atom has eight fluorine neighbours arranged in a slightly distorted antiprism and there is a basic repeating unit of five uranium atoms arranged in a distorted pyramid with four atoms forming a rhomb-shaped base and the fifth comprising the apex of the pyramid. The structure, viewed down the positive b_0 axis is shown in Figure 2.10. The U–F distances range between 2.249 and 2.318 Å.

Properties. The tetrafluorides are high melting solids; some vapour pressure studies have been reported[228–232] and the results are summarized in Table 2.16. The tetrafluorides are monomeric in the vapour phase.

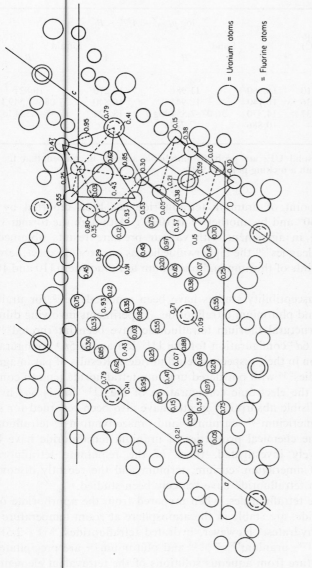

Figure 2.10 The structure of UF_4 viewed down the positive b_0 axis.[277] Numbers represent the positional parameter y as a fraction of the unit cell edge. The slightly distorted antiprisms about U(1) ($y = 0.21$) and U(2) ($y = 0.43$) are also shown. (*After* A. C. Larson, R. B. Roof, Jr. and D. T. Cromer, *Acta Cryst.*, **17**, 555 (1964))

○ = Uranium atoms

○ = Fluorine atoms

5

TABLE 2.16

Vapour Pressure Data for the Actinide Tetrafluorides[228-232]

Compound	m.p. (°C)	$\log p_{mm} - A/T + B$			
		Solid		Liquid	
		A	B	A	B
ThF$_4$	1110	16,860	11.986	15,270	10.821
UF$_4$[a]	1036	16,140	12.945	16,840	37.086–7.549 log T
PuF$_4$	1037	14,370	36.07–7.554 log T	18,124	37.97–7.554 log T
AmF$_4$	—	10,886	7.727	—	—

[a] The data for solid UF$_4$ are probably unreliable since it is claimed that they are consistent with a melting point of 960°.

The melting point of uranium tetrafluoride has been reported by one group[233] as 960° and by another[229] as 1036°. Although the former value appears to be in agreement with vapour pressure results obtained by Popov and colleagues[230], the higher value lies, as one would expect, between the melting points of thorium and plutonium tetrafluoride, 1110 and 1037° respectively.

Magnetic susceptibility studies have been reported[234,235] for uranium tetrafluoride and plutonium tetrafluoride. Experiments involving dilution with the isostructural thorium tetrafluoride have been interpreted[235] on the basis of a $6d^2$ configuration for the U^{4+} ion and a $5f^4$ configuration for the Pu^{4+} ion in their respective tetrafluorides. Results of paramagnetic resonance studies[236] with powdered uranium tetrafluoride tend to confirm the view that the electronic configuration of the U^{4+} ion is $6d^2$ in this compound. Visible absorption spectra have also been recorded for solid uranium[237], americium[238], curium[238] and praseodymium[226] tetrafluoride.

Although the chemical properties of uranium tetrafluoride have been fairly extensively investigated, those of the remaining tetrafluorides, particularly of americium, curium, terbium and the recently discovered praseodymium tetrafluoride have scarcely been studied.

The actinide tetrafluorides, when prepared from the appropriate oxide or lower fluoride, are stable in the atmosphere at room temperature and do not form hydrates. However, hydrated tetrafluorides, MF$_4 \cdot 2.5H_2O$, of thorium[177,178], uranium[179,239,240] and plutonium[241] are precipitated at room temperature from aqueous solutions of the tetravalent elements by hydrofluoric acid. Complete dehydration of, for example, UF$_4 \cdot 2.5H_2O$, yields a phase which retains a structure similar to that of the original hydrate and this phase readily rehydrates to form $UF_4 \cdot 2.5H_2O$. The preparation of the above hydrates and their thermal degradation to lower hydrates[177–179,239,242], e.g. $ThF_4 \cdot 0.5H_2O$ and $UF_4 \cdot 0.4H_2O$, or to the corresponding anhydrous tetrafluorides have been studied in detail. The various results are discussed in other reviews[4,5,9,14] and it suffices to say here that the stable crystal hydrates for thorium, uranium and plutonium tetrafluoride are of the type $MF_4 \cdot 2.5H_2O$ and that, at least in the case of uranium (IV), all hydrates of lower water content are gradually converted to this form.

The composition of the protactinium (IV) fluoride hydrate which precipitates from aqueous solution[243] has not been determined and although $NpF_4 \cdot 2.5H_2O$ should exist it has not yet been reported. Americium (IV) and curium (IV) are unstable in water in the absence of high concentrations of fluoride ion and consequently hydrated tetrafluorides of these elements cannot be obtained from aqueous solution. Similarly terbium and praseodymium tetrafluoride hydrates are unlikely to be prepared.

Cerium, terbium and the actinide tetrafluorides (Th → Pu inclusive) are only slightly soluble in water and aqueous hydrofluoric acid: americium tetrafluoride reacts with water[222] evolving unidentified gases and being reduced to americium (III). Uranium tetrafluoride is attacked by oxidizing media such as fuming perchloric acid, nitric acid–boric acid mixtures (the boron being necessary to preferentially complex the fluoride) and sulphuric acid–ammonium persulphate mixtures, dissolving to give uranium (VI) solutions. Protactinium tetrafluoride, the most stable tetravalent protactinium halide known, similarly requires strong oxidizing conditions to effect dissolution. The tetrafluorides of cerium and of the elements thorium to plutonium inclusive are decomposed by steam at high temperatures (>500°) with the formation of the oxide and liberation of hydrogen fluoride. This pyrohydrolysis is obviously of analytical importance. The trivalent terbium oxyfluoride, TbOF, has been observed[218] as the product when terbium tetrafluoride is heated in air at 400°. Cerium tetrafluoride is stable in dry oxygen[223] up to 700° as is plutonium tetrafluoride[39] at 600° but the more easily oxidized uranium compound is converted to a mixture of uranyl fluoride and uranium hexafluoride (p. 23). Protactinium tetrafluoride will probably react to form the stable diprotactinium oxyoctafluoride (V), Pa_2OF_8, when heated in dry oxygen.

Owing to the ease with which protactinium, uranium, neptunium and plutonium attain oxidation states greater than 4+, the action of oxidizing agents on their tetrafluorides is of interest. For obvious reasons this aspect of the chemistry of thorium, americium, curium, cerium, terbium and praseodymium tetrafluoride has attracted little attention. However, the recent characterization of americyl (VI) oxyfluoride, AmO_2F_2, could well lead to renewed interest in the possibility of oxidizing americium tetrafluoride to a higher fluoride. Fluorine oxidation of the tetrafluorides of protactinium to plutonium inclusive (pp. 21 and 36), yields PaF_5, UF_6 NpF_6 and PuF_6 respectively; the conversion of uranium tetrafluoride to the hexafluoride by chlorine trifluoride, bromine trifluoride and sulphur tetrafluoride (p. 23) has already been discussed. In view of the fact that bromine trifluoride and sulphur tetrafluoride reduce plutonium hexafluoride it is unlikely that they will be capable of oxidizing americium or curium tetrafluoride but their reactions with protactinium tetrafluoride, and in particular neptunium tetrafluoride, are worthy of investigation. It is possible that one or other of these reagents is capable of oxidizing neptunium tetrafluoride to the presently unknown pentafluoride.

As described earlier (p. 49) uranium tetrafluoride reacts with uranium hexafluoride to yield intermediate fluorides such as U_2F_9, U_4F_{17} and the pentafluoride, UF_5. Although it is reported that no reaction occurs in the analogous neptunium and plutonium systems, the possible existence of Pu_4F_{17} (p. 50) suggests that these systems merit further attention. Chlorine does not react with uranium tetrafluoride but it may be capable of oxidizing the protactinium (IV) compound, possibly with the formation of protactinium (V) mixed fluoro–chloro compounds.

The reduction of thorium and protactinium tetrafluoride to lower fluorides has not been reported and it is likely that it will only be achieved by an investigation of the metal–metal tetrafluoride systems. Uranium tetrafluoride can be reduced to the trifluoride only with difficulty, for example by hydrogen[244] at 1000°, aluminium[245] at 900° or by finely divided uranium metal[246] in an argon atmosphere at 1050°. The reaction with magnesium[247] at 560° also yields uranium trifluoride but at a slightly higher temperature (600°) uranium metal is produced. The remaining tetrafluorides are more easily reduced, for example cerium[224] and plutonium[248] trifluoride are obtained in hydrogen at moderate temperatures. In view of the oxidizing conditions necessary to prepare the tetrafluorides of neptunium, americium, curium and terbium these compounds should all be reduced by hydrogen at only moderate temperatures and, like cerium tetrafluoride[223], by gaseous ammonia.

The reduction of thorium, protactinium, uranium and plutonium tetrafluorides to their respective metals constitutes an important source of these elements, in particular of uranium, whereas the remaining actinide and the lanthanide metals are usually obtained by reduction of the more easily accessible trifluorides (p. 87). The large-scale reduction of thorium, uranium and plutonium tetrafluoride, which will not be discussed here, can be achieved by heating them with more strongly electropositive metals such as calcium, magnesium and sodium (for reviews of the methods available see references 249–253 and 353). Protactinium tetrafluoride has been reduced to the metal on the milligram scale with barium vapour using the double crucible technique described later (p. 87) but gram amounts are best reduced[431] by a 10% magnesium in zinc alloy at 800°.

Thorium tetrafluoride and thoria react at 900° to form the oxydifluoride, $ThOF_2$ (p. 78). The analogous protactinium reaction has not yet been studied but uranium (IV) oxydifluoride cannot be prepared in this way. The plutonium tetrafluoride–plutonium dioxide reaction has been shown to be reversible[39], the products being plutonium trifluoride and oxygen,

$$PuO_2 + 3PuF_4 \underset{600°}{\overset{\text{vacuum}}{\rightleftharpoons}} 4PuF_3 + O_2$$

Dawson and co-workers[241] have measured equilibrium oxygen pressures over PuF_4/PuO_2 mixtures and confirmed the reversibility of the system. Cerium tetrafluoride[223] apparently undergoes a similar reaction with cerium dioxide.

Tetravalent Fluoro Complexes

Many tetravalent fluoro complexes are known for the lanthanide elements cerium, praseodymium and terbium and for the actinide elements thorium to curium inclusive. There is some evidence for the existence of analogous neodymium (IV) and dysprosium (IV) salts[254,437] and further studies will be of interest. The tetravalent fluoro complexes have been prepared in a variety of ways, including fluorine oxidation of lower valence compounds, hydrogen reduction of higher valence fluoro complexes, fusion of stoicheiometric amounts of the component fluorides and, in a few instances only, by precipitation from aqueous solution. Many of the reported complexes have, in fact, only been identified during detailed studies of binary fluoride fused-salt systems. Some of the presently known alkali fluoride–metal (IV) fluoro complexes are listed in Table 2.17, together with the cation radius ratios M^+/M^{4+} as reported by Thoma[255]. From a consideration of these ratios Thoma has predicted the existence of many fluoro complexes in the systems not yet studied and such complexes are denoted by the letter *a* in Table 2.17. The stoicheiometry reported in the literature for certain of the actinide (IV) fluoro complexes is incorrect. For example the complex originally reported[256–258] as $Li_7U_6F_{31}$ has recently been shown[259] by a single crystal structure determination to have the composition $LiUF_5$ and since the reported $Li_7Th_6F_{31}$

TABLE 2.17

Alkali Metal Fluoride–Metal (IV) Fluoro Complexes

| Tetrafluoride | Lithium fluoride | | | | Sodium fluoride | | | | | |
| | M^+/M^{4+} ratio | Stoicheiometry | | | M^+/M^{4+} ratio | Stoicheiometry | | | | |
		3:1	1:1	1:4		3:1	2:1	7:6	1:1	1:2
ThF_4	0.69	Li_3ThF_7	$LiThF_5$	$LiTh_4F_{17}$	0.99	Na_3ThF_7	Na_2ThF_6	$Na_7Th_6F_{31}$	$NaThF_5{}^b$	$NaTh_2F_9$
PaF_4	0.71	a	$LiPaF_5$	a	1.02	Na_3PaF_7		$Na_7Pa_6F_{31}$	c	a
UF_4	0.73	$Li_3UF_7{}^b$	$LiUF_5$	LiU_4F_{17}	1.05	Na_3UF_7	Na_2UF_6	$Na_7U_6F_{31}$	c	NaU_2F_9
NpF_4	0.74	a	$LiNpF_5$	a	1.06	a	Na_2NpF_6	$Na_7Np_6F_{31}$	c	
CeF_4	0.74	a	a	a	1.06	Na_3CeF_7	Na_2CeF_6	a	c	
PrF_4	0.76	a	a	a	1.09	Na_3PrF_7	Na_2PrF_6	$Na_7Pr_6F_{31}$	c	
PuF_4	0.76	a	$LiPuF_5$	a	1.09	a	Na_2PuF_6	$Na_7Pu_6F_{31}$	c	
AmF_4	0.76	a	$LiAmF_5$	a	1.10	a	a	$Na_7Am_6F_{31}$	c	
CmF_4	0.77	a	$LiCmF_5$	a	1.11	a	a	$Na_7Cm_6F_{31}$	c	
BkF_4	0.77	a	a	a	1.11	a	a	a	c	
TbF_4	0.86	a	a	a	1.24	Na_3TbF_7	††	a	?	

TABLE 2.17 contd.

Tetrafluoride	M^+/M^{+4} ratio	Potassium fluoride Stoicheiometry						
		3:1	2:1	7:6	1:1	1:2	1:3	1:6
ThF_4	1.34	$K_3ThF_7{}^b$	K_2ThF_6	$K_7Th_6F_{31}$	c	KTh_2F_9	KTh_3F_{13}	KTh_6F_{25}
PaF_4	1.38	K_3PaF_7	d	$K_7Pa_6F_{31}$	c	a	a	a
UF_4	1.43	K_3UF_7	K_2UF_6	$K_7U_6F_{31}$	c	KU_2F_9	KU_3F_{13}	KU_6F_{25}
NpF_4	1.44	a	a	$K_7Np_6F_{31}$	c	KNp_2F_9		
CeF_4	1.44	a	K_2CeF_6	a	c	a		
PrF_4	1.48	K_3PrF_7	K_2PrF_6	a	c	a		
PuF_4	1.48	a	K_2PuF_6	$K_7Pu_6F_{31}$	c	KPu_2F_9		
AmF_4	1.49	a	a	$K_7Am_6F_{31}$	c			
CmF_4	1.51	a	a	$K_7Cm_6F_{31}$	c			
BkF_4	1.51	a	a		c			
TbF_4	1.68	K_3TbF_7	††	a	$KTbF_5$			

TABLE 2.17 contd.

Tetrafluoride	Rubidium fluoride							Caesium fluoride			
	M^+/M^{4+} ratio	Stoichiometry						M^+/M^{4+} ratio	Stoichiometry		
		3:1	2:1	7:6	1:1	1:3	1:6		3:1	2:1	1:1
ThF_4	1.49	Rb_3ThF_7	Rb_2ThF_6	$Rb_7Th_6F_{31}$	—	$RbTh_3F_{13}$	$RbTh_6F_{25}$	1.69	Cs_3ThF_7	Cs_2ThF_6	$CsThF_5$
PaF_4	1.54	a		$Rb_7Pa_6F_{31}$		a	a	1.74	a	a	a
UF_4	1.59	Rb_3UF_7	Rb_2UF_6	$Rb_7U_6F_{31}$	$RbUF_5$	RbU_3F_{13}	RbU_6F_{25}	1.80	Cs_3UF_7	Cs_2UF_6	$CsUF_5$
NpF_4	1.61	a	Rb_2NpF_6	$Rb_7Np_6F_{31}$	a			1.82	Cs_3NpF_7		a
CeF_4	1.61	Rb_3CeF_7	Rb_2CeF_6	a	a			1.82	Cs_3CeF_7	Cs_2CeF_6	$CsCeF_5$
PrF_4	1.64	Rb_3PrF_7	Rb_2PrF_6	a	a			1.86	Cs_3PrF_7	Cs_2PrF_6	$CsPrF_5$
PuF_4	1.64	a	Rb_2PuF_6	$Rb_7Pu_6F_{31}$	a			1.86	a	a	a
AmF_4	1.66	a	Rb_2AmF_6		a			1.88	a	a	a
CmF_4	1.68	a	Rb_2CmF_6		a			1.90	a	a	a
BkF_4	1.68	a	a		a			1.90	a	a	a
TbF_4	1.87	Rb_3TbF_7	††		$RbTbF_5$			2.12	Cs_3TbF_7	††	$CsTbF_5$

a System not studied but complex formation is predicted.
b Metastable phase.
c It is likely that any reported 1:1 compounds are actually 7:6 complexes (page 59).
d Hydrogen reduction of K_2PaF_7 leads to the formation of $K_7Pa_6F_{31}$ not K_2PaF_6.[272]
†† Attempted preparation has given a mixture of 3:1 and 1:1 complexes.[437]

complex is isostructural[257,260] this is undoubtedly $LiThF_5$. The $7:6$ stoicheiometry does however persist throughout the tetravalent series of fluoro complexes where M^+/M^{4+} lies between 0.99 and 1.64 and such sodium, potassium, ammonium and rubidium salts $M_7^I M_6^{IV} F_{31}$ ($M^I = Na$, K, NH_4 and Rb) all possess rhombohedral symmetry, space group $R\bar{3}$. Thoma originally predicted that the upper limit of M^+/M^{4+} for the $7:6$ complexes was 1.68 but recent studies[455] have shown that $Rb_7Am_6F_{31}$ and $Rb_7Cm_6F_{31}$ cannot be prepared and these results indicate a limit of 1.64. It is highly probable that the plutonium complexes reported by various authors[261-263] as $NaPuF_5$, $KPuF_5$ and $RbUF_5$, and shown by Zachariasen[264] to be rhombohedral, are in fact $Na_7Pu_6F_{31}$, $K_7Pu_6F_{31}$ and $Rb_7Pu_6F_{31}$ respectively.

This unusual stoicheiometry was recently confirmed by a single crystal structure determination[265] on the isostructural zirconium (IV) salt $Na_7Zr_6F_{31}$. Similarly the complexes identified by x-ray power diffraction analysis[264] as $NaM^{IV}F_5$ and $KM^{IV}F_5$ ($M^{IV} = $ Th and U) possess rhombohedral symmetry and therefore are likely to have the composition $M_7^I M_6^{IV} F_{31}$. In the light of the present evidence the existence of genuine, stable $1:1$ complexes $M^I M^{IV} F_5$ ($M^I = $ Na or K; $M^{IV} = $ an actinide or lanthanide element) where the cation radius ratio is $\leqslant 1.51$ is questionable and in the following discussion such phases, although referred to as $1:1$ complexes in the literature, will be considered as the $7:6$ complexes. $KTbF_5(M^+/M^{4+} = 1.68)$ is known, however,[437] and there is some evidence in favour of the existence of $NaTbF_5$ but the purity of the latter is in doubt.

The situation with the larger ammonium and rubidium cations is quite different. Stable $7:6$ complexes of rhombohedral symmetry have been identified in each case[266-268] but in addition the $1:1$ uranium (IV) complexes NH_4UF_5 and $RbUF_5$ have been isolated. These compounds possess a different, unidentified structure; support for the existence of the ammonium complex, NH_4UF_5, is provided[267] by the reported thermal decomposition of $(NH_4)_4UF_8$ which proceeds,

$$(NH_4)_4UF_8 \xrightarrow[\text{air}]{130°} (NH_4)_2UF_6 \xrightarrow[\text{air}]{180°} (NH_4)_7U_6F_{31} \xrightarrow[\text{air}]{260°}$$

$$NH_4UF_5 \xrightarrow[\text{vacuum}]{290°} NH_4U_3F_{13} \xrightarrow[\text{vacuum}]{400°} UF_4$$

The plutonium (IV) complex $(NH_4)_4PuF_8$ decomposes in a similar fashion[267] and the results are shown diagrammatically in Figure 2.11. This investigation led to the suggestion that, on the basis of x-ray powder data, the previously reported[262] compounds $(NH_4)_2PuF_6$ and NH_4PuF_5 were actually $(NH_4)_4PuF_8$ and $(NH_4)_2PuF_6$ respectively. Thoma[255] has predicted that with rubidium, stable $7:6$ *and* $1:1$ complexes are capable

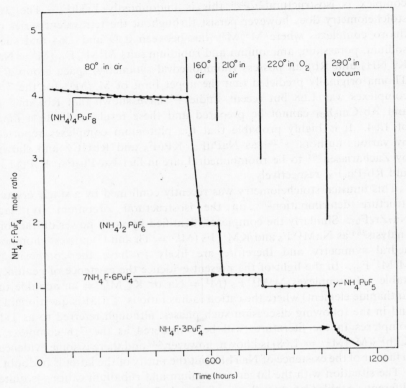

Figure 2.11 Thermal decomposition of compounds in the NH₄F–PuF₄ system.[267] (*After* R. Benz, R. M. Douglass, F. H. Kruse and R. A. Penneman, *Inorg. Chem.*, **2**, 799 (1963))

of existence where M^+/M^{4+} lies between 1.59 and 1.68 (Table 2.17) and since NH_4^+ and Rb^+ are very similar in size it is likely that both 7:6 and 1:1 ammonium salts will also be stable for the same elements (as discussed above the upper limit for the existence of the rubidium 7:6 complexes is more probably 1.64).

The various tetravalent fluoro complexes are most conveniently compared by considering the range of compounds formed with a given univalent cation and this procedure is used in the following discussion. Details of the numerous binary fluoride fused salt systems have been collected together in two recent compilations[269,270], which also include information on ternary fluoride systems and systems such as ThF_4–UF_4 and ZrF_4–UF_4 which will not be dealt with here. The x-ray powder data for many of the phases observed have also been published[260] in a single report. The few infrared data available are listed in Table B.3.

Lithium fluoride–metal (IV) fluoride complexes. Several of the known complexes are listed in Table 2.17. In addition to those shown for uranium (IV) and thorium (IV), the compounds Li_4UF_8 and $LiTh_2F_9$ have been identified during phase studies[256-258]. All the thorium (IV) and

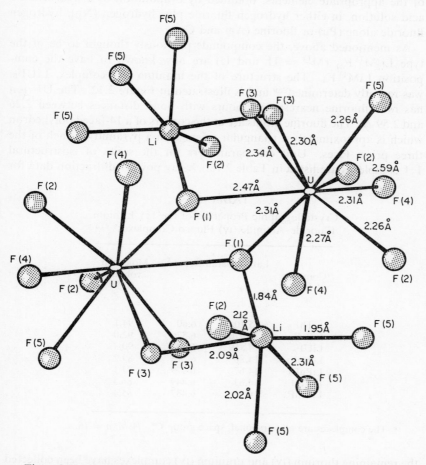

Figure 2.12 Two centrosymmetrically related asymmetric units of $LiUF_5$[259]. (In order to get maximum separation of the atoms, the unit was rotated 60° counter-clockwise around y and 6° clockwise around x. The pivot point was the symmetry centre ($\frac{1}{2}$, $\frac{1}{2}$, $\frac{1}{2}$) and the starting orientation was the conventional cartesian system with x and y in the plane of the illustration and z towards the viewer.) (*After* G. D. Brunton, *Acta Cryst.*, **21**, 814 (1964))

uranium (IV) complexes can be prepared[256-258] by heating together the required amounts of the component fluorides at high temperatures. The only known protactinium (IV) complex, $LiPaF_5$, is made[271,272] in a similar manner or by hydrogen reduction of $LiF–PaF_5$ mixtures at 400–500°. The neptunium (IV), plutonium (IV), americium (IV) and curium (IV) 1:1 complexes have been prepared[273] by heating a stoicheiometric mixture of the appropriate elements, obtained by evaporation of a hydrochloric acid solution, in either hydrogen fluoride plus hydrogen (Np), hydrogen fluoride alone (Pu) or fluorine (Am and Cm).

As mentioned above, the compounds previously thought to be of the type $Li_7M_6^{IV}F_{31}$ (M^{IV} = Th and U) are now known to have the composition $LiM^{IV}F_5$. The structure of the uranium (IV) complex, $LiUF_5$, was recently determined[259] and is illustrated in Figure 2.12. The U^{4+} ion has nine fluorine nearest neighbours with bond distances between 2.26 and 2.59 Å. The fluorine atoms are at the corners of a 14-faced polyhedron which is approximately a triangular prism with a pyramid on each of the three prism faces. Unit cell parameters for the series of isostructural 1:1 complexes are listed in Table 2.18. X-ray powder diffraction data for

TABLE 2.18

Crystallographic Properties of the 1:1 Lithium Fluoride–Actinide (IV) Fluoro Complexes[a] [273]

Complex	Lattice parameters (Å)		Molecular volume (Å³)
	a_0	c_0	
$LiThF_5$	15.10	6.60	94.1
$LiPaF_5$	14.970	6.576	92.0
$LiUF_5$	14.859	6.543	90.3
$LiNpF_5$	14.80	6.516	89.2
$LiPuF_5$	14.67	6.479	87.1
$LiAmF_5$	14.63	6.449	86.3
$LiCmF_5$	14.57	6.437	85.4

[a] The complexes are all tetragonal, space group $C_{4h}^6–I4_1/a$, n = 16.

the remaining thorium (IV) and uranium (IV) complexes have been collected together[260] in a recent publication. The octafluorouranate (IV) salt Li_4UF_8 is orthorhombic[274] with a_0 = 9.960, b_0 = 9.883, c_0 = 5.986 Å, space group *Pnma*; each uranium atom has eight fluorine neighbours at distances between 2.21 and 2.39 Å with a ninth at 3.30 Å.

Sodium fluoride–metal (IV) fluoride complexes. Some of the regularly occurring sodium complexes are listed in Table 2.17. The detailed phase diagrams for the NaF–ThF_4[257-277] and NaF–UF_4[256,276] systems have been reported and other phases identified in these studies include Na_4ThF_8, $Na_7Th_2F_{15}$, $Na_3Th_2F_{11}$, $Na_7U_2F_{15}$, $Na_5U_3F_{17}$ and the metastable complex $NaThF_5$. As mentioned earlier (p. 59) those complexes previously considered to be of the type $NaM^{IV}F_5$ (M^{IV} = Th, U and Pu) and found to be rhombohedral are actually of the type $Na_7M_6^{IV}F_{31}$ and this applies also to the praseodymium (IV) complex originally reported[278] as $NaPrF_5$. Others have reported[262,279] the preparation of $NaUF_5$ and $NaPuF_5$ from aqueous solution but provide no x-ray powder diffraction data for their products. It is likely, however, that these were either the 7:6 complexes or possibly the monohydrates $NaM^{IV}F_5 \cdot H_2O$ (M^{IV} = U and Pu) since $NaThF_5 \cdot H_2O$ and $NaUF_5 \cdot H_2O$ have been identified[280,281] as the products

TABLE 2.19

Crystallographic Properties of Certain Sodium Fluoride–Metal (IV)
Fluoro Complexes

Compound	Symmetry	Space group	Lattice parameters (Å)			Reference
			a_0	b_0	c_0	
Na_4ThF_8	Cubic	—	12.706	—	—	264
Na_4ThF_8	Cubic	—	11.04	—	—	275
Na_3ThF_7	Hexagonal	—	12.713	—	10.377	260
Na_3UF_7	Tetragonal	D_{4h}^{17}–$I4/mmm$	5.448	—	10.896	264
Na_3CeF_7	Tetragonal	D_{4h}^{17}–$I4/mmm$	5.40	—	10.80	282
Na_3PrF_7	Tetragonal	D_{4h}^{17}–$I4/mmm$	5.44	—	10.81	283
Na_3TbF_7	Tetragonal	D_{4h}^{17}–$I4/mmm$	5.39	—	10.74	437
β_2-Na_2ThF_6	Hexagonal	$C32$	5.977	—	3.827	264
δ-Na_2ThF_6	Hexagonal	—	6.14	—	7.36	260
α-Na_2UF_6	Cubic	O_h^5–$Fm3m$	5.565	—	—	264
β_2-Na_2UF_6	Hexagonal	$C32$	5.94	—	3.74	264
δ-Na_2UF_6	Hexagonal	—	6.11	—	7.25	260
γ-Na_2UF_6	Orthorhombic	D_{2h}^{25}–$Immm$	5.56	4.01	11.64	264
Na_2NpF_6	Hexagonal	—	6.074	—	7.167	284
$Na_2PuF_6{}^a$	Hexagonal	—	6.059	—	7.130	284
$Na_2PuF_6{}^b$	Hexagonal	$C32$	6.055	—	3.571	262
Na_2CeF_6	Hexagonal	—	5.93	—	3.69	282
Na_2PrF_6	Orthorhombic	—	5.54	3.41	11.57	278
$NaTh_2F_9$	Cubic	T_d^3–$I\bar{4}3m$	8.705	—	—	264

[a] Prepared by hydrofluorination of NaF/PuO_2 mixtures.
[b] Prepared from aqueous solution.

of the reactions between sodium fluoride and either thorium or uranium tetrafluoride in aqueous solution.

Zachariasen[264] first prepared many of the thorium (IV) and uranium (IV) complexes by heating together varying amounts of the component fluorides. He identified the various products by x-ray powder diffraction analysis, showing that Na_2ThF_6 and Na_2UF_6 exist in several different crystal forms. It has recently been suggested[256] that his phase, α-Na_2UF_6, has the composition $Na_5U_3F_{17}$. In addition, efforts to establish temperature stability ranges for the phases designated γ-Na_2UF_6 and β-Na_2UF_6 by Zachariasen have failed although phases with the x-ray properties described by Zachariasen for γ-Na_2UF_6 and β_2-Na_2UF_6 were observed in a few quenched samples. A hexagonal phase, designated β_3-Na_2UF_6*, was found to be the only stable form of this compound between 273° and the incongruent melting point 650°. The crystallographic properties of the various forms of Na_2UF_6 and Na_2ThF_6 are listed in Table 2.19 together with similar information concerning certain other sodium fluoride-metal (IV) fluoro complexes.

The known tetravalent protactinium, neptunium, plutonium, americium and curium complexes are conveniently prepared by the reactions listed for the lithium fluoro complexes of these elements (p. 65). Sodium hexafluoroplutonate (IV) has also been prepared[262] from aqueous solution. The cerium (IV) praseodymium (IV) and terbium (IV) salts[278,282,283,437] have been made by heating together stoicheiometric amounts of sodium chloride and respectively CeO_2, Pr_6O_{11} or Tb_4O_7 in fluorine at about 400°. The visible spectra[226] and magnetic properties[283,437] of these and other praseodymium (IV) and terbium (IV) complexes have been recorded and the results confirm the $4f^1$ configuration of the Pr^{4+} ion and the $4f^7$ configuration for terbium (IV) in these complexes.

The rhombohedral $Na_7M_6^{IV}F_{31}$ complexes are isostructural with $Na_7Zr_6F_{31}$. The basic structural unit of the latter[265] is an approximately square antiprism formed by eight fluorine atoms around a zirconium atom as illustrated in Figure 2.13a. Six such antiprisms share corners to form an octahedral array (Figure 2.13b, right) which encloses a cavity containing one additional fluorine atom. The twelve fluorine atoms at the shared corners form a tetrakisdecahedron having six square and eight triangular faces, an approximately regular cuboctahedron, as shown in Figure 2.13b, left. Lattice parameters for the isostructural sodium, potassium, rubidium and ammonium 7:6 complexes are given in Tables 2.20–2.23.

---

* This phase was later shown to be isostructural with δ-Na_2ThF_6 and is now referred to as δ-Na_2UF_6 rather than β_3-Na_2UF_6 (see reference 302).

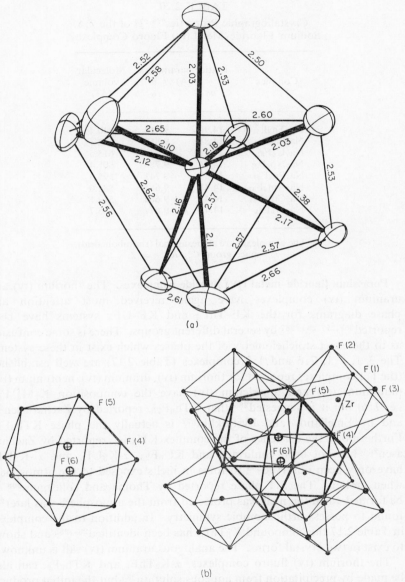

Figure 2.13 The structure of Na₇Zr₆F₃₁. (*a*) The square antiprism arrangement in Na₇Zr₆F₃₁. (*b*) *Right*; Six antiprisms sharing corners to form an octahedral array which encloses a cavity containing the 'extra' fluorine (F6). (*b*) *Left*; The approximately regular cuboctahedron cavity showing the two sites over which (F6) is statistically distributed. (*After* J. H. Burns, R. D. Ellison and H. A. Levy, *Acta Cryst.*, **B24**, 230 (1968))

TABLE 2.20

Crystallographic Properties[226,273] of the 7:6
Sodium Fluoride–Metal (IV) Fluoro Complexes[a]

Complex	Lattice parameters (Å)		Molecular volume (Å³)
	a_0	c_0	
$Na_7Th_6F_{31}$	14.96	9.912	640.4
$Na_7Pa_6F_{31}$	14.81	9.850	623.7
$Na_7U_6F_{31}$	14.72	9.84	615.5
$Na_7Np_6F_{31}$	14.64	9.785	605.4
$Na_7Pu_6F_{31}$	14.55	9.741	595.3
$Na_7Am_6F_{31}$	14.48	9.665	585.0
$Na_7Cm_6F_{31}$	14.41	9.661	579.1
$Na_7Pr_6F_{31}$	14.48	9.677	585.7

[a] The complexes are all hexagonal (rhombohedral),
space group $C_{3i}^2 - R\bar{3}$, $n = 2$.

Potassium fluoride–metal (IV) fluoride complexes. The thorium (IV) and uranium (IV) complexes have again received most attention and phase diagrams for the $KF-ThF_4$ and $KF-UF_4$ systems have been reported[268,277,285-287] by several different groups. There is some confusion as to the exact stoicheiometry of the phases which exist in these systems. The 3:1, 2:1, 7:6 and 1:2 complexes (Table 2.17) are well established (the 1:1 complexes reported for thorium (IV), uranium (IV), neptunium (IV), plutonium (IV) and americium (IV) have the composition $K_7M_6^{IV}F_{31}$, i.e. 7:6) but it is suggested[285] that KTh_6F_{25}, reported by Zachariasen[264] and by Emeljanov and Evstjukhin[277] is actually the phase KTh_3F_{13}. Furthermore, the existence of the complex KU_3F_{13}, reported by Zachariasen[264] has not been confirmed and KU_6F_{25}, $\alpha-K_3UF_7$ and $\alpha'-K_3UF_7$ have only been observed[268] to form in melts exposed to the atmosphere when molten. The 3:1 phase reported by Thoma and colleagues[268] to be formed when the melt was protected from the atmosphere was later[260] found to possess orthorhombic symmetry. In addition to the complexes in Table 2.17 the compound K_5ThF_9 has been identified[264,285] and shown to exist in two crystal forms. The analogous uranium (IV) salt is unknown.

The thorium (IV) fluoro complexes, $\alpha-K_2ThF_6$ and KTh_2F_9, can also be made by precipitation from aqueous solution[285] but the initial products contain water which is only completely removed at about 300° in an inert atmosphere. The compounds KNp_2F_9, K_2PuF_6, $K_7Pu_6F_{31}$ (originally reported as $KPuF_5$) and KPu_2F_9 have also been prepared[262,288,289]

TABLE 2.21
Crystallographic Properties for Some Potassium Fluoride–Metal (IV) Fluoro Complexes

Complex	Symmetry	Space group	Lattice parameters (Å)			Reference
			a_0	b_0	c_0	
β-K_5ThF_9	Orthorhombic	D_{2h}^{17}–$Ccmm$	12.87	7.90	10.83	264
α-K_3UF_7	Cubic	O_h^5–$Fm3m$	9.21	—	—	264
α_1-K_3UF_7	Tetragonal	D_{4h}^{14}–$I4/amd$	9.20	—	18.40	264
K_3UF_7	Orthorhombic	$Pnmm$ or $Pnm2$	6.59	8.30	7.20	260
K_3TbF_7	Cubic	O_h^5–$Fm3m$	9.085	—	—	437
α-K_2ThF_6	Cubic	O_h^5–$Fm3m$	5.994	—	—	264
β_1-K_2ThF_6	Hexagonal	$C\bar{6}2m$	6.565	—	3.815	264
α-K_2UF_6	Cubic	O_h^5–$Fm3m$	5.934	—	—	264
β_1-K_2UF_6	Hexagonal	$C\bar{6}2m$	6.54	—	3.76	264
β_2-K_2UF_6	Hexagonal	$C32$	6.53	—	4.04	264
K_2CeF_6	Hexagonal	$C\bar{6}2m$	6.52	—	3.71	282
$K_7Th_6F_{31}$	Rhombohedral	C_{3i}^2–$R\bar{3}$	9.510 ($\alpha = 107° \, 17'$)	—	—	264
$K_7Pa_6F_{31}$	Rhombohedral	C_{3i}^2–$R\bar{3}$	9.44 ($\alpha = 107° \, 9'$)	—	—	272
$K_7U_6F_{31}$	Rhombohedral	C_{3i}^2–$R\bar{3}$	9.36 ($\alpha = 107° \, 10'$)	—	—	455
$K_7Np_6F_{31}$	Rhombohedral	C_{3i}^2–$R\bar{3}$	9.31 ($\alpha = 107° \, 12'$)	—	—	284
$K_7Pu_6F_{31}$	Rhombohedral	C_{3i}^2–$R\bar{3}$	9.27 ($\alpha = 107° \, 2'$)	—	—	292
$K_7Am_6F_{31}$	Rhombohedral	C_{3i}^2–$R\bar{3}$	9.27 ($\alpha = 107° \, 35'$)	—	—	290
$K_7Cm_6F_{31}$	Rhombohedral	C_{3i}^2–$R\bar{3}$	9.25 ($\alpha = 107° \, 15'$)	—	—	455
KTh_2F_9	Orthorhombic	D_{2h}^{16}–$Pnam$	8.85	7.16	11.62	264
KU_2F_9	Orthorhombic	D_{2h}^{16}–$Pnam$	8.68	7.02	11.44	264
KNp_2F_9	Orthorhombic	D_{2h}^{16}–$Pnam$	8.63	7.01	11.43	292
KPu_2F_9	Orthorhombic	D_{2h}^{16}–$Pnam$	8.56	6.95	11.33	292
KU_3F_{13}	Orthorhombic	D_{2h}^5–$Pmcm$	8.03	7.25	8.53	264
KTh_6F_{25}[a]	Hexagonal	D_{6h}^4–$C6_3/mmc$	8.32	—	16.78	264
KU_6F_{25}	Hexagonal	D_{6h}^4–$C6_3/mmc$	8.18	—	16.42	264

[a] Asker and co-workers[285], believe that the phase possessing the lattice parameters quoted has the composition KTh_3F_{13}.

6

from aqueous solution but the protactinium (IV), cerium (IV), praseodymium (IV), terbium (IV) and americium (IV) complexes have only been prepared by high temperature reactions involving, for example, hydrogen reduction of a higher valence fluoro complex[271,272] (Pa (IV)), fluorination of chloride–oxide mixtures[282,283,437] (Ce (IV), Pr (IV) and Tb (IV)) or fluorination of a carbonato complex[290] (Am (IV)). An unidentified precipitate believed to be a fluoro complex[291], is obtained when americium (IV) hydroxide is treated with a 12 M potassium fluoride solution.

The 7:6 complexes $K_7M_6^{IV}F_{31}$ (M^{IV} = Th, Pa, U, Np, Pu and Am) are isostructural with $Na_7Zr_6F_{31}$ (p. 68). Available lattice parameters for the various potassium salts, many of which are polymorphic, are listed in Table 2.21.

Rubidium fluoride–metal (IV) fluoride complexes. The $RbF–ThF_4$ and $RbF–UF_4$ phase diagrams have been reported[268,287,293,294]; some of the regularly occurring phases observed are listed in Table 2.17, others include Rb_5ThF_9, $Rb_7Th_2F_{15}$, and $Rb_2U_3F_{14}$. It is interesting to note that in contrast to the analogous sodium and potassium fluoride–uranium tetrafluoride systems the existence of both 7:6 and 1:1 complexes is observed[268] in the $RbF–UF_4$ system (Figure 2.14). Like the sodium, potassium and ammonium complexes $Rb_7U_6F_{31}$ is rhombohedral but the x-ray powder

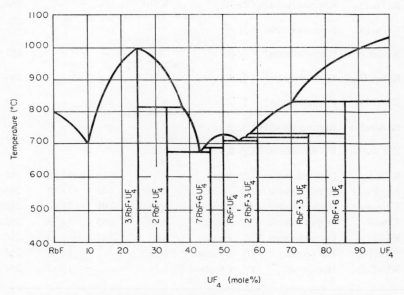

Figure 2.14   The system $RbF–UF_4$[268]. (*After* R. E. Thoma, H. Insley, B. S. Landau, H. A. Friedman and W. R. Grimes, *J. Am. Ceram. Soc.*, **41**, 538 (1958))

data for $RbUF_5$ have not been interpreted. The only protactinium (IV) complex known[143,271,272] is $Rb_7Pa_6F_{31}$ which is made by hydrogen reduction of the pentavalent complex $RbPaF_6$. The cerium (IV) and praseodymium (IV) complexes $Rb_3M^{IV}F_7$ and $Rb_2M^{IV}F_6$ (M^{IV} = Ce and Pr) have only been prepared[282,283] by fluorination of rubidium chloride–lanthanide oxide mixtures at about 400° whereas the uranium (IV)[295], plutonium

TABLE 2.22

Crystallographic Properties of Some Rubidium
Fluoride–Metal (IV) Fluoro Complexes

Compound	Symmetry	Space group	Lattice parameters (Å)			Reference
			a_0	b_0	c_0	
Rb_3ThF_7	Cubic	O_h^5–$Fm3m$	9.62	—	—	287
Rb_3UF_7	Cubic	O_h^5–$Fm3m$	9.567	—	—	268
Rb_3CeF_7	Cubic	O_h^5–$Fm3m$	9.52	—	—	282
Rb_3PrF_7	Cubic	O_h^5–$Fm3m$	9.54	—	—	283
Rb_3TbF_7	Cubic	O_h^5–$Fm3m$	9.49	—	—	437
Rb_2ThF_6	Hexagonal	C–$62m$	6.85	—	3.83	293
Rb_2UF_6	Orthorhombic	D_{2h}^{17}–$Cmcm$	6.998	12.098	7.669	456
Rb_2NpF_6	Orthorhombic	D_{2h}^{17}–$Cmcm$	6.986	12.068	7.628	456
Rb_2PuF_6	Orthorhombic	D_{2h}^{17}–$Cmcm$	6.971	12.033	7.602	456
Rb_2AmF_6	Orthorhombic	D_{2h}^{17}–$Cmcm$	6.962	12.001	7.579	456
Rb_2CmF_6	Orthorhombic	D_{2h}^{17}–$Cmcm$	6.931	11.996	7.567	456
$Rb_2CeF_6{}^a$	Hexagonal	—	6.90	—	7.49	282
$Rb_2PrF_6{}^a$	Hexagonal	—	6.80	—	7.50	283
$Rb_7Th_6F_{31}$	Rhombohedral	C_{3i}^2–$R\bar{3}$	9.58 ($\alpha = 106.9°$)	—	—	260
$Rb_7Pa_6F_{31}$	Rhombohedral	C_{3i}^2–$R\bar{3}$	9.64 ($\alpha = 107°$)	—	—	271
$Rb_7U_6F_{31}$	Rhombohedral	C_{3i}^2–$R\bar{3}$	9.595 ($\alpha = 107.7°$)	—	—	260
$Rb_7Np_6F_{31}$	Rhombohedral	C_{3i}^2–$R\bar{3}$	9.47 ($\alpha = 107.2°$)	—	—	284
$Rb_7Pu_6F_{31}$	Rhombohedral	C_{3i}^2–$R\bar{3}$	9.46 ($\alpha = 106.9°$)	—	—	284
$RbTh_6F_{25}$	Hexagonal	—	8.330	—	25.40	260
RbU_6F_{25}	Hexagonal	D_{6h}^4–$C6_3/mmc$	8.195	—	16.37	268

a The hexagonal parameters reported probably refer to a pseudo-cell ($\equiv \delta$-Na_2ThF_6) and the complexes are likely to be orthorhombic like their heavier actinide (IV) analogues.

(IV)[261,262] and americium (IV)[295] compounds $Rb_2M^{IV}F_6$ ($M^{IV} = U$, Pu and Am) and $Rb_7Pu_6F_{31}$ can be crystallized from aqueous solution. With plutonium (IV) the initial precipitate is the green compound $Rb_7Pu_6F_{31}$ (reported[261,262] as $RbPuF_5$) which is slowly converted[262] to the pink hexafluoro complex, Rb_2PuF_6, on standing in contact with the supernatant solution. The rhombohedral neptunium (IV) complex $Rb_7Np_6F_{31}$ has been made[284] by heating together rubidium fluoride and neptunium dioxide in hydrogen fluoride above 400°. The analogous plutonium (IV) complex can be prepared in a similar manner provided oxygen is present to prevent reduction to plutonium (III). The curium complex Rb_2CmF_6 is obtained[456] when a 2:1 mixture (Rb:Cm) is evaporated to dryness from aqueous hydrofluoric acid and then fluorinated at 350°.

The crystallographic properties of the rubidium salts are given in Table 2.22. The actinide hexafluoro complexes, $Rb_2M^{IV}F_6$ ($M^{IV} = U$ — Cm) possess pseudohexagonal symmetry[295] and the fact that Rb_2CeF_6 and Rb_2PrF_6 have been assigned hexagonal unit cells[282,283] ('similar' to δ-Na_2ThF_6) suggests that these complexes are really orthorhombic like their tetravalent actinide analogues. Since Rb_2ThF_6 is reported to be hexagonal[293] and isostructural with β_1-K_2ThF_6 and not δ-Na_2ThF_6 it would be interesting to have information on the protactinium (IV) analogue to see whether this follows Rb_2ThF_6 or Rb_2UF_6.

Ammonium fluoride–metal (IV) fluoride complexes. Relatively few ammonium fluoro complexes have been reported and only the uranium (IV) and plutonium (IV) systems have been studied in detail. Octafluoro complexes of the type $(NH_4)_4M^{IV}F_8$ ($M^{IV} = $ Pa, U, Np, Pu and Am) have been characterized. The protactinium (IV) complex is formed[271,296] merely by grinding together protactinium tetrafluoride and an excess of ammonium fluoride at room temperature. $(NH_4)_4UF_8$ and $(NH_4)_4PuF_8$ are conveniently prepared by sealed tube reactions[267] between the component fluorides at 80–130° or by precipitation[266] from aqueous solution. The red americium (IV) complex has been made[291] by reacting americium (IV) hydroxide with 13M NH_4F solution. The conversion of $(NH_4)_4UF_8$ to complexes containing less ammonium fluoride has been mentioned (p. 63) and the thermal decomposition[267] of $(NH_4)_4PuF_8$ is illustrated in Figure 2.11. X-ray powder diffraction data are available for these isostructural octafluoro complexes[267]; the results have not all been interpreted but a single crystal structure analysis is currently being carried out on $(NH_4)_4UF_8$. On the basis of x-ray powder results Benz and colleagues[267] suggest that the pink complex reported by Alenchikova and co-authors[262] as $(NH_4)_2PuF_6$ would appear to be $(NH_4)_4PuF_8$ while their green 1:1 complex 'NH_4PuF_5' is $(NH_4)_2PuF_6$.

No 3:1 complexes are known (cf. the RbF–MF$_4$ systems, p. 72) but $(NH_4)_2UF_6$ and $(NH_4)_2PuF_6$ have been prepared[267] by controlled thermal decomposition of the respective octafluoro complexes, by heating together stoicheiometric quantities of the component fluorides at 80–130° or, see references 262 and 266, by precipitation from aqueous solution. The rhombohedral complexes $(NH_4)_7M_6^{IV}F_{31}$ (M^{IV} = U and Pu) which are isostructural with $Na_7Zr_6F_{31}$ (p. 68) can be prepared[267] by similar reactions. Ammonium hexafluorouranate (IV), $(NH_4)_2UF_6$ exists in four crystal modifications none of which is isostructural with the rubidium analogue (p. 73). Unit-cell parameters are available only for the γ-phase (Table 2.23).

TABLE 2.23

Crystallographic Properties of Some Ammonium
Fluoride–Actinide (IV) Fluoro Complexes

Compound	Symmetry	Space group	Lattice parameters (Å)			Reference
			a_0	b_0	c_0	
γ-$(NH_4)_2UF_6$	Orthorhombic	$Pmc2_1$ or $Pmcm$	4.05	7.03	11.76	266
$(NH_4)_7U_6F_{31}$	Rhombohedral	C_{3i}^2–$R\bar{3}$	9.55 (α = 107.4°)	—	—	267
$(NH_4)_7Pu_6F_{31}$	Rhombohedral	C_{3i}^2–$R\bar{3}$	9.42 (α = 107.4°)	—	—	267
$(NH_4)_4PaF_8$	Monoclinic	—	13.18 (β = 117.16°)	6.71	13.22	272

Although the existence of stable 1:1 fluoro complexes with the sodium and potassium cations, $M^IM^{IV}F_5$ (M^I = Na and K; M^{IV} = a tetravalent lanthanide or actinide element), is doubtful, ammonium pentafluorouranate (IV) and plutonate (IV) have been characterized[267] and analytical results suggest the existence[223] of NH_4CeF_5. The uranium (IV) complex was not identified during a study of the NH_4F–UF_4–H_2O system[266] but both it and NH_4PuF_5 have been prepared[267] in three different crystalline forms by the controlled thermal decomposition of the respective ammonium octa- or hexafluoro complexes. One form, the α-phase, must be regarded as an ammonium fluoride deficient form of the 7:6 complex from which it cannot be distinguished by x-ray powder diffraction analysis, but the β- and γ-phases each give rise to unique x-ray powder patterns.

The ammonium pentafluoro complexes, NH_4ThF_5, NH_4UF_5 and NH_4PuF_5, have also been reported as intermediates in the conversion of actinide dioxides to the tetrafluorides by heating with ammonium fluoride or ammonium bifluoride (p. 51). An ammonium fluoroneptunate (IV) precipitated from aqueous ammonium fluoride by the addition of hydrofluoric acid[297] was identified[298] crystallographically as NH_4NpF_5 since it was isostructural with 'RbPuF₅'. However, since the latter was said to be rhombohedral it was probably $Rb_7Pu_6F_{31}$ and the neptunium complex was therefore more probably $(NH_4)_7Np_6F_{31}$.

Thermal decomposition of NH_4UF_5 and NH_4PuF_5 above 290° *in vacuo*[267] leads to the formation of $NH_4U_3F_{13}$ and $NH_4Pu_3F_{13}$ respectively; these in turn decompose at higher temperatures to yield the respective tetrafluorides. Similarly $N_2H_5UF_5$ and $(NH_3OH)UF_5$ decompose[449] to yield uranium tetrafluoride above 400°. The cerium (IV) complex $NH_4Ce_2F_9$, which results[223] from the interaction between cerium dioxide and ammonium fluoride at 200°, decomposes not to the tetrafluoride but to the trifluoride at 400° in a vacuum. Apart from $NH_4Ce_2F_9$ and NH_4CeF_5, tetravalent lanthanide fluoro complexes with the ammonium cation are unknown.

Caesium fluoride–metal (IV) fluoride complexes. Phase diagrams for the $CsF–ThF_4$ and $CsF–UF_4$ systems have only been reported[294,299] recently. In addition to those complexes listed in Table 2.17, phases having the composition $Cs_2M_3^{IV}F_{14}$, $CsTh_2F_9$, $CsTh_3F_{13}$ and $CsM_6^{IV}F_{25}$ ($M^{IV} = Th$

TABLE 2.24

Crystallographic Properties for Some Caesium
Fluoride–Metal (IV) Fluoro Complexes

Compound	Symmetry	Space group	Lattice parameters (Å)			Reference
			a_0	b_0	c_0	
Cs_3ThF_7	Cubic	—	10.04	—	—	260
Cs_3ThF_7	Cubic	—	9.659	—	—	299
Cs_3CeF_7	Cubic	—	9.93	—	—	282
Cs_3PrF_7	Cubic	—	9.92	—	—	283
Cs_3TbF_7	Cubic	—	9.80	—	—	300, 437
Cs_2CeF_6	Hexagonal	—	7.20	—	7.60	282
Cs_2PrF_6	Hexagonal	—	7.14	—	4.09	283
$Cs_2U_3F_{14}$	Monoclinic	$P2_1m$ or $P2_1$	8.39	8.46	20.88 ($\beta = 119.9°$)	260
$CsTh_6F_{25}$	Hexagonal	$D_{6h}^4–C6_3/mmc$	8.31	—	16.91	260
CsU_6F_{25}	Hexagonal	$D_{6h}^4–C6_3/mmc$	8.19	—	16.63	260

and U) have been characterized. Caesium heptafluorouranate (IV), Cs_3UF_7, has been crystallized from aqueous solution and has also been observed during the high temperature phase studies; crystals obtained in the former manner are not isostructural with high temperature preparations of Cs_3ThF_7[260]. Analogous protactinium (IV) complexes are at present unknown but will undoubtedly be capable of existence as will be complexes of plutonium (IV), americium (IV) and curium (IV). The cerium (IV), praseodymium (IV) and terbium (IV) complexes (Table 2.17) have all been prepared[282,283,300] by fluorination of caesium chloride–lanthanide oxide mixtures at about 400°. The addition of caesium fluoride to aqueous acid solutions of plutonium (IV) results only in the formation[262] of the hydrated complex $CsPu_2F_9 \cdot 3H_2O$.

TABLE 2.25

Crystallographic Properties[a] for Tetravalent
Fluoro Complexes of the Type $M^{II}M^{IV}F_6$[303]

Compound	Lattice parameters (Å)[b]	
	a_0	c_0
$CaThF_6$	6.994	7.171
$CaUF_6$	6.928	7.127
$CaNpF_6$	6.918	7.100
$CaPuF_6$	6.918	7.097
$SrCeF_6$	7.065	7.242
$SrThF_6$	7.150	7.313
$SrUF_6$	7.124	7.271
$SrNpF_6$	7.093	7.242
$SrPuF_6$	7.060	7.236
$BaThF_6$	7.419	7.516
$BaUF_6$	7.403	7.482
$BaNpF_6$	7.374	7.450
$PbCeF_6$	7.214	7.287
$PbThF_6$	7.280	7.404
$PbUF_6$	7.245	7.355
$PbNpF_6$	7.212	7.360
$CdThF_6$	6.963	7.109
$EuThF_6$	7.124	7.360

[a] The complexes all possess the LaF_3-type structure.
[b] Data for the strontium complexes refer to crystals prepared from aqueous solution. Slightly different parameters are reported for high temperature preparations.

Lattice parameters for certain of the caesium fluoro complexes are shown in Table 2.24; powder data have been reported, but not interpreted, for the remaining thorium (IV) and uranium (IV) compounds. It is interesting to note that 7:6 complexes are not known with the caesium cation and in fact Thoma[255] predicts that the M^+/M^{4+} cation radius ratio is too great for such complexes to be capable of existence.

Miscellaneous tetravalent fluoro complexes. Compounds of the type $M^{II}M^{IV}F_6$ (M^{II} = Pb, Ca, Sr, Ba, Cd and Eu; M^{IV} = Th, U, Np, Pu and Ce) have been prepared[105,303] by melting together the component fluorides or by precipitation from aqueous solution. The magnesium complex $MgTh_2F_{10}$ is also known[301]. Some crystallographic properties are given in Table 2.25.

Tetravalent Oxydifluorides

The only known tetravalent oxydifluoride is the thorium compound $ThOF_2$. This can be prepared by heating the hydrated tetrafluoride above 800° in air[304] or better by heating together stoicheiometric amounts of thoria and thorium tetrafluoride[305] at 900°. It was reported by Zachariasen[105] to possess the lanthanum trifluoride-type structure but later studies by D'Eye[305] failed to reproduce this phase. The latter preparations possessed orthorhombic symmetry with $a_0 = 14.07$, $b_0 = 4.041$, and $c_0 = 7.25$ Å, the cell being related to that reported by Zachariasen. D'Eye suggests that the deviation from hexagonal symmetry is due to the ordering of the oxygen and fluorine atoms. Attempts to prepare the uranium (IV) analogue by reacting uranium dioxide with uranium tetrafluoride have been unsuccessful and the reaction between antimony trioxide and uranium tetrafluoride has yielded[389] only a mixture of uranium dioxide and the tetrafluoride. Analogous reactions involving protactinium tetrafluoride have not yet been reported.

The plutonium dioxide–plutonium tetrafluoride reaction has been discussed earlier (p. 59).

TRIVALENT

Stable trifluorides are known for scandium, yttrium and all the lanthanide elements (lanthanum to lutetium inclusive). Trifluorides of the actinide elements uranium to curium inclusive become progressively easier to prepare with increasing atomic number and elements beyond curium should readily form trifluorides. The fact that such compounds are presently unknown merely reflects the difficulty of preparing and handling the appropriate elements. Actinium trifluoride has been reported but thorium and protactinium trifluoride are unknown and, should they be capable of existence, powerful reducing conditions will be necessary for their preparation. A few trivalent lanthanide fluoro complexes of the types $M^IM^{III}F_4$ and $M^I_3M^{III}F_6$ (M^I = variously Li, Na, K, Rb, Cs and NH_4; M^{III} = lanthanide) are known but analogous actinide complexes have been reported only for trivalent uranium, plutonium and americium.

Oxyfluorides of the type MOF are easy to prepare for the elements scandium, yttrium and the majority of the lanthanides, but in the $5f$ series of elements only AcOF and PuOF are presently known.

Trifluorides

Scandium, yttrium and the lanthanide trifluorides can be prepared by several relatively simple methods and fluorination of the metal or carbide, reported by Moissan[306] at the end of the last century, now finds little application. The three most widely used methods, all of which are similar in nature, are:

(1) direct hydrofluorination of the oxide[307-309],

$$M_2O_3 + 6HF \xrightarrow{700°} 2MF_3 + 3H_2O$$

(2) reaction of the oxide with ammonium bifluoride[307,308,310-314],

$$M_2O_3 + 6NH_4HF_2 \xrightarrow{300°} 2MF_3 + 6NH_4F + 3H_2O$$

(3) dehydration of the trifluoride hydrate obtained by precipitation from aqueous solution[307-309,315-318],

$$MF_3 \cdot xH_2O \ (x = 0.5 \text{ to } 1) \xrightarrow[\text{600° in HF gas}]{\text{300° in vacuum or}} MF_3 + xH_2O$$

Spedding and Daane[308], who first reported the ammonium bifluoride method, have assessed the above reactions for the preparation of high-grade lanthanide trifluorides suitable for conversion to metals of low oxygen content. They find that direct hydrofluorination of the oxides yields superior-grade trifluorides and that the most reactive oxides are obtained by decomposition of the trivalent lanthanide oxalates at 700–800°. Although a static-bed apparatus is quite satisfactory[307] for hydrofluorination of small amounts of oxide (0.5 to 1 kg batches), larger amounts (25 kg batches) are more efficiently converted by a rotary-batch method. It has also been suggested[307] that a fluidized-bed method may be useful for preparative work on this scale.

Although the reaction between excess ammonium bifluoride and the lanthanide oxides is highly suitable[307,308] for the preparation of small amounts (up to 1 kg) of the trifluorides, complete removal of the excess ammonium bifluoride is troublesome when larger amounts are employed.

By dissolving lanthanide nitrates in hydrofluoric acid in the presence of ammonium carbonate, fluoro complexes of the type $(NH_4)_3M^{III}F_6$ are said to be formed. Thermal decomposition of the products[319,320] at 300–400° leads to the formation of the anhydrous trifluorides.

The main disadvantages associated with the dehydration of the lanthanide trifluoride hydrates appear to be the difficulty of obtaining hydrate precipitates which filter easily[307,308] and the slightly higher oxygen content of the resulting anhydrous trifluorides. However, this method has been frequently used and the trifluoride hydrates, usually obtained by the addition of hydrofluoric acid to hydrochloric acid solutions of the trivalent lanthanides, can be dehydrated by vacuum drying above 1000°, by heating them in anhydrous hydrogen fluoride at about 600° or by heating them in helium at 400°.

The reaction between yttrium trichloride or sesquioxide and fluorine is reported[321] to yield inferior quality trifluoride. Chlorine trifluoride reacts[322] only to a slight extent with the lanthanide oxides under anhydrous conditions, even at 800°, but in the presence of moisture the trifluorides of lanthanum to samarium inclusive can be obtained from this reaction. Even under these conditions, however, the heavier lanthanide oxides $(Eu_2O_3–Er_2O_3)$ are incompletely converted to the trifluorides and those of thulium to lutetium do not react at all.

It has recently been shown[204] that scandium, yttrium, the lanthanide elements lanthanum to samarium inclusive and gadolinium react with anhydrous hydrogen fluoride at 225° in sealed containers to yield their respective trifluorides. Under similar conditions thorium and uranium metal react to yield tetrafluorides but it is likely that in the presence of the hydrogen produced during the reaction metals of the higher actinide elements will be converted to their trifluorides.

The trifluorides of actinium, americium and curium are not readily oxidized and can be prepared in a similar manner to the lanthanide trifluorides; for example, by hydrofluorination[323–325] of the hydrated oxide, sesquioxide or dioxide at high temperatures or[222,326–329] by dehydration of the solid precipitated from aqueous hydrofluoric acid. The oxide–ammonium bifluoride reaction has only been employed[429] for the preparation of americium and curium trifluoride but will doubtless be useful for the preparation of higher actinide trifluorides. Neptunium and plutonium trifluoride require only mild reducing conditions for their formation, for example[213,330–332] the reaction of the appropriate dioxide or, in the case of plutonium, $PuF_3 \cdot 0.75H_2O$, with a mixture of hydrogen and hydrogen fluoride at about 500°. Plutonium trifluoride has also been made by heating other plutonium compounds[330], such as the trivalent oxalate, in hydrogen–hydrogen fluoride mixtures, by reacting plutonium (III) or (IV) oxalate[333] with Freon-12 at 400–450° and by vacuum thermal decomposition of $PuF_4 \cdot 2.5H_2O$. The last reaction probably proceeds by way of partial hydrolysis to the dioxide which then reacts with the tetrafluoride as mentioned earlier (p. 51),

$$3PuF_4 + PuO_2 \rightleftharpoons 4PuF_3 + O_2$$

The more easily oxidized uranium trifluoride is only obtained under strongly reducing conditions. Thus, uranium tetrafluoride is reduced by hydrogen[244] at 1000°, by aluminium[245] at 900° and by finely divided uranium metal[246,334] at 1050°. During reduction by aluminium the monofluoride AlF, which is capable of existence only at high temperature, sublimes out of the reaction zone,

$$UF_4 + Al \rightarrow UF_3 + AlF \uparrow$$

Magnesium can also be employed[247] to reduce uranium tetrafluoride to the trifluoride. The reaction proceeds smoothly at 560° but at 600° reduction to uranium metal occurs. Temperatures in excess of those mentioned above are not recommended for the preparation of uranium trifluoride since it is thermally unstable and disproportionates[347] to a slight extent (0.1 %/h) even at 800°,

$$4UF_3 \rightleftharpoons 3UF_4 + U$$

Crystal structures. Lanthanum trifluoride has been found to exist in only one crystal modification, possessing hexagonal (trigonal) symmetry, space group $D_{3d}^4 - P\bar{3}cl$. Following the work of Oftedal[335], who stated on the basis of faint reflections on single crystal photographs of the mineral tysonite, $(La, Ce)F_3$, that the cell was hexamolecular, Schylter[336] suggested that the unit cell was actually bimolecular with $a^s = a^0/\sqrt{3}$ and Zachariasen[337] indexed the powder patterns of lanthanum trifluoride and the isostructural actinide trifluorides on the basis of the smaller unit cell. The results of electron spin resonance studies[338] involving neodymium (III) in lanthanum trifluoride crystals and the optical absorption spectrum[339,340] of praseodymium (III) in lanthanum trifluoride have been reported to be in accord with the C_{2v} point symmetry required for the metal atom in a hexamolecular cell. On the other hand the spectrum[339] of neodymium (III) in lanthanum trifluoride and the polarization spectrum[341] of erbium (III) in the same matrix are said to be incompatible with C_{2v} symmetry and the infrared and Raman spectra of lanthanum trifluoride are reported[342] to indicate that the cell is bimolecular. However, Zalkin and Templeton[309]

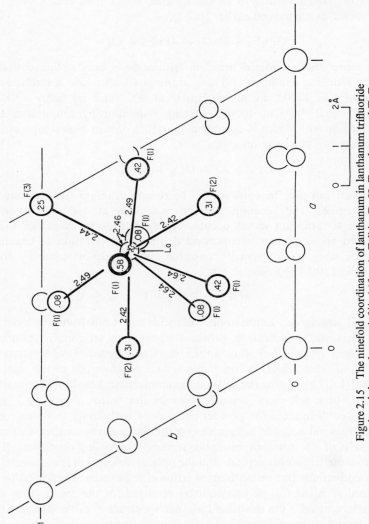

Figure 2.15   The ninefold coordination of lanthanum in lanthanum trifluoride as viewed down the *c* axis.[344] (*After* A. Zalkin, D. H. Templeton and T. E. Hopkins, *Inorg. Chem.*, **5**, 1466 (1966))

report that Weissenberg photographs of single crystals of synthetic cerium trifluoride indicate that the true cell is hexamolecular and this is confirmed by the recent structural studies[343,344] on lanthanum trifluoride. In view of these last results the unit cell data reported for the actinide trifluorides on the basis of powder results alone have been corrected and the values listed in Table 2.26 refer to the larger unit cell.

TABLE 2.26

Some Physical and Crystallographic Properties of
the Actinide Trifluorides

Compound	Colour	m.p. (°C)	Lattice parameters (Å)[a]		Reference
			a_0	c_0	
AcF_3	White	—	7.41	7.55	337
UF_3	Black	d	7.181	7.348	337
NpF_3	Purple	1425	7.129	7.288	337
PuF_3	Purple	—	7.092	7.240	337
AmF_3	Pink	1395	7.044	7.225	329, 429
CmF_3	White	1406	6.999	7.179	329, 429

[a] All possess the hexagonal LaF_3-type structure, space group $D_{3d}^4-P\bar{3}cl$.

The recent structural determinations, reported independently by Mannsman[343] and by Zalkin and co-workers[344], are in good agreement. The lanthanum atoms have almost exactly the positions reported by Oftedal[335] but the fluorine positions are changed as a consequence of the change in point symmetry. Thus, there are nine fluorines around each lanthanum atom at distances between 2.416 and 2.640 Å (Figure 2.15) and the lanthanum atoms lie on a two-fold axis of symmetry.

The known actinide trifluorides[329,337] (Table 2.26) and the trifluorides of lanthanum to neodymium inclusive have been observed to crystallize only with hexagonal symmetry[309,314] but yttrium trifluoride and the trifluorides of samarium to lutetium inclusive are dimorphic. At room temperature these latter trifluorides generally possess orthorhombic symmetry (with the yttrium trifluoride structure[309]) but hexagonal phases have also been found to be stable at room temperature for samarium, holmium and thulium trifluoride. The transition temperatures[314], orthorhombic → hexagonal, are plotted in Figure 2.16 which also shows that the hexagonal

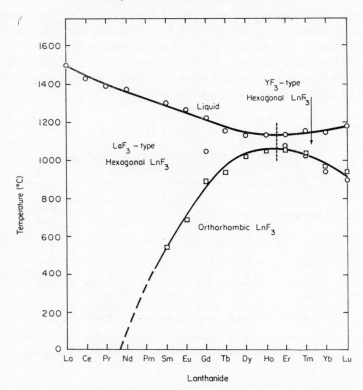

Figure 2.16   Dimorphism among the rare earth trifluorides.[314]   (*After* R. E. Thoma and G. D. Brunton, *Inorg. Chem.*, **5**, 1937 (1966))

phase observed for the trifluorides of samarium to holmium inclusive is of the lanthanum trifluoride type whereas that for the trifluorides of erbium to lutetium inclusive is the same as the high-temperature yttrium trifluoride structure.   The assignment of hexagonal symmetry to the high-temperature modification of yttrium trifluoride and the related phases is only tentative owing to the restricted diffraction data obtained.

The room temperature modification of yttrium trifluoride possesses[309] orthorhombic symmetry, space group D_{2h}^{16}–*Pnma*, with four molecules per unit cell.   Each yttrium atom has eight fluorine neighbours at distances between 2.25 and 2.32 Å with a ninth at 2.60 Å.   The arrangement of these neighbours is shown in Figure 2.17.   Unit cell parameters for the various crystal forms of the lanthanide trifluorides are provided in Table 2.27.

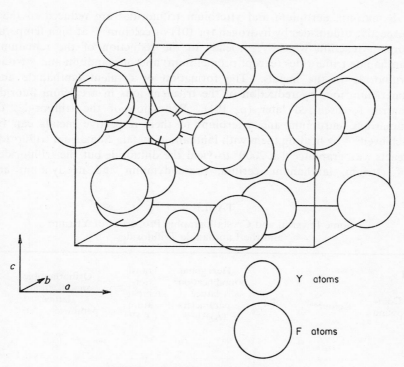

Figure 2.17   The structure of YF$_3$[309]
Y–F bond distances are in the range 2.25–2.32 Å for eight neighbours with a
ninth at 2.60 Å. (*After* A. Zalkin and D. H. Templeton, *J. Am. Chem. Soc.*,
75, 2543 (1953))

Scandium trifluoride (m.p. 1515°) is reported to possess rhombohedral
symmetry with $a_0 = 4.023$ Å and $\alpha = 89° 34'$. The structure is considered[345,346] to be a slightly distorted ReO$_3$-type. The trifluorides
of niobium, tantalum and molybdenum possess the latter type of
structure.

Properties. The trifluorides are stable, high-melting solids which are
only slightly soluble in water and hydrofluoric acid. Recently determined
melting points for yttrium fluoride and the lanthanide trifluorides, except
promethium trifluoride, are given in Table 2.27. Unlike uranium trifluoride the higher actinide trifluorides do not disproportionate and
vapour pressure data have been recorded for plutonium[348] and americium
trifluoride[349] (Table 2.28). Similar data are available for the majority of
the lanthanide trifluorides (Table 2.28).

Samarium, europium and ytterbium trifluorides are reduced to their respective difluorides by hydrogen (p. 101) or calcium[308] at high temperature, a reaction which[308,316] leads to the reduction of the remaining lanthanide trifluorides (except promethium) and of scandium and yttrium trifluoride to the metals. The formation of divalent lanthanide and americium ions by reduction of the trivalent ions in a calcium fluoride matrix is discussed later (p. 101). Reduction of the trifluorides of samarium, europium and ytterbium to their respective metals can be achieved[308] by heating them with lanthanum metal. Samarium trifluoride reacts with graphite[350] at 2000° to yield the difluoride but the trifluorides of yttrium, lanthanum, cerium, praseodymium and neodymium are

TABLE 2.27

Some Physical and Crystallographic Properties of Yttrium
and the Lanthanide Trifluorides

Compound	Colour[307]	m.p. (°C)[307]	Hexagonal modification[a] lattice parameters (Å)[314,318]		Transition temperature (°C)[314]	Orthorhombic modification[b] lattice parameters (Å)[309]		
			a_0	c_0		a_0	b_0	c_0
LaF₃	White	1493	7.186	7.352	—	—	—	—
CeF₃	White	1430	7.112	7.279	—	—	—	—
PrF₃	Green	1395	7.075	7.238	—	—	—	—
NdF₃	Violet	1374	7.030	7.200	—	—	—	—
PmF₃	Purplish-pink	—	6.970	7.190	—	—	—	—
SmF₃	White	1306	7.07	7.24	555	6.669	7.059	4.405
EuF₃	White	1276	7.04	7.26	700	6.622	7.019	4.396
GdF₃	White	1231	7.06	7.20	900	6.570	6.984	4.393
TbF₃	White	1172	7.03	7.10	950	6.513	6.949	4.384
DyF₃	Light green	1154	7.01	7.05	1030	6.460	6.906	4.376
HoF₃	Brownish-pink	1143	7.01	7.08	1070	6.404	6.875	4.379
ErF₃	Pink	1140	6.97	8.27	1075	6.354	6.846	4.380
TmF₃	White	1158	7.03	8.35	1030	6.283	6.811	4.408
YbF₃	White	1157	6.99	8.32	985	6.216	6.786	4.434
LuF₃	White	1182	6.96	8.30	945	6.151	6.758	4.467
YF₃	White	1152	7.13	8.45	1052	6.353	6.850	4.393

[a] The trifluorides of lanthanum to holmium inclusive possess the hexagonal LaF₃-type structure, space group D_{3d}^4–$P\bar{3}c1$. The remaining trifluorides possess a different unidentified structure.

[b] The trifluorides of samarium to lutetium inclusive possess the orthorhombic YF₃-type structure, space group D_{2h}^{16}–$Pnma$.

unaffected at this temperature. Actinide trifluorides are also reduced by electropositive elements but no stable divalent fluorides have been reported, reduction to the metal taking place. For example, actinium trifluoride[328] is reduced by lithium vapour at 1250°, uranium trifluoride by calcium[351] at 900°, and by magnesium at 600°, plutonium trifluoride[331,352] by calcium at 1000°, aluminium at 900° and by magnesium at 800° and the trifluorides of neptunium[213], americium[325] and curium[326] are reduced by barium at 1200°. The first preparations of many of the actinide metals, by reduction of the trifluorides, were performed on the submilligram scale and the apparatus used[213] for this work is illustrated in Figure 2.18. The barium reductant is vaporized by rapidly raising the temperature to 1200° and the actinide trifluoride reduced *in situ* in the inner crucible by the barium vapour.

The oxidation of cerium, terbium, plutonium and americium trifluoride by fluorine has already been discussed (p. 53). Uranium trifluoride is converted to soluble uranium (VI) salts by oxidizing acids and reacts rapidly with silver perchlorate, giving a silver mirror. Like plutonium

TABLE 2.28

Vapour Pressure Data for Certain Actinide
and Lanthanide Trifluorides

Compound	Temperature range (°K)	$\log P_{atm} = -(A/T) + B$		Reference
		A	B	
ScF₃	1172–1402	19,380	9.43	438
YF₃	1256–1434	21,850	9.77	438
LaF₃	1340–1650	21,730	9.608	448
LaF₃	1200–1434	20,200	8.20	438
CeF₃	—	19,830	8.816	390
CeF₃	1373–1634	20,460	9.205	439
NdF₃	1383–1520	18,730	8.03	440
DyF₃	1426–1622	18,420	7.538	441
HoF₃	1278–1429	18,470	7.333	441
ErF₃	1374–1521	19,300	7.777	441
TmF₃	1353[a]	19,600	8.240	442
YbF₃	1362[a]	18,670	7.750	442
LuF₃	1368[a]	21,000	9.410	442
PuF₃ (s)	1200–1440	21,120	9.587	348
PuF₃ (l)	1440–1770	19,400	8.392	348
AmF₃	1126–1469	24,628	34.007−7.048 log T	349

[a] Average temperature.

7

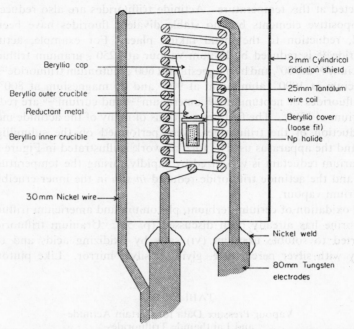

Beryllia cap

2 mm Cylindrical
radiation shield

Beryllia outer crucible

25mm Tantalum
wire coil

Reductant metal

Beryllia inner crucible

Beryllia cover
(loose fit)

Np halide

30mm Nickel wire

Nickel weld

80mm Tungsten
electrodes

Figure 2.18   Apparatus for metal trifluoride reductions.[213]   (*After* S. Fried
and N. Davidson, *J. Am. Chem. Soc.*, **70**, 3539 (1948))

trifluoride, it is stable in the atmosphere at room temperature but at 900°
it is oxidized to U_3O_8; plutonium trifluoride is converted to the dioxide
at 300° under similar conditions.   Chlorine, bromine and iodine oxidize
uranium trifluoride at moderate temperatures to produce[345,355] the mixed
halides UF_3X (X = Cl, Br or I) but analogous reactions involving other
readily oxidized trifluorides have not been reported.

The absorption spectra of certain trivalent lanthanide elements in a
lanthanum trifluoride matrix have already been mentioned (p. 81).   Other
investigations have included details of the spectra of solid neodymium
trifluoride[317], terbium trifluoride[327], americium trifluoride[327,238] and
curium trifluoride[238], the spectrum of Nd^{3+} in lanthanum trifluoride
crystals[394] and the spectra of several trivalent lanthanide ions[395] in
cadmium difluoride.   During an investigation of the absorption spectra
of certain lanthanide trifluorides in the latter matrix Mandel and colleagues[356] made the first observation of the forbidden $^2F_{7/2} \leftarrow\, ^2F_{5/2}$
electronic transition in cerium (III).   The excited state ($^2F_{7/2}$) lies as predicted 2250 cm^{-1} above the ground state ($^2F_{5/2}$).   The paramagnetic

resonance of many trivalent lanthanide ions in a calcium fluoride matrix has been reported[357,396-403] and Gosh and co-workers[236] have investigated the resonance of powdered uranium trifluoride samples. Although these investigators concluded that the electronic configuration of the U^{3+} ion involved both $5f$ and $6d$ orbitals, later calculations by O'Brien[358] showed that the configuration was more probably $5f^3$. The paramagnetic resonance of U^{3+} in single crystals of calcium fluoride is similar[359] to that of Nd^{3+} in the same matrix; hyperfine structure in the former case suggests that the $5f$ electrons have more extended wave functions than the $4f$ electrons.

Magnetic susceptibility studies have been reported for cerium trifluoride by several groups[360-362]. The effective magnetic moment[362] is 2.5 B.M. with a Weiss constant of 61°. The effective moment of uranium trifluoride is variously reported to be 3.67 B.M. ($\theta = -110°$)[430], 3.66 B.M. ($\theta = -98°$)[334] or 3.50 B.M. ($\theta = -32.2°$)[138], the first two values being slightly higher than those found by Dawson[363] for other uranium trihalides and close to the predicted moment (3.62 B.M.) for a $5f^3$ electronic configuration. The dependence of the magnetic susceptibility of plutonium trifluoride[364] on temperature is similar to that of samarium (III) compounds (cf. $PuCl_3$, p. 157) with a minimum in the χ versus T plot between 500 and 550°K which is considered to be good qualitative evidence that plutonium (III) has the $5f^5$ electronic configuration. The molar magnetic susceptibility of curium trifluoride, measured using a 10% solution in the isostructural lanthanum trifluoride[365], is 26,500 ± 700 c.g.s. units at 296°K. This result is strong evidence in favour of a $5f^7$ electronic configuration for curium (III) since, for such a configuration with Russell–Saunders coupling, a theoretical value of 26,000 c.g.s. units is calculated.

Trivalent Fluoro Complexes

The lanthanide elements form stable trivalent complexes of the types $M^IM^{III}F_4$ (M^I = Li, Na and K) and $M_3^IM^{III}F_6$ (M^I = K, Rb, Cs and NH_4). Examples of both types have not, however, been reported for each of the lanthanide elements with the cations listed. In fact Thoma[255] predicted a few years ago that where the cation radius ratio M^+/M^{3+} lies between 0.77 and 1.40, complexes of the type $M^IM^{III}F_4$ can exist, and where M^+/M^{3+} is greater than 1.43, congruently-melting complexes of the type $M_3^IM^{III}F_6$ will be formed. His predictions concerning trivalent lanthanide and actinide complexes, modified to include recently reported studies, are shown in tabular form in Figure 2.19. Thoma stated that insufficient data were available to fix the exact radius ratio for the transition in system type and this is still the case because relatively few systems with a large

Figure 2.19 (*for caption see page 91*)

radius ratio have been thoroughly studied. However, the existence[391] of $LiEuF_4$ indicates a slightly lower value than 0.77 as the lower limit for 1:1 complexes. Both 1:1 and 3:1 complexes of trivalent cerium, $KCeF_4$ and K_3CeF_6, have recently been reported[320] and this may indicate a lower limit than 1.43 for the existence of the 3:1 complexes and in conjunction with the fact that $RbPrF_4$ is now known, may also give an indication of the region where both types of complex will be formed. However, the same authors also report the existence of Na_3YbF_6 in addition to $NaYbF_4$ whereas the results of the exhaustive investigations of $NaF-MF_3$ phase systems reported by Thoma and his colleagues[311-313] indicate that 3:1 complexes are not formed. At the time of Thoma's predictions no more than one intermediate compound had been observed in any $M^IF-M^{III}F_3$ (M^I = alkali metal; M^{III} = lanthanide and actinide element) system. Subsequently, in addition to the 1:1 complexes $NaM^{III}F_4$, phases of the type $Na_5M^{III}_9F_{32}$ (M^{III} = Y, Pr, Nd and Sm–Lu inclusive) have been observed[311,312] during $NaF-M^{III}F_3$ phase studies. However, with the exception of $Na_5Lu_9F_{32}$, these phases are unstable at room temperature. Both 1:1 and 3:1 complexes are known with certain cations for trivalent scandium (p. 98), lanthanum, cerium, praseodymium and uranium (p. 96) but there are still too few data to permit generalizations concerning those systems where several stable intermediate compounds may occur. It will be apparent from Figure 2.19 that there is scope for further investigation of the alkali metal–metal trifluoride phase systems, particularly those concerning the actinide trifluorides.

The majority of the complexes have been identified in phase studies[311-313,319,320,366-369] or by melting together[264,370-372] stoicheiometric amounts of the component fluorides. Many equilibrium phase diagrams are illustrated in a recent review by Thoma[436]. In a few instances they have been prepared by fluorination of 3:1 mixtures of an alkali metal chloride and a lanthanide halide at 300–400°[373], by hydrofluorination

---

Figure 2.19   Cation radius ratios, M^+/M^{3+}, and predicted complex formation for certain $MF-MF_3$ systems

Underlined values and those not enclosed indicate systems not yet investigated. In 1962 Thoma[255] made the following predictions which are illustrated as indicated;

(1) Ratios $0 \rightarrow 0.67$; No complex formed: in [   ]
(2) Ratios $0.77 \rightarrow 1.40$; 1:1 complex formed: in [////]
(3) Ratios > 1.43; 3:1 complex formed: in [\\\\]

These limits are now slightly changed because

a, K_3CeF_6 and K_3UF_6 now known
b, $RbPrF_4$ now known
c, $LiEuF_4$, $LiGdF_4$ and $LiTbF_4$ now known.

of sodium fluoride–lanthanide sesquioxide mixtures[371] or by heating sodium fluoride–metal dioxide (metal = Ce, Pr, Tb, Pu and Am) mixtures at 450–650° in a mixture of hydrogen and hydrogen fluoride. The scandium complexes $M_3^IScF_6$ (M^I = K and NH_4) and $NaYF_4$ have also been isolated from aqueous solution[373,374].

Lithium fluoride–metal trifluoride complexes. Equilibrium phase diagrams reported for lithium fluoride with lanthanum[375], cerium[376], uranium[377] and plutonium[378] trifluoride show that no complex formation occurs in these systems. Dergunov[366] has reported that the lithium fluoride–yttrium trifluoride phase system is similar, with a eutectic at 744° (18 mole % YF_3), but a more recent investigation[367] has shown that the 1:1 salt, $LiYF_4$ is formed. This white solid which melts incongruently at 819° and possesses tetragonal symmetry, space group $C_{4h}^6–I4_1/a$, is isostructural with $CaWO_4$. X-ray powder diffraction data have been reported[391] for the analogous complexes of Eu–Lu inclusive (Table 2.29)

TABLE 2.29

Crystallographic Properties[a] for Some Lanthanide
Tetrafluoro Complexes, $LiM^{III}F_4$ [391]

Complex	Lattice parameters (Å)	
	a_0	c_0
$LiEuF_4$	5.228	11.03
$LiGdF_4$	5.219	10.97
$LiTbF_4$	5.200	10.89
$LiDyF_4$	5.188	10.83
$LiHoF_4$	5.175	10.75
$LiErF_4$	5.162	10.70
$LiTmF_4$	5.145	10.64
$LiYbF_4$	5.132	10.59
$LiLuF_4$	5.124	10.54
$LiYF_4$	5.175	10.74

[a] All are tetragonal, space group
$C_{4h}^6–I4_1/a$.

which were prepared by heating the appropriate lanthanide trifluoride or sesquioxide with lithium fluoride in an atmosphere of hydrogen fluoride.

Sodium fluoride–metal trifluoride complexes. Complete phase diagrams have been reported by Thoma and his colleagues for the $NaF–ScF_3$[313], $NaF–YF_3$[311] and $NaF–MF_3$[312] (M = lanthanide elements except cerium and promethium) systems and the results of similar studies of the $NaF–CeF_3$[368], $NaF–UF_3$[369] and $Na–PuF_3$[368] systems have been published by others. In these last three and the NaF–LaF$_3$ phase systems a single equilibrium compound of formula NaM^{III}F$_4$ (M^{III} = La, Ce, U and Pu) is formed. However, two equilibrium complexes of the types NaM^{III}F$_4$ and Na$_5$M$_9^{III}$F$_{32}$ (M^{III} = Y and Pr–Lu inclusive) are observed for each of the other systems[312], apart from the NaF–ScF$_3$ system. The 1:1 complexes, NaM^{III}F$_4$, (M^{III} = Y and Pr–Lu inclusive) possess hexagonal symmetry at low temperature but above about 700° they are converted[312] to disordered fluorite-like cubic phases of variable composition. The upper composition limit of the cubic phases, determined from refractive index, lattice constant and phase transition data, corresponds to the composition Na$_5$M$_9^{III}$F$_{32}$ whereas the lower limit extends progressively from 55.5 mole % MF$_3$ at SmF$_3$ to 39 mole % MF$_3$ at LuF$_3$. These differences are illustrated in Figure 2.20. The cubic, solid solution phases are unstable below temperatures varying from 800° to 530° and on cooling they transform to a variety of products depending on the composition of the decomposing phases. For example, at equimolar NaF–MF$_3$ compositions partial ordering occurs and the hexagonal NaM^{III}F$_4$ phase is formed. At the Na$_5$M$_9^{III}$F$_{32}$ phase boundary in the NaF–PrF$_3$ to NaF–TbF$_3$ systems hexagonal NaM^{III}F$_4$ and MF$_3$ are formed, whereas in the systems NaF–DyF$_3$ to NaF–LuF$_3$ the cubic phase initially transforms to an orthorhombic phase of the same composition. Apart from Na$_5$Lu$_9$F$_{32}$, this orthorhombic phase is also unstable with respect to NaM^{III}F$_4$ and MF$_3$ at lower temperatures. Although size factors alone indicate that fluorite-like phases should be formed in the actinide systems NaF–UF$_3$ and NaF–PuF$_3$, no such phases have been observed. On the assumption that polarizability of the heavy-metal ions influences formation of the fluorite-like phase (as indicated by the trends throughout the NaF–MF$_3$ phase systems) it has been predicted[312] that such phases will occur with the heavier actinides.

The NaF–ScF$_3$ phase system has been studied by several groups but the most reliable data are probably those of Thoma and Karraker[313]. Two complexes NaScF$_4$ and Na$_3$ScF$_6$ are formed; the former possesses hexagonal symmetry with $a_0 = 12.97$, $c_0 = 9.27$ Å and the latter exists in a high-temperature and a low-temperature modification each of which is isostructural with the corresponding cryolite phase. The low-temperature phase is monoclinic, space group C_{2h}^5–$P2_1/n$, with $a_0 = 5.60$, $b_0 = 5.81$, $c_0 = 8.12$ Å and $\beta = 90° 45'$. These results confirm that cation size and polarizability are the principal controlling factors in determining the nature of interactions between pairs of ionic salts.

The remaining 1:1 complexes discussed above possess hexagonal symmetry at room temperature. Although they were previously considered to be isostructural with β_2–Na$_2$ThF$_6$, the recently determined structure

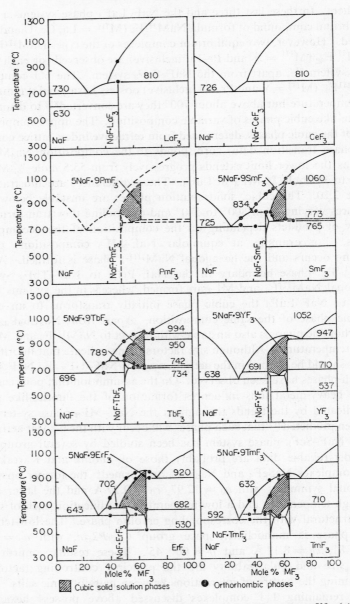

Figure 2.20 Equilibrium diagrams for the NaF–MF₃ systems.[312] (*After* R. E. Thoma, H. Insley and G. M. Herbert, *Inorg. Chem.*, **5**, 1222 (1966))

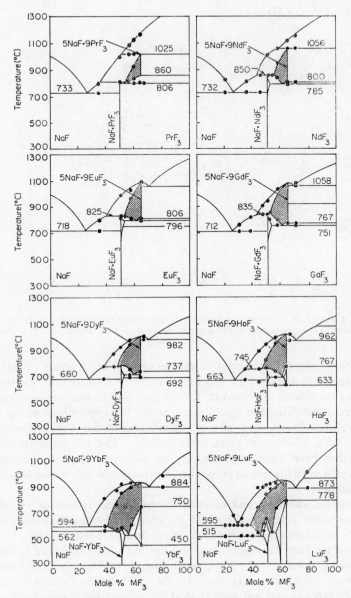

Figure 2.20 Equilibrium diagrams for the NaF–MF₃ systems.[312] (*After* R. E. Thoma, H. Insley and G. M. Herbert, *Inorg. Chem.*, **5**, 1222 (1966))

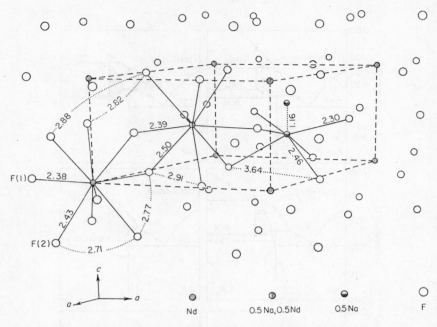

Figure 2.21   Perspective drawing of one unit cell of NaNdF₄ with its
fluoride ion neighbours[379] (distances in Å). (*After* J. H. Burns, *Inorg. Chem.*,
4, 881 (1965))

of NaNdF$_4$ shows[379] that this is not the case. The structure of the latter
is illustrated in Figure 2.21 and unit cell dimensions for the isostructural
series of complexes are listed in Table 2.30. Lattice parameters have also
been recorded[312] for the limiting 1:1 and 5:9 compositions of the high-
temperature, fluorite-like phases. Similar data for the orthorhombic
Na$_5$M$^{III}_9$F$_{32}$ phases are given in Table 2.31. The composition of such
phases has been established entirely by empirical methods and cannot
be accepted until the results of the current structure analysis[312] of Na$_5$Lu$_9$F$_{32}$
are available.

Potassium, rubidium and caesium fluoride–metal trifluoride complexes.
Relatively few complexes formed between potassium fluoride and the
actinide and lanthanide trifluorides have been reported. The 1:1 complexes of trivalent lanthanum and cerium have been observed to exist in
two crystal forms[105,264,372,434], as do many of the analogous sodium
complexes. β-KLaF$_4$ was recently shown[434] to be isostructural with
NaNdF$_4$ (above) and not with β$_1$–K$_2$UF$_6$ as previously reported[372]. The
3:1 complexes of cerium, samarium and erbium[319,320] have been recorded

although their existence has not been substantiated. The $3:1$ uranium (III) complex K_3UF_6 is known[164] and both $1:1$ and $2:1$ complexes have been reported, but full details of their preparation and properties are lacking at present. K_3UF_6 possesses face-centred cubic symmetry and is isomorphous with α-K_3UF_7 (p. 71) and K_3UF_8 (p. 46); all three complexes possess almost identical cell parameters. On the other hand the trivalent scandium and yttrium complexes, K_3ScF_6 and K_3YF_6 respectively, are reported[373] to possess different structures; the latter is tetragonal (Table 2.32).

Other alkali metal salts which have been reported include the $1:1$ lanthanide compounds[260,366], $RbM^{III}F_4$ ($M^{III} = La$, Ce and Pr) and the $3:1$ compounds[319,320,373,437] $Rb_3M^{III}F_6$ ($M^{III} = Y$, La, Ce, Pr, Sm, Tb and Er) and $Cs_3M^{III}F_6$ ($M^{III} = Y$, La, Ce, Pr, Sm, Gd, Dy, Ho and Er). Some crystallographic properties for these complexes are listed in Table 2.32.

TABLE 2.30

Crystallographic Data[a] for the Low-Temperature $NaM^{III}F_4$ Phases[260,312,371]

Complex	Lattice parameters (Å)	
	a_0	c_0
$NaLaF_4$	6.157	3.822
$NaCeF_4$	6.131	3.776
$NaPrF_4$	6.123	3.743
$NaNdF_4$	6.100	3.711
$NaPmF_4$[b]	(6.056)	(3.670)
$NaSmF_4$	6.051	3.640
$NaEuF_4$	6.044	3.613
$NaGdF_4$	6.020	3.601
$NaTbF_4$	6.008	3.580
$NaDyF_4$	5.985	3.554
$NaHoF_4$	5.981	3.528
$NaErF_4$	5.959	3.514
$NaTmF_4$	5.953	3.494
$NaYbF_4$	5.929	3.471
$NaLuF_4$	5.912	3.458
$NaYF_4$	5.967	3.523
$NaUF_4$	6.167	3.770
$NaPuF_4$	6.119	3.752
$NaAmF_4$	6.109	3.731

[a] All possess hexagonal symmetry, space group C_{3h}^1-$P\bar{6}$.

[b] Values in parentheses are interpolated.

It has been stated that $1:1$, $2:1$ and $3:1$ trivalent uranium complexes $M_x^I UF_{3+x}(M^I = Rb$ and Cs; $x = 1$, 2 and 3) can be made[164] but full details of their preparation and properties have yet to be published.

Miscellaneous complexes and phase studies. The ammonium complexes NH_4ScF_4 and $(NH_4)_3ScF_6$ can be prepared[373,374,450] from aqueous solution. The latter exists in two crystal forms; α-$(NH_4)_3ScF_6$, the high-temperature modification, possessing[373] face-centred cubic symmetry.

TABLE 2.31

Crystallographic Data for the Orthorhombic $Na_5M_9^{III}F_{32}$ Complexes[312]

Complex	Lattice parameters (Å)		
	a_0	b_0	c_0
$Na_5Dy_9F_{32}$	5.547	39.23	7.845
$Na_5Ho_9F_{32}$	5.525	39.07	7.814
$Na_5Er_9F_{32}$	5.514	38.99	7.798
$Na_5Tm_9F_{32}$	5.493	38.84	7.768
$Na_5Yb_9F_{32}$	5.480	38.75	7.750
$Na_5Lu_9F_{32}$	5.463	38.63	7.725

TABLE 2.32

Crystallographic Data for Certain Trivalent Fluoro Complexes

Compound	Symmetry	Space group	Lattice parameters (Å)		Reference
			a_0	c_0	
α-$KLaF_4$	Cubic	O_h^5–$Fm3m$	5.932	—	105
α-$KCeF_4$	Cubic	O_h^5–$Fm3m$	5.894	—	105
β_1-$KLaF_4$	Hexagonal	C_{3h}^1–$P\bar{6}$	6.530	3.800	434
β_1-$KCeF_4$	Hexagonal	C_{3h}^1–$P\bar{6}$	6.496	3.750	372, 434
K_3UF_6	Cubic	O_h^5–$Fm3m$	9.20	—	164
K_3YF_6	Tetragonal	D_{4h}^{17}–$I4/mmm$	6.20	9.10	373
Rb_3YF_6	Tetragonal	D_{4h}^{17}–$I4/mmm$	6.55	9.40	373
Rb_3CeF_6	Cubic	O_h^5–$Fm3m$	9.42	—	373
Rb_3PrF_6	Cubic	O_h^5–$Fm3m$	9.48	—	373
Cs_3YF_6	Tetragonal	D_{4h}^{17}–$I4/mmm$	6.89	9.78	373
α-$(NH_4)_3ScF_6$	Cubic	O_h^5–$Fm3m$	9.26	—	373

The existence of analogous trivalent lanthanide fluoro complexes was mentioned earlier (p. 80). Hydrated fluoro complexes have also been reported[380]. Complex formation is not observed in the systems $M^{II}F_2$–$M^{III}F_3$ (M^{II} = Ca, Sr and Ba; M^{III} = variously Y, La, Ce, U, Pu and Am).

Trivalent Oxyfluorides

A study of the thermal decomposition of certain lanthanide trifluoride hydrates has indicated[381] that oxyfluoride formation occurs above 600°. Oxyfluorides of the type MOF (M = La, Ce, Pr, Nd, Pm, Sm, Eu, Gd, Dy, Sc and Y) have been prepared[318,382-384,392] by partial hydrolysis of the trifluorides in moist air or ammonia vapour at 800°. Hydrolysis of actinium trifluoride requires a temperature of 1200°. Lanthanum[385], scandium[392] and yttrium[386] oxyfluoride have also been prepared by heating together stoicheiometric amounts of the appropriate trifluoride and sesquioxide. Terbium oxyfluoride is formed[387] when the tetrafluoride is heated at 400° in air and has been prepared accidentally[383] by fluorination of Tb_4O_7 followed by hydrolysis in a stream of undried hydrogen. Although the fluorite-like holmium oxyfluoride phase has been reported[309], its existence remains questionable in view of the recent work on high-temperature fluorite-like phases of the general formula $NaM^{III}F_4$ (p. 92). However, a tetragonal form of this and other lanthanide oxyfluorides which have not been characterized analytically (M^{III} = Dy, Er, Tm, Yb and Lu) has been observed[428] on the exterior of pellets used for the preparation of the lithium complexes $LiM^{III}F_4$.

Crystal structures. Three related structure types have been observed for the lanthanide and actinide oxyfluorides; lanthanum oxyfluoride is reported to exist in all three forms. Cubic, fluorite-like phases have been reported for certain of the oxyfluorides (Table 2.33) and in a thorough investigation of the hydrolysis of lanthanum and yttrium trifluoride, Zachariasen[382] has observed that a tetragonal phase is stabilized by the presence of excess fluoride over the ideal composition MOF. Although plutonium oxyfluoride was originally reported to be cubic, a re-examination[382] of the powder data showed that the phase was actually tetragonal. This tetragonal structure is a small distortion of the cubic fluorite arrangement. At the stoicheiometric composition, MOF, Zachariasen[382] found that the tetragonal phase was unstable with respect to a rhombohedral phase which is again a small distortion of the cubic fluorite arrangement. This structure has been reported for the majority of the lanthanide oxyfluorides (Table 2.33). The interatomic distances for lanthanum and yttrium oxyfluorides in the rhombohedral modification[382] are La–4F

TABLE 2.33

Crystallographic Properties for the Trivalent Oxyfluorides

Compound	Colour	Symmetry	Space group	Lattice parameters (Å; α in °)			Reference
				a_0	c_0	α	
α-LaOF	White	Cubic	O_h^5–$Fm3m$	5.756	—	—	385
β-LaOF		Rhombohedral	D_{3d}^5–$R\bar{3}m$	7.132	—	33.01	382
γ-LaOF		Tetragonal	D_{4h}^7–$P4/nmm$	4.091	5.852	—	382
CeOF	Black	Cubic	O_h^5–$Fm3m$	5.703	—	—	383
α-PrOF	Brown	Cubic	O_h^5–$Fm3m$	5.644	—	—	388
β-PrOF		Rhombohedral	D_{3d}^5–$R\bar{3}m$	7.016	—	33.03	383
α-NdOF	Purple	Cubic	O_h^5–$Fm3m$	5.595	—	—	388
β-NdOF		Rhombohedral	D_{3d}^5–$R\bar{3}m$	6.953	—	33.04	383
α-PmOF	Pink	Cubic	O_h^5–$Fm3m$	5.560	—	—	318
γ-PmOF		Tetragonal	D_{4h}^7–$P4/nmm$	3.980	5.58	—	318
α-SmOF	Grey-green	Cubic	D_{4h}^7–$P4/nmm$	5.519	—	—	388
β-SmOF		Rhombohedral	D_{3d}^5–$R\bar{3}m$	6.865	—	33.07	383
EuOF	White	Rhombohedral	D_{3d}^5–$R\bar{3}m$	6.827	—	33.05	383
GdOF	Yellow	Rhombohedral	D_{3d}^5–$R\bar{3}m$	6.800	—	33.05	383
TbOF	White	Rhombohedral	D_{3d}^5–$R\bar{3}m$	6.758	—	33.02	383
DyOF	White	Rhombohedral	D_{3d}^5–$R\bar{3}m$	6.716	—	33.07	428
HoOF	Yellow	Rhombohedral	D_{3d}^5–$R\bar{3}m$	6.647	—	33.15	428
ErOF	Pink	Rhombohedral	D_{3d}^5–$R\bar{3}m$	6.628	—	33.14	428
AcOF	White	Cubic	O_h^5–$Fm3m$	5.931	—	—	382
PuOF		Tetragonal	D_{4h}^7–$P4/nmm$	4.05	5.72	—	382
α-YOF	White	Cubic	O_h^5–$Fm3m$	5.363	—	—	386
β-YOF		Rhombohedral	D_{3d}^5–$R\bar{3}m$	6.697	—	33.2	382
γ-YOF		Tetragonal	D_{4h}^7–$P4/nmm$	3.938	5.47	—	382
ScOF	White	Cubic	O_h^5–$Fm3m$	5.575	—	—	392

= 2.42 Å;   La–4O = 2.58 Å;   Y–4F = 2.28 Å;   and   Y–4O = 2.44 Å. The designation of the various forms of the lanthanide oxyfluorides (α, β or γ) follows the notation used by Wyckoff[389] who has discussed the relationships between the three structure-types.

Recently a few lanthanide thiofluorides, MSF (M = La, Ce and Eu) have been made by heating together the trifluorides and sesquisulphides at 450–500°. These are reported to possess tetragonal symmetry with a PbClF-type of structure (Table 2.34).

TABLE 2.34

Crystallographic Data[a] for Some Lanthanide Thiofluorides[393]

Compound	Lattice parameters (Å)	
	a_0	c_0
LaSF	4.02	6.99
CeSF	4.01	6.95
EuSF	3.87	6.73

[a] All possess tetragonal symmetry, space group D_{4h}^7–$P4/nmm$.

DIVALENT

Difluorides

Only samarium, ytterbium and europium form stable difluorides. These are obtained either by hydrogen reduction[404–409] of the corresponding trifluorides at elevated temperatures or, less satisfactorily[409], by metathesis of the divalent lanthanide sulphates with ammonium fluoride. Reduction of the appropriate trifluoride by calcium metal also yields samarium, europium and ytterbium difluoride whereas the remaining lanthanum trifluorides are reduced directly to the metals. Partial reduction of samarium trifluoride by graphite at 2000° has been observed[350] and EuF_2 has been prepared in a similar manner[435]. Variable lattice constants have been obtained for the fluorite-like unit cells of the difluorides and the analytical results of Asprey and co-workers[409] indicate that this is probably due to the formation of a solid solution of the trifluoride in the difluoride. The presence of trivalent europium may also explain the low magnetic moment reported[404] for europium difluoride.

Recent interest has centred on the formation of divalent lanthanide and actinide ions in a host lattice such as calcium fluoride and it has been demonstrated, for example, see references 410–416, that divalent ions of all the lanthanides and of americium can be formed by reduction of the trivalent ions using ionizing radiations such as x-rays or γ-rays, and by ultraviolet light. Only a fraction (5–10%) of the trivalent ions are converted to the divalent state by these methods and this fraction reconverts to the trivalent state fairly rapidly, depending on the temperature and illumination. It has been shown that more extensive reduction to the divalent state can be achieved[417] by heating the fluoride crystal (e.g. CaF_2, BaF_2 or SrF_2) containing the trivalent element with the appropriate alkaline-earth metal vapour. Similar results have been obtained[416,418–420] by solid state electrolysis at temperatures between 300° and 700° and in each case the resultant divalent ions are relatively stable in comparison with those produced by irradiation. The differences in the extent of reduction and the stability of the divalent ions obtained by the different techniques appear to be due to the fact that only those trivalent ions in cubic sites are reduced by the ionizing radiations and the divalent ions so produced are unstable owing to the presence of recombination hole-centres. However, reduction by the other methods leads to the formation of divalent ions in non-cubic sites as well as in cubic sites and also results in the elimination of the recombination centres. Some information is available[418,420,458] for corresponding lanthanide chloride and bromide systems.

Several spectral[411–414,417,421–424,457] and paramagnetic resonance absorption studies[410,415,416,425,457] have been reported for the divalent ions in calcium, barium or strontium fluoride crystals.

Properties. Few physical or chemical properties have been reported for the stable divalent lanthanide fluorides. They are high-melting solids which possess the cubic fluorite-type structure (Table 2.35). Asprey and

TABLE 2.35

Some Physical and Crystallographic[a] Properties of the
Lanthanide Difluorides

Compound	Colour	m.p. (°C)[427]	Lattice parameters (Å)[409] a_0
SmF_2	Yellow	1417	5.81
EuF_2	—	1416	5.84
YbF_2	—	1407	5.57

[a] All possess the cubic, fluorite-type structure.

colleagues[409] experienced difficulty in obtaining truly stoicheiometric phases, MF_2, and their cell constants quoted in Table 2.35 refer to those phases nearest to the required 1:2 stoicheiometry. Magnetic susceptibility studies have led to a value of 7.4 B.M. for the effective moment of europium difluoride. This is lower than that expected for a $4f^7$ configuration and lower than the value (7.9 B.M.) found for europium dichloride, dibromide and diiodide; this probably indicates that the difluoride was not pure. At lower temperatures[426] europium difluoride, unlike the other europium dihalides, becomes antiferromagnetic.

REFERENCES

1. J. J. Katz and E. Rabinowitch, 'The Chemistry of Uranium', *Nat. Nucl. Energy Ser., Div. VIII*, Vol. 5, McGraw-Hill, New York, 1951.
2. 'The Chemistry of the Transuranium Elements' (G. T. Seaborg, J. J. Katz and W. M. Manning, Eds.), *Nat. Nucl. Energy Ser., Vol. IV*, McGraw-Hill, New York, **14B**, 1949.
3. 'The Chemistry of the Actinide Elements' (G. T. Seaborg and J. J. Katz, Eds.), *Nat. Nucl. Energy Ser., Vol. IV*, McGraw-Hill, New York, **14A**, 1954.
4. P. Pascal (Ed.), *Nouveau Traite de Chimie Minérale*, Vol. IX, Masson et Cie, Paris, 1963.
5. P. Pascal (Ed.), *Nouveau Traite de Chimie Minérale*, Vol. XV, Part 2, Masson et Cie, Paris, 1961.
6. P. Pascal (Ed.), *Nouveau Traite de Chimie Minérale*, Vol. XV, Part 3, Masson et Cie, Paris, 1961.
7. P. Pascal (Ed.), *Nouveau Traite de Chimie Minérale*, Vol. VII, Masson et Cie, Paris, 1959.
8. A. G. Sharpe in *Advances in Fluorine Chemistry* (M. Stacey, J. C. Tatlow and A. G. Sharpe, Eds.), Vol. I, Butterworths, London, 1960, p. 29.
9. N. Hodge in *Advances in Fluorine Chemistry* (M. Stacey, J. C. Tatlow and A. G. Sharpe, Eds.), Vol. II, Butterworths, London, 1961, p. 138.
10. K. W. Bagnall in *Halogen Chemistry* (V. Gutmann, Ed.), Vol. 3, Academic Press, London, 1967, p. 303.
11. J. H. Simons in *Fluorine Chemistry* (J. H. Simons, Ed.), Vol. 5, Academic Press, New York, 1964, p. 1.
12. M. J. Steindler, U.S. Report ANL-6753 (1963).
13. R. DeWitt, U.S. Report GAT-280 (1960).
14. I. V. Tananaev, N. S. Nikolaev, Yu. A. Luk'yanchev and A. A. Opalovskii, *Russ. Chem. Rev.* (English Transl.), **30**, 654 (1961).
15. S. Tsujimura, D. Cohen, C. Chevnik and B. Weinstock, *J. Inorg. Nucl. Chem.*, **25**, 226 (1963).
16. O. Rüff and A. Heinzlemann, *Z. Anorg. Chem.*, **72**, 63 (1911).
17. Reference 1, p. 398.
18. B. Weinstock and J. G. Malm, *J. Inorg. Nucl. Chem.*, **2**, 380 (1956).
19. A. E. Florin, I. A. Tannenbaum and J. F. Lemons, *J. Inorg. Nucl. Chem.*, **2**, 368 (1956).
20. C. J. Mandleberg, H. K. Rae, R. Hunt, G. Long, D. Davies and K. E. Francis, *J. Inorg. Nucl. Chem.*, **2**, 358 (1956).
21. J. G. Malm, B. Weinstock and E. E. Weaver, *J. Phys. Chem.*, **62**, 1506 (1958).
22. M. J. Steindler, D. V. Steidl and R. K. Steunenberg, U.S. Report ANL-5875 (1958).
23. M. D. Adams, R. K. Steunenberg and R. C. Voge, U.S. Report ANL-5796 (1957).
24. J. G. Malm, B. Weinstock and H. H. Claasen, U.S. Pat. 2,893,826 (1959).
25. T. A. O'Donnell, D. F. Stewart and P. Wilson, *Inorg. Chem.*, **5**, 1438 (1966).
26. M. J. Steindler, D. V. Steidl and R. K. Steunenberg, *Nucl. Sci. Eng.*, **6**, 333 (1959).
27. H. J. Emeléus, A. G. Maddock, G. L. Miles and A. G. Sharpe, *J. Chem. Soc.*, **1948**, 1991.

28. H. J. Emeléus and A. A. Woolf, *J. Chem. Soc.*, **1950**, 164.
29. G. J. Vogel and R. W. Vogel, U.S. Report K-727 (1951).
30. V. Y. Labaton, *J. Inorg. Nucl. Chem.*, **10**, 86 (1959).
31. J. F. Ellis and C. W. Forrest, *J. Inorg. Nucl. Chem.*, **16**, 150 (1960).
32. M. J. Steindler and S. Vogler, U.S. Report ANL-5422, p. 35 (1955).
33. A. L. Oppegard, W. C. Smith, E. L. Muetterties and V. A. Englehardt, *J. Am. Chem. Soc.*, **82**, 3835 (1960).
34. C. E. Johnson, J. Fischer and M. J. Steindler, *J. Am. Chem. Soc.*, **83**, 1620 (1961).
35. M. J. Steindler, U.S. Pat. 3,046,089 (1962).
36. L. M. Ferris, *J. Am. Chem. Soc.*, **79**, 5419 (1957); *Ind. Eng. Chem.*, **51**, 200 (1959).
37. S. Fried and N. R. Davidson, U.S. Report TID-5290, p. 688 (1958).
38. N. J. Hawkins and J. W. Codding, U.S. Report KAPL-1536, p. 57 (1956).
39. S. Fried and N. R. Davidson, reference 2, p. 784.
40. J. K. Dawson, R. M. Elliot, R. Hurst and A. E. Truswell, *J. Chem. Soc.*, **1954**, 558.
41. R. Rosen, U.S. Pat. 2,894,811 (1959).
42. Reference 1, p. 399.
43. J. F. Ellis, L. H. Brooks and K. D. B. Johnson, *J. Inorg. Nucl. Chem.*, **6**, 199 (1958).
44. W. H. Mears, R. V. Townend, R. D. Broadley, A. D. Turissini and R. F. Stahl, *Ind. Eng. Chem.*, **50**, 1771 (1958).
45. G. I. Cathers, M. R. Bennett and R. L. Jolley, *Ind. Eng. Chem.*, **50**, 1709 (1958).
46. J. L. Hoard and J. D. Stroup, reference 1, page 439.
47. J. G. Malm and B. Weinstock, *Proc. Intern. Conf. Peaceful Uses At. Energy, 2nd Geneva*, **28**, 125 (1958).
48. V. Schomaker and R. Glauber, *Phys. Rev.*, **91**, 1182 (1953); *Nature*, **170**, 290 (1952).
49. B. B. Cunningham and J. C. Hindman. Reference 3, p. 471.
50. E. H. Claassen, B. Weinstock and J. G. Malm, *J. Chem. Phys.*, **25**, 426 (1956).
51. D. W. Magnusson, *J. Chem. Phys.*, **24**, 344 (1955).
52. B. Weinstock and G. L. Goodman, *Advan. Chem. Phys.*, **IX**, 169 (1965).
53. N. J. Hawkins, H. C. Mattraw and B. Weinstock, *J. Chem. Phys.*, **23**, 2191 (1955).
54. J. G. Malm, B. Weinstock and E. H. Claassen, *J. Chem. Phys.*, **23**, 2192 (1955).
55. G. D. Oliver, H. T. Milton and J. W. Grysard, *J. Am. Chem. Soc.*, **75**, 2827 (1953).
56. E. E. Weaver, J. G. Malm and B. Weinstock, *J. Inorg. Nucl. Chem.*, **11**, 104 (1959).
57. V. P. Henkel and W. Klemm, *Z. Anorg. Chem.*, **222**, 70 (1935).
58. D. M. Gruen, J. G. Malm and B. Weinstock, *J. Chem. Phys.*, **24**, 905 (1956).
59. B. Weinstock and J. G. Malm, *J. Chem. Phys.*, **27**, 594 (1957).
60. C. A. Hutchinson and B. Weinstock, *J. Chem. Phys.*, **32**, 56 (1960).
61. C. L. Goodman and M. Fred, *J. Chem. Phys.*, **30**, 849 (1959).
62. M. H. Popov and Yu. V. Gagarinskii, *Russ. J. Inorg. Chem.*, **2**, 9 (1959).

63. M. J. Steindler, U.S. Report ANL-6287 (1961).
64. M. J. Steindler, U.S. Report ANL-6753, pp. 25–27 (1963).
65. J. F. Ellis, L. H. Brooks and K. D. B. Johnson, *J. Inorg. Nucl. Chem.*, **6**, 194 (1958).
66. J. Fischer and R. C. Vogel, *J. Am. Chem. Soc.*, **76**, 4829 (1954).
67. J. K. Dawson, D. W. Ingram and L. L. Bircumshaw, *J. Chem. Soc.*, **1950**, 1421.
68. A. D. Tevebaugh and F. Vaslov, U.S. Pat. 2,638,406 (1953).
69. A. S. Wolf, W. E. Hobbs and K. E. Rapp, *Inorg. Chem.*, **4**, 755 (1965).
70. L. E. Trevorrow, J. Fischer and W. H. Gunther, *Inorg. Chem.*, **2**, 1281 (1963).
71. J. S. Nairn, D. A. Collins and J. C. Taylor, *Proc. Intern. Conf. Peaceful Uses At. Energy, 2nd Geneva*, **4**, 191 (1958).
72. B. Cohen, A. J. Edwards, M. Mercer and R. D. Peacock, *Chem. Commun.*, **1965**, 322.
73. T. A. O'Donnell and D. F. Stewart, *Inorg. Chem.*, **5**, 1434 (1966).
74. M. Michallet, M. Chevreton, P. Plurien and D. Massingnon, *Compt. Rend.*, **249**, 691 (1959).
75. G. A. Rampy, U.S. Report GAT-T-697 (1959).
76. N. P. Galkin, B. N. Sudarikov and V. A. Zaitsev, *Soviet J. At. Energy* (English Transl.), **11**, 1210 (1960).
77. I. B. Johns, A. D. Tevebaugh, E. Gladrow, K. Walsh, P. Chiotti, B. Ayres, F. Vaslov and R. W. Fischer, U.S. Report TID-5290, p. 39 (1958).
78. J. R. Geichmann, E. A. Smith and P. R. Ogle, *Inorg. Chem.*, **2**, 1012 (1963).
79. J. R. Geichmann, P. R. Ogle and L. R. Swaney, U.S. Report GAT-T-809 (1961).
80. H. Martin and A. Albers, *Naturwiss.*, **33**, 370 (1946).
81. H. Martin, A. Albers and H. P. Dust, *Z. Anorg. Chem.*, **265**, 128 (1951).
82. R. E. Worthington, U.K.A.E.A. Report IGR-R/CA-200 (1957).
83. F. E. Massoth and W. E. Hensel, *J. Phys. Chem.*, **62**, 479 (1958); **63**, 697 (1959).
84. G. I. Cathers, M. R. Bennett and R. L. Jolly, *Ind. Eng. Chem.*, **50**, 1709 (1958).
85. I. Sheft, H. H. Hyman, R. M. Adams and J. J. Katz, *J. Am. Chem. Soc.*, **83**, 291 (1961).
86. S. Katz, *Inorg. Chem.*, **3**, 1598 (1964).
87. J. G. Malm, H. Selig and S. Siegel, *Inorg. Chem.*, **5**, 130 (1966).
88. N. S. Nikolaev and V. F. Sukhoverkhov, *Dokl. Akad. Nauk SSSR.*, **136**, 621 (1961).
89. B. Volavsek, *Croat. Chem. Acta*, **35**, 61 (1963).
90. B. Frlec, B. S. Brcic and J. Slivnik, *Croat. Chem. Acta*, **36**, 173 (1964).
91. J. J. Berzelius, *Ann. Phys.*, **1**, 34 (1842).
92. S. Fried, according to reference 3, page 471.
93. T. Keenan, unpublished observations.
94. K. O Johnson and G. H. Klewett, British Report AECD-4182 (1946).
95. G. O. Morriss, U.S. Report, BDDA-217 (1944).
96. Reference 1, p. 566.
97. K. W. Bagnall, D. Brown and J. F. Easey, *J. Chem. Soc.*, (A), **1968**, in press.

98. K. W. Bagnall and J. B. Laidler, *J. Chem. Soc.*, **1964**, 2693.
99. Reference 1, p. 565.
100. H. H. Anderson, reference 2, page 825.
101. I. F. Alenchikova, L. L. Zaitseva, L. V. Lipis, N. S. Nikolaev, V. V. Fomin and N. T. Chebotarev, *Zh. Neorgan. Khim.*, **3**, 951 (1958).
102. L. H. Brooks, E. V. Garner and E. Whitehead, British Report IGR/TN/CA-277 (1956).
103. C. A. Kraus, U.S. Report CC-1717 (1944).
104. W. H. Zachariasen, *Acta Cryst.*, **1**, 277 (1948).
105. W. H. Zachariasen, *Acta Cryst.*, **2**, 388 (1949).
106. W. L. Marshall, W. S. Gill and C. H. Secoy, *J. Am. Chem. Soc.*, **76**, 4279 (1954).
107. R. Kunin, U.S. Report A-3255 (1945), according to reference 1, p. 570.
108. I. F. Alenchikova, L. V. Lipis and N. S. Nikolaev, *At. Energ.* (*USSR*), **10**, 592 (1961).
109. L. M. Ferris and F. G. Baird, *J. Electrochem. Soc.*, **107**, 305 (1960).
110. L. M. Ferris and R. P. Gardner, U.S. Report ORNL-2690 (1959).
111. G. A. Rampy, U.S. Report GAT-T-961 (1961).
112. A. von Unruh, Dissertation, Rostock, 1909, according to reference 1, p. 576.
113. Reference 1, p. 572.
114. L. L. Laitseva, L. V. Lipis, V. V. Fomin and N. T. Chebotarev, *Russ. J. Inorg. Chem.*, **7**, 795 (1962).
115. L. M. Ferris, *J. Chem. Eng. Data*, **5**, 241 (1960).
116. B. Sahoo and K. C. Satapathy, *J. Inorg. Nucl. Chem.*, **26**, 1379 (1964).
117. K. W. Bagnall, D. Brown and J. F. Easey, unpublished observations.
118. E. A. Ippolitova and L. N. Kovba, *Dokl. Akad. Nauk SSSR*, **138**, 605 (1961).
119. G. Mitra, *Z. Anorg. Chem.*, **326**, 98 (1963).
120. W. H. Zachariasen, *Acta Cryst.*, **7**, 783 (1954).
121. E. Staritzky, D. T. Cromer and D. I. Walker, *Anal. Chem.*, **28**, 1355 (1956).
122. N. P. Galkin, V. D. Veryatin and V. I. Karpov, *Russ. J. Inorg. Chem.*, **7**, 1043 (1962).
123. Reference 2, p. 1115.
124. L. Brewer, L. A. Bromley, P. W. Gilles and N. L. Lofgren, U.S. Report UCRL-633 (1950).
125. G. A. Rampy, U.S. Report GAT-T-265 (1959).
126. L. Stein, *Inorg. Chem.*, **3**, 995 (1964).
127. D. Brown and J. F. Easey, unpublished observations.
128. L. B. Asprey, R. A. Penneman and F. H. Kruse, 'Physico-chimie du Protactinium', Centre Nat. de la Recherche Scientifique, Pub. No. 154, Orsay, Paris, p. 113 (1966).
129. A. von Grosse, *Proc. Roy. Soc.* (*London*), *Ser. A*, **150**, 365 (1935); *J. Am. Chem. Soc.*, **56**, 2501 (1934).
130. A. von Grosse, U.S. Report TID-5290, 315 (1958).
131. A. S. Wolf, J. C. Posey and K. E. Rapp, *Inorg. Chem.*, **4**, 751 (1965).
132. P. Agron, A. Grenall, R. Kunin and S. Weller, U.S. Report TID-5290, 652 (1958).
133. Reference 1, p. 386.

134. W. E. Hobbs, U.S. Patent 3,035,894 (1962).
135. J. R. Geichman, L. R. Swaney and P. R. Ogle, U.S. Report GAT-T-808 (1962).
136. W. H. Zachariasen, *Acta Cryst.*, **2**, 296 (1949).
137. H. F. Priest, U.S. Report TID-5290, 738 and 742 (1958).
138. H. Nguyen-Nghi, H. Marquet-Ellis and A. J. Dianoux, *Compt. Rend.*, **259**, 4683 (1964).
139. L. B. Asprey and R. A. Penneman, *Science*, **145**, 924 (1964).
140. D. Brown and J. F. Easey, *J. Chem. Soc.* (A), **1966**, 254.
141. L. O. Keller, Jnr., and A. Chetham-Strode, Jnr., reference 128, p. 119.
142. L. B. Asprey, F. H. Kruse, A. Rosenzweig and R. A. Penneman, *Inorg. Chem.*, **5**, 659 (1966).
143. L. B. Asprey, F. H. Kruse and R. A. Penneman, *J. Am. Chem. Soc.*, **87**, 3518 (1965).
144. L. B. Asprey, T. K. Keenan and R. A. Penneman, *Inorg. Nucl. Chem. Letters*, **2**, 19 (1966).
145. R. A. Penneman, G. D. Sturgeon, L. B. Asprey and F. H. Kruse, *J. Am. Chem. Soc.*, **87**, 5803 (1965).
146. L. B. Asprey and R. A. Penneman, *Inorg. Chem.*, **3**, 727 (1964).
147. G. B. Sturgeon, R. A. Penneman, F. H. Kruse and L. B. Asprey, *Inorg. Chem.*, **4**, 748 (1965).
148. L. B. Asprey and R. A. Penneman, *J. Am. Chem. Soc.*, **89**, 172 (1967).
149. P. R. Ogle, J. R. Geichman and S. S. Trond, U.S. Report GAT-T-552 (1959).
150. J. R. Geichman, E. A. Smith, S. S. Trond and P. R. Ogle, *Inorg. Chem.*, **1**, 661 (1962).
151. J. R. Geichman, L. R. Swaney and P. R. Ogle, U.S. Report GAT-T-971 (1962).
152. R. A. Penneman, L. B. Asprey and G. Sturgeon, *J. Am. Chem. Soc.*, **84**, 4608 (1962).
153. R. A. Penneman, G. D. Sturgeon and L. B. Asprey, *Inorg. Chem.*, **3**, 126 (1964).
154. H. Nguyen-Nghi, A. J. Dianoux, H. Marquet-Ellis and P. Plurien, *Compt. Rend.*, **260**, 1963 (1965).
155. R. Bougon and P. Plourien, *Compt. Rend.*, **260**, 4217 (1965).
156. P. Charpin, *Compt. Rend.*, **260**, 1914 (1965).
157. J. H. Burns, H. A. Levy and O. L. Keller, U.S. Report ORNL-4146 (1967); *Acta Cryst.*, in press (1968).
158. M. Reisfeld and G. A. Crosby, *Inorg. Chem.*, **4**, 65 (1965).
159. D. Brown and J. F. Easey, *Nature*, **205**, 589 (1965).
160. M. N. Bukhsh, J. Flegenheimer, F. M. Hall and A. G. Maddock, *J. Inorg. Nucl. Chem.*, **28**, 421 (1966).
161. D. Brown and A. J. Smith, *Chem. Commun.*, 554 (1965); D. Brown, S. F. A. Kettle and A. J. Smith, *J. Chem. Soc.* (A), **1967**, 1429.
162. W. Rüdorff and H. Leutner, *Ann. Chemie.*, **632**, 1 (1960).
163. K. W. Bagnall, D. Brown and J. F. Easey, to be published.
164. R. E. Thoma, H. A. Friedman and R. A. Penneman, *J. Am. Chem. Soc.* **88**, 2046 (1966).

165. S. S. Kirslis, T. S. McMillan and H. A. Bernhardt, U.S. Report K-567 (1956).
166. L. B. Asprey, F. H. Ellinger and W. H. Zachariasen, *J. Am. Chem. Soc.*, **76**, 5235 (1954).
167. T. Keenan, *Inorg. Chem.*, **4**, 1500 (1965).
168. R. Livingston and W. Burns, U.S. Report CN-982 (1943), according to reference 1, p. 383.
169. P. Agron, A. Grenall, R. Kunin and S. Weller, U.S. Report MDDC-1588 (1948); U.S. Report TID-5290, p. 652 (1958).
170. J. S. Broadley and P. B. Longton, British Report R and D.B. (C) TN-60 (1954).
171. V. Y. Labaton, British Report IGR-R/CA-193 (1956).
172. Nguyen-Hoang-Nghi, French Report CEA 1976 (1961).
173. P. Agron, U.S. Report TID-5290, p. 610 (1958).
174. W. H. Zachariasen, *Acta Cryst.*, **2**, 390 (1949).
175. L. Stein, reference 128, p. 101.
176. Reference 3, p. 378.
177. R. W. M. D'Eye and G. W. Booth, *J. Inorg. Nucl. Chem.*, **1**, 326 (1955); *J. Inorg. Nucl. Chem.*, **4**, 13 (1957).
178. Yu. V. Gagarinskii and V. P. Mashirev, *Zh. Neorgan. Khim.*, **4**, 1246 (1959).
179. J. K. Dawson, R. W. M. D'Eye and A. E. Truswell, *J. Chem. Soc.*, **1954**, 3922.
180. N. Barson and M. Smutz, U.S. Report AECD-3705 (1954).
181. Reference 1, page 360.
182. A. S. Newton, H. Lipkind, W. H. Keller and J. E. Iliff, U.S. Report TID-5223 (1952).
183. R. W. Fisher, *Progr. Nucl. Energy Ser. III* (F. R. Bruce, J. M. Fletcher and H. H. Hyman, Eds.), Pergamon Press, London, **2**, 149 (1958).
184. S. H. Smiley and D. C. Brater, *Progr. Nucl. Energy Ser. III* (F. R. Bruce, J. M. Fletcher and H. H. Hyman, Eds.), Pergamon Press, London, **2**, 171 (1958).
185. B. A. Lister and G. B. Gilles, *Progr. Nucl. Energy Ser. III* (F. R. Bruce, J. M. Fletcher and H. H. Hyman, Eds.), Pergamon Press, London, **1**, 19 (1956).
186. N. M. Levitz, E. J. Petkus, M. M. Katz and A. A. Jonke, *Chem. Eng. Progr.*, **53**, 199 (1957).
187. *Proc. Intern. Conf. Peaceful Uses At. Energy*, 2nd Geneva, **4**, 44, 133, 139, 165 (1958).
188. Reference 1, p. 362.
189. Reference 1, p. 365.
190. J. Sanlaville, *Proc. Intern. Conf. Peaceful Uses At. Energy*, 2nd Geneva, **4**, 102 (1958).
191. J. D. Pedregal and F. Aquilar, *Proc. Intern. Conf. Peaceful Uses At. Energy*, 2nd Geneva, **4**, 88 (1958).
192. J. R. Long, U.S. Report AECD-3203 (1948).
193. J. D. Pedregal and F. Aquilar, *Energia Nucl.* (*Madrid*), **3**, 39 (1959).
194. T. R. Braddock and D. Copenhafer, U.S. Reports N-17 and N-24, according to reference 1, p. 365.

195. H. S. Booth, W. Krasny Ergen and R. Heath, *J. Am. Chem. Soc.*, **68,** 1969 (1946).
196. A. Cacciari, R. de Leone, C. Fizzotti and M. Gabaglio, *Energia Nucl.* (*Milan*), **3,** 462 (1956).
197. A. Cacciari, C. Fizzotti, G. M. Gabaglio and R. de Leone, *Energia Nucl.* (*Madrid*), **1,** 11 (1957).
198. V. V. Dapade and N. S. Krishna Prasad, *Proc. Intern. Conf. Peaceful Uses At. Energy*, *2nd Geneva*, **4,** 125 (1958).
199. L. Grainger, *Proc. Intern. Conf. Peaceful Uses At. Energy*, *1st Geneva*, **8,** 149 (1955).
200. B. Goldschmidt and P. Vertes, *Proc. Intern. Conf. Peaceful Uses At. Energy*, *1st Geneva*, **8,** 152 (1955).
201. *Proc. Intern. Conf. Peaceful Uses At. Energy*, *2nd Geneva*, Collected papers in Volume **4** (1958).
202. J. S. Nairn, D. A. Collins and J. C. Taylor, *Proc. Intern. Conf. Peaceful Uses At. Energy*, *2nd Geneva*, **4,** 191 (1958).
203. S. H. Smiley and D. C. Brater, *Proc. Intern. Conf. Peaceful Uses At. Energy*, *2nd Geneva*, **4,** 196 (1958); reference 183, p. 107.
204. E. L. Muetterties and J. E. Castle, *J. Inorg. Nucl. Chem.*, **18,** 148 (1961).
205. H. Lipkind and A. S. Newton, U.S. Report TID-5290, 398 (1952).
206. Reference 1, p. 364.
207. F. H. Spedding, A. S. Newton, J. C. Warf, O. Johnson, R. W. Nottorf, I. B. Johns and A. H. Daane, *Nucleonics*, **4,** 4 (1949).
208. Reference 3, p. 138.
209. H. Moissan and M. Martinsen, *Compt. Rend.*, **104,** 1510 (1905).
210. E. Chauvenet, *Compt. Rend.*, **146,** 973 (1908).
211. B. Sahoo and D. Patnaik, *Current Sci.* (*India*), **28,** 401 (1959).
212. P. A. Sellers, S. Fried, R. E. Elson and W. H. Zachariasen, *J. Am. Chem. Soc.*, **76,** 5935 (1954).
213. S. Fried and N. Davidson, *J. Am. Chem. Soc.*, **70,** 3539 (1948).
214. A. E. Florin and R. E. Heath, U.S. Report CK-1372 (1944).
215. F. Meyer and H. Zvolner, U.S. Report CK-1763 (1944).
216. J. Maly, I. Peka, M. Talas and M. Tympl, *Radiokhimiya*, **3,** 195 (1961).
217. W. B. Tolley, U.S. Report H.W.-31211 (1954).
218. B. B. Cunningham, D. C. Freay and M. A. Rollier, *J. Am. Chem. Soc.*, **76,** 3361 (1954).
219. L. B. Asprey, F. H. Ellinger, S. Fried and W. H. Zachariasen, *J. Am. Chem. Soc.*, **79,** 5825 (1957).
220. L. B. Asprey and T. Keenan, *J. Inorg. Nucl. Chem.*, **7,** 27 (1958).
221. C. von Wartenberg, *Z. Anorg. Chem.*, **244,** 339 (1940).
222. L. B. Asprey, *J. Am. Chem. Soc.*, **76,** 2019 (1954).
223. W. J. Asher and A. W. Wylie, *Aust. J. Chem.*, **18,** 959 (1965).
224. W. Klemm and P. Henkel, *Z. Anorg. Chem.*, **220,** 180 (1934).
225. J. Soriano, M. Givon and J. Shamir, *Inorg. Nucl. Chem. Letters*, **2,** 13 (1966).
226. L. B. Asprey, J. S. Coleman and M. J. Reisfeld in *Lanthanide/Actinide Chemistry*, Advances in Chemistry Series 71 (R. E. Gould, Ed.), Amer. Chem. Soc. Publ., Washington, 1967, p. 122.
227. A. C. Larson, R. B. Roof, Jnr., and D. T. Cromer, *Acta Cryst.*, **17,** 555 (1964).

228. A. Darnell and F. J. Keneshea, *J. Phys. Chem.*, **62**, 143 (1958).
229. S. Langar and F. E. Blankenship, *J. Inorg. Nucl. Chem.*, **14**, 26 (1960).
230. M. M. Popov, F. A. Kostleyer and N. V. Zubova, *Zh. Neorgan. Khim.*, **4**, 1708 (1959).
231. J. J. Katz and I. Sheft in *Advances in Inorganic and Radiochemistry* (H. J. Emeleus and A. G. Sharpe, Eds.), Vol. 2, Academic Press, New York, 1960, p. 226.
232. G. N. Yakovlev and V. N. Kosylkov, *Proc. Intern. Conf. Peaceful Uses At. Energy, 2nd Geneva*, **28**, 373 (1958).
233. Reference 1, p. 366.
234. N. Elliott, *Phys. Rev.*, **76**, 431 (1949).
235. J. K. Dawson, *J. Chem. Soc.*, **1951**, 429, 2884; **1952**, 1185, 1882.
236. S. N. Gosh, W. Goody and D. Hill, *Phys. Rev.*, **96**, 36 (1954).
237. D. M. Gruen and M. Fred, *J. Am. Chem. Soc.*, **76**, 3850 (1954).
238. L. B. Asprey and T. K. Keenan, *J. Inorg. Nucl. Chem.*, **7**, 27 (1958).
239. Yu. V. Gagarinskiĭ and V. P. Mashirev, *Zh. Neorgan. Khim.*, **4**, 1253 (1959).
240. M. M. Popov and Yu. V. Gagarinskiĭ, *Zh. Neorgan. Khim.*, **2**, 3 (1957).
241. K. J. Dawson, R. M. Elliott, R. Hurst and A. E. Truswell, *J. Chem. Soc.*, **1954**, 558.
242. G. L. Gal'chenko, Yu. V. Gagarinskiĭ and M. M. Popov, *Zh. Neorgan. Khim.*, **5**, 1631 (1960).
243. M. Haissinsky and G. Bouissières, *Bull. Soc. Chim. France*, **18**, 146 (1951).
244. H. J. Spencer-Palmer, Report BR-422 (1944), according to reference 1, p. 531.
245. O. J. C. Runnalls, *Can. J. Chem.*, **31**, 694 (1953).
246. J. C. Warf, U.S. Report AECD-2523 (1949).
247. C. M. Schwarz and D. A. Vaughan, U.S. Report BMI-266 (1953).
248. Reference 3, p. 378.
249. H. Huet, reference 5, Part I, p. 180; R. Pascard, reference 6, p. 492.
250. J. H. Gittus, *Uranium*, Butterworths, London (1963).
251. H. A. Wilhelm (Ed.), *The Metal Thorium*, American Society for Metals, Cleveland, 1958.
252. A. S. Coftinberry and W. N. Miner (Eds.), *The Metal Plutonium*, University of Chicago Press, Chicago, 1961.
253. W. D. Wilkinson (Ed.), *Extraction and Physical Metallurgy of Plutonium and its Alloys*, Interscience, New York, 1960.
254. L. B. Asprey and B. B. Cunningham in *Progress in Inorganic Chemistry*, (F. A. Cotton, Ed.) Vol. 2, Interscience, New York, 1960, p. 270; L. B. Asprey, in *Rare Earth Research* (E. V. Kleber, Ed.), Macmillan, New York, 1961, p. 58.
255. R. E. Thoma, *Inorg. Chem.*, **1**, 220 (1962).
256. C. J. Barton, H. A. Friedman, W. R. Grimes, H. Insley, R. E. Moore and R. E. Thoma, *J. Am. Ceram. Soc.*, **41**, 63 (1958).
257. L. A. Harris, G. D. White and R. E. Thoma, *J. Phys. Chem.*, **63**, 1974 (1959).
258. C. F. Weaver, R. E. Thoma, H. Insley and H. A. Friedman, *J. Am. Ceram. Soc.*, **43**, 213 (1960).
259. G. D. Brunton, U.S. Report ORNL-3913, p. 10 (1965); *Acta Cryst.*, **21**, 814 (1966).

260. G. D. Brunton, H. Insley, T. N. McVay and R. E. Thoma, U.S. Report ORNL-3761 (1965).
261. H. H. Anderson, reference 2, p. 775.
262. I. F. Alenchikova, L. L. Zaitseva, L. V. Lipis, V. V. Fomin and N. T. Chebotarev, *Proc. Intern. Conf. Peaceful Uses At. Energy*, *2nd Geneva*, **28**, 309 (1958).
263. E. N. Diechmann and I. V. Tananaev, *Radiokhimyia*, **4**, 66 (1961).
264. W. H. Zachariasen, *J. Am. Chem. Soc.*, **70**, 2147 (1948); reference 2, p. 1464.
265. J. H. Burns, R. D. Ellison and H. A. Levy, U.S. Report ORNL-3913, p. 17 (1965); *Acta Cryst.*, **B24**, 230 (1968).
266. R. A. Penneman, F. H. Kruse, R. S. George and J. S. Coleman, *Inorg. Chem.*, **3**, 309 (1964).
267. R. Benz, R. M. Douglass, F. H. Kruse and R. A. Penneman, *Inorg. Chem.*, **2**, 799 (1963).
268. R. E. Thoma, H. Insley, B. S. Landau, H. A. Friedman and W. R. Grimes, *J. Am. Ceram. Soc.*, **41**, 538 (1958).
269. E. M. Levin, C. R. Robbins and H. F. McMurdie (Eds.), *Phase Diagrams for Ceramists*, The American Ceramic Society, 1964.
270. R. E. Thoma, U.S. Report ORNL-2548 (1959).
271. R. A. Penneman, T. K. Keenan and L. B. Asprey, ref. 226 p. 248.
272. L. B. Asprey, F. H. Kruse and R. A. Penneman, *Inorg. Chem.*, **6**, 544 (1967).
273. T. K. Keenan, *Inorg. Nucl. Chem. Letters*, **2**, 153 (1966); **2**, 211 (1966).
274. J. H. Burns, D. R. Sears and G. D. Brunton, U.S. Report ORNL-3913, p. 17 (1965); G. D. Brunton, U.S. Report ORNL-4076, p. 7 (1966); *J. Inorg. Nucl. Chem.* **29**, 1631 (1967).
275. R. E. Thoma, H. Insley, B. S. Landau, H. A. Friedman and W. R. Grimes, *J. Phys. Chem.*, **63**, 1266 (1959).
276. R. E. Thoma, H. Insley, G. M. Herbert, H. A. Friedman and C. F. Weaver, *J. Am. Ceram. Soc.*, **46**, 37 (1963).
277. V. S. Emelyanov and A. I. Evstyukhin, *J. Nucl. Energy*, **5**, 108 (1957).
278. L. B. Asprey and T. K. Keenan, *J. Inorg. Nucl. Chem.*, **16**, 260 (1961).
279. M. Brodsky and P. Pagny, *Proc. Intern. Conf. Peaceful Uses At. Energy*, *2nd Geneva*, **4**, 69 (1958).
280. I. V. Tananaev and Lu Chzhao-Da, *Zh. Neorgan. Khim.*, **4**, 2116 (1959).
281. I. V. Tananaev, N. P. Galkin, G. S. Savchenko and V. M. Sutyagin, *Zh. Neorgan. Khim.*, **7**, 1675 (1962).
282. R. Hoppe and K. M. Rödder, *Z. Anorg. Chem.*, **313**, 154 (1961).
283. R. Hoppe and W. Liebe, *Z. Anorg. Chem.*, **313**, 221 (1961).
284. C. Keller and H. Schmutz, *Inorg. Nucl. Chem. Letters*, **2**, 355 (1966).
285. W. J. Asker, E. R. Segnit, and A. W. Wylie, *J. Chem. Soc.*, **1952**, 4470.
286. G. E. Kaplan, *Proc. Intern. Conf. Peaceful Uses At. Energy*, *1st Geneva*, **8**, 184 (1956).
287. A. G. Dergunov and E. P. Bergman, *Dokl. Akad. Nauk. SSSR*, **60**, 391 (1948).
288. Reference 3, p. 472.
289. Reference 3, p. 380.
290. L. B. Asprey, *J. Am. Chem. Soc.*, **76**, 2019 (1954).
291. L. B. Asprey and R. A. Penneman, *Inorg. Chem.*, **1**, 134 (1962).

292. W. H. Zachariasen, reference 2, p. 1469.
293. L. A. Harris, *Acta Cryst.*, **13**, 502 (1960).
294. R. E. Thoma, U.S. Report ORNL-2548, p. 76 (1959).
295. L. B. Asprey and R. A. Penneman, *Inorg. Chem.*, **1**, 137 (1962).
296. L. B. Asprey and R. A. Penneman. Reference 128, p. 109.
297. T. J. LaChapelle, L. B. Magnusson and J. C. Hindman, reference 2, p. 1097.
298. R. C. Mooney, U.S. Report ANL-4082 (1947).
299. R. E. Thoma and T. S. Carlton, *J. Inorg. Nucl. Chem.*, **17**, 88 (1961).
300. R. Hoppe and K. M. Rödder, *Z. Anorg. Chem.*, **312**, 277 (1961).
301. J. O. Blomeke, U.S. Report ORNL-1030 (1951).
302. R. E. Thoma, Oak Ridge National Laboratory Central Files Memorandum No. 58-12-40 (1958).
303. M. Salzer, German Report KFK-385 (1966).
304. E. Chauvenet, *Compt. Rend.*, **23**, 425 (1911).
305. R. W. M. D'Eye, *J. Chem. Soc.*, **1958**, 196.
306. H. Moissan, *Compt. Rend.*, **123**, 148 (1896); *Compt. Rend.*, **131**, 597 (1900); H. Moissan and A. Étard, *Compt. Rend.*, **122**, 153 (1896).
307. O. N. Carlson and F. A. Schmidt in *The Rare Earths* (F. H. Spedding and A. H. Daane, Eds.), Wiley, New York, 1961, Chap. 6, p. 77.
308. F. H. Spedding and A. H. Daane in *Progr. Nucl. Energy Ser.* V, **1**, 413 (H. M. Finniston and J. R. Howe Eds.), Pergamon, London (1956).
309. A. Zalkin and D. H. Templeton, *J. Am. Chem. Soc.*, **75**, 2453 (1953).
310. J. Walker and E. Olsen, U.S. Report IS-2 (1959).
311. R. E. Thoma, G. M. Herbert, H. Insley and C. F. Weaver, *Inorg. Chem.*, **2**, 1005 (1963).
312. R. E. Thoma, H. Insley and G. M. Herbert, *Inorg. Chem.*, **5**, 1222 (1966).
313. R. E. Thoma and R. H. Karraker, *Inorg. Chem.*, **5**, 1933 (1966).
314. R. E. Thoma and G. D. Brunton, *Inorg. Chem.*, **5**, 1937 (1966).
315. J. Popovici, *Chem. Ber.*, **41**, 634 (1908).
316. A. H. Daane and F. H. Spedding, *J. Electrochem. Soc.*, **100**, 442 (1953).
317. E. Staritzky and L. B. Asprey, *Anal. Chem.*, **29**, 855, 856 (1957).
318. F. Weigel and V. Scherrer, *Radiochim. Acta*, **7**, 40 (1967).
319. E. P. Dergunov, *Dokl. Akad. Nauk SSSR*, **85**, 1025 (1952).
320. G. A. Bukhalov and E. P. Babaeva, *Russ. J. Inorg. Chem.*, **11**, 337 (1966).
321. R. L. Tischer and G. Burnet, U.S. Report IS-8 (1959).
322. A. I. Popov and G. E. Kundson, *J. Am. Chem. Soc.*, **76**, 3921 (1954).
323. S. Fried, F. Hagemann and W. H. Zachariasen, *J. Am. Chem. Soc.*, **72**, 771 (1950).
324. S. Fried, *J. Am. Chem. Soc.*, **73**, 416 (1951).
325. E. F. Westrum and L. Eyring, *J. Am. Chem. Soc.*, **73**, 3396 (1951).
326. J. C. Wallman, W. W. T. Crane and B. B. Cunningham, *J. Am. Chem. Soc.*, **73**, 493 (1951).
327. D. C. Feay, U.S. Report UCRL-2547 (1954).
328. J. G. Stites, M. L. Salutsky and B. D. Stone, *J. Am. Chem. Soc.*, **77**, 237 (1955).
329. L. B. Asprey, T. K. Keenan and F. H. Kruse, *Inorg. Chem.*, **4**, 985 (1965).
330. Reference 3, p. 377.
331. F. Anselin, P. Fangeras and E. Guison, *Compt. Rend.*, **242**, 1996 (1952).
332. J. K. Dawson and A. E. Truswell, Report AERE C/R 662 (1951).

333. L. L. Burger and W. E. Roake, U.S. Report HW 26022 (1952); U.S. Pat. 2,992,066 (1961).

334. H. Nguyen-Nghi, A. J. Dianoux and H. Marquet-Ellis, *Compt. Rend.*, **259**, 811 (1964).

335. I. Oftdel, *Z. Physik. Chem.*, **B5**, 272 (1929); **B13**, 190 (1931).

336. K. Schylter, *Arkiv Kemi*, **5**, 73 (1953).

337. W. H. Zachariasen, reference 3, p. 788.

338. J. M. Baker and R. S. Rubins, *Proc. Phys. Soc. (London)*, **78**, 1353 (1961).

339. F. Wong, O. Stafsudd and D. Jonston, *J. Chem. Phys.*, **39**, 786 (1963); *Phys. Rev.*, **131**, 990 (1963).

340. E. V. Sayre and S. Freed, *J. Chem. Phys.*, **23**, 2066 (1955).

341. W. Kruphe and J. Gruder, *J. Chem. Phys.*, **39**, 1024 (1963).

342. H. H. Caspers, R. A. Buchanan and H. R. Marlin, *J. Chem. Phys.*, **41**, 94 (1964).

343. M. Mansmann, *Z. Anorg. Chem.*, **331**, 98 (1964).

344. A. Zalkin, D. H. Templeton and T. E. Hopkins, *Inorg. Chem.*, **5**, 1467 (1966).

345. A. F. Wells, *Structural Inorganic Chemistry*, Clarendon Press, Oxford, 1962, p. 341.

346. R. W. G. Wyckoff, *Crystal Structures*, Vol. II, Interscience, New York, 1965, p. 52.

347. L. O. Gilpatrick, R. Baldock and J. R. Stites, U.S. Report ORNL-1376 (1952).

348. T. E. Phipps, G. W. Sears, R. L. Siefert and O. C. Simpson, *J. Chem. Phys.*, **18**, 713 (1950).

349. S. C. Carniglia and B. B. Cunningham, *J. Am. Chem. Soc.*, **77**, 1451 (1955).

350. A. D. Kirshenbaum and J. A. Cahill, *J. Inorg. Nucl. Chem.*, **14**, 148 (1960).

351. Reference 1, p. 354.

352. D. J. C. Runnals, U.S. Report AECL-543 (1958).

353. E. Grison, W. B. H. Lord and R. D. Fowler (Eds.), *Plutonium* 1960, Cleaver Hume Press, London, 1961, Chap. 4, p. 171.

354. J. C. Warf and F. Edwards, U.S. Report CC-1496 (1944).

355. Reference 1, p. 541.

356. G. Mandel, R. P. Baumen and E. Banks, *J. Chem. Phys.*, **33**, 192 (1960).

357. W. Low, *Ann. N.Y. Acad. Sci.*, **72**, 69 (1958).

358. M. C. M. O'Brien, *Proc. Phys. Soc.*, **68A**, 351 (1955).

359. B. Bleaney, P. M. Llewellyn and D. A. Jones, *Proc. Phys. Soc.*, **69B**, 858 (1956).

360. W. J. de Haas and E. C. Wiersma, *Congr. Intern. Froid, 6th Buenos Aires, Comm. Kammerlingh Ormes Lab. Univ. Leiden Suppl.*, **74**, 36 (1932), according to *Chem. Abs.*, **28**, 12⁸.

361. W. J. de Haas and C. J. Garter, *Proc. Acad. Sci. Amsterdam*, **33**, 949, 1930.

362. H. Kubicka, *Roczniki Chem.*, **35**, 385 (1961), according to *Chem. Abs.*, **55**, 19407i.

363. J. K. Dawson, *J. Chem. Soc.*, **1951**, 429.

364. J. K. Dawson, C. J. Mandleberg and D. Davies, *J. Chem. Soc.*, **1951**, 2047.

365. W. W. T. Crane, U.S. Report UCRL-1220 (1951); W. W. T. Crane, B. B. Cunningham and J. Wallman, U.S. Report UCRL-846 (1950).

366. E. P. Dergunov, *Dokl. Akad. Nauk SSSR*, **60**, 1185 (1948), according to *Chem. Abs.*, **42**, 6696d.

367. R. E. Thoma, C. F. Weaver, H. A. Friedman, H. Insley, L. A. Harris and H. A. Yakel, Jnr., *J. Phys. Chem.*, **65**, 1096 (1961).
368. C. J. Barton, J. D. Redman and R. A. Strehlow, *J. Inorg. Nucl. Chem.*, **20**, 45 (1961).
369. Reference 270, p. 87.
370. D. M. Roy and R. Roy, *J. Electrochem. Soc.*, **111**, 421 (1964).
371. C. Keller, *Z. Naturforsch.*, **19b**, 1080 (1964).
372. W. H. Zachariasen, *Acta Cryst.*, **1**, 265 (1948).
373. H. Bode and E. Voss, *Z. Anorg. Chem.*, **290**, 1 (1957).
374. R. J. Meyer, *Z. Anorg. Chem.*, **86**, 275 (1914).
375. L. A. Khripin, *Izv. Sibirsk. Otd. Akad. Nauk SSSR, Ser. Khim. Nauk.*, 107 (1963), according to *Chem. Abs.*, **60**, 2378e.
376. Reference 270, p. 69.
377. Reference 270, p. 84.
378. C. J. Barton and R. A. Strehlow, *J. Inorg. Nucl. Chem.*, **18**, 143 (1961).
379. J. H. Burns, *Inorg. Chem.*, **4**, 881 (1965).
380. L. R. Batsanova, *Izv. Sibirsk. Otd. Akad. Nauk SSSR, Ser. Khim. Nauk.*, 83 (1963), according to *Chem. Abs.*, **59**, 13578f.
381. W. Wendlandt, *Science*, **129**, 842 (1959).
382. W. H. Zachariasen, *Acta Cryst.*, **4**, 231 (1951).
383. N. C. Baenziger, J. R. Holden, G. E. Knudsen and A. I. Popov, *J. Am. Chem. Soc.*, **76**, 4734 (1954).
384. L. R. Batsanova and G. N. Kustova, *Zh. Neorgan. Khim.*, **9**, 330 (1964).
385. W. Klemm and H. A. Klein, *Z. Anorg. Chem.*, **248**, 167 (1941).
386. F. Hund, *Z. Anorg. Chem.*, **265**, 62 (1951).
387. D. H. Templeton and C. H. Dauben, *J. Am. Chem. Soc.*, **76**, 5237 (1954).
388. L. Mazza and A. Iandelli, *Atti accord. ligures sci e lettere (Pavia)*, **7**, 44 (1951), according to *Chem. Abs.*, **47**, 4194a.
389. D. Brown and J. F. Easey, unpublished observations.
390. M. J. Lim, U.S. Report UCRL-16150 (1965).
391. C. Keller and H. Schmutz, *J. Inorg. Nucl. Chem.*, **27**, 900 (1965).
392. F. Kutek, *Zh. Neorgan. Khim.*, **9**, 2784 (1964).
393. H. Hahn and R. Schmid, *Naturwiss.*, **52**, 475 (1965).
394. H. Caspers, H. E. Rast and R. A. Buchanan, *J. Chem. Phys.*, **42**, 3214 (1965).
395. K. Recker and J. Lierbertz, *Naturwiss.*, **49**, 391 (1962).
396. J. M. Baker, B. Bleaney and W. Hayes, *Proc. Roy. Soc. (London), Ser. A*, **247**, 141 (1958).
397. M. Dvir and W. Low, *Proc. Phys. Soc. (London)*, **75**, 136 (1960).
398. G. Wincow and W. Low, *Phys. Rev.*, **122**, 1390 (1961).
399. W. Low, *J. Phys. Soc. Japan, Suppl. B-I*, **17**, 440 (1962).
400. W. Low and U. Rosenberger, *Compt. Rend.*, **254**, 1771 (1962).
401. W. Low, *Phys. Rev.*, **134**, 265 (1964).
402. B. Bleaney, *Proc. Roy. Soc. (London), Ser. A*, **277**, 289 (1964).
403. U. Ranon, *Phys. Rev.*, **132**, 1609 (1963).
404. W. Klemm and W. Döll, *Z. Anorg. Chem.*, **241**, 233 (1939).
405. W. Döll and W. Klemm, *Z. Anorg. Chem.*, **241**, 236 (1939).
406. G. Beck and W. Nowacki, *Naturwiss.*, **26**, 495 (1938).
407. W. Klemm and H. Serff, *Z. Anorg. Chem.*, **241**, 359 (1939).

408. W. Nowacki, *Z. Krist.*, **99**, 339 (1938).
409. L. B. Asprey, F. H. Ellinger and E. Staritzky in *Rare Earth Research* (K. S. Vorres, Ed.), Vol. 2, Gordon and Breach, New York, 1964, p. 11.
410. W. Hayes and J. W. Twidell, *J. Chem. Phys.*, **35**, 1521 (1961).
411. D. S. McLure and Z. J. Kiss, *J. Chem. Phys.*, **39**, 3251 (1963).
412. Z. J. Kiss, *Phys. Rev.*, **127**, 708 (1961).
413. F. K. Fong, *J. Chem. Phys.*, **41**, 245 (1964).
414. E. S. Sabisky, *J. Chem. Phys.*, **41**, 892 (1964).
415. E. S. Sabisky, *J. Appl. Phys.*, **36**, 1788 (1965).
416. N. Edelstein, W. Easley and R. McLaughlin, U.S. Report UCRL-16608 (1966). *J. Chem. Phys.* **44**, 3130 (1966); Ref. 226, p. 203.
417. Z. J. Kiss and P. N. Yocum, *J. Chem. Phys.*, **41**, 1511 (1964).
418. F. K. Fong, *R.C.A. Rev.*, **25**, 303 (1964); *J. Chem. Phys.*, **41**, 2291 (1964).
419. H. Guggenheim and J. V. Kane, *Appl. Phys. Letters*, **4**, 172 (1964).
420. F. K. Fong in *Rare Earth Research* (K. S. Vorres, Ed.), Vol. 3, Gordon and Breach, New York, 1965, p. 373.
421. P. P. Sorokin, M. J. Stevensen, J. R. Lankard and G. D. Pettit, *Phys. Rev.*, **127**, 503 (1962).
422. J. D. Kingsley and J. S. Prener, *Phys. Rev.*, **126**, 458 (1962).
423. D. L. Wood and W. Kaiser, *Phys. Rev.*, **126**, 2079 (1962).
424. B. P. Zakharchenya and A. Ya. Ryskin, *Opt. i Spektroskopiya*, **13**, 875 (1962).
425. V. M. Vinokurov, M. M. Zaripov, V. G. Stepanov, G. K. Chirkin and L. Ya. Shekum, *Fiz. Tverd. Tela*, **5**, 1936 (1963).
426. T. R. McGure and M. W. Shafer, *J. Appl. Phys.*, **35**, 984 (1964).
427. F. H. Spedding and A. H. Daane, *Met. Rev.*, **5**, 297 (1960).
428. K. S. Vorres and R. Riviello in *Rare Earth Research* (L. Eyring, Ed.), Vol. 3, Gordon and Breach, New York, 1965, p. 521.
429. J. Burnett, U.S. Report AECD/65/332-9 (1965).
430. M. Berger and M. J. Sienko, *Inorg. Chem.*, **6**, 324 (1967).
431. J. Lee, personal communication (1966).
432. D. Brown, J. F. Easey and D. G. Holah, *J. Chem. Soc.*, (A) **1967**, 1979.
433. A. J. Dianoux, H. Marquet-Ellis, Nguyen-Nghi and P. Plurien, paper presented at the 3rd International Symposium on Fluorine Chemistry, Munich, Sept., 1965.
434. D. R. Sears, U.S. Report ORNL-4076, p. 11 (1966).
435. D. G. Brunton, unpublished observations.
436. R. E. Thoma in *Progress in the Science and Technology of The Rare Earths* (L. Eyring, Ed.), Vol. 2, Pergamon Press, London, 1966, p. 90.
437. R. Hoppe, personal communication (1967); K. Rödder, Thesis, Westfälischen, Wilhelms-Universität, Münster (1963).
438. R. A. Kent, K. F. Zmbov, A. S. Kana'an, G. Besenbruch, J. D. McDonald and J. L. Margrave, *J. Inorg. Nucl. Chem.*, **28**, 1419 (1966).
439. M. Lim and A. W. Searcy, *J. Phys. Chem.*, **70**, 1762 (1966).
440. K. F. Zmbov and J. L. Margrave, *J. Chem Phys.*, **45**, 3167 (1966).
441. G. Besenbruch, T. V. Charlu, K. F. Zmbov and J. L. Margrave, *J. Less-Common Metals*, **12**, 375 (1967).
442. K. F. Zmbov and J. L. Margrave, *J. Less-Common Metals*, **12**, 494 (1967).
443. N. S. Nikolaev and Yu. D. Shishkov, *Dokl. Akad. Nauk SSSR*, **143**, 130 (1962).

444. R. C. Shrewsberry and E. L. Williamson, USAEC Report KY-L-362, Part I (1964).
445. R. L. Jarry and M. J. Steindler, *J. Inorg. Nucl. Chem.*, **29**, 1591 (1967).
446. I. Peka and J. Vachuska, *Collection Czech. Chem. Commun.*, **32**, 426 (1967).
447. M. G. Otey and R. A. LeDoux, *J. Inorg. Nucl. Chem.*, **29**, 2249 (1967).
448. R. W. Mar and A. W. Searcy, *J. Phys. Chem.*, **71**, 888 (1967).
449. B. Frlec, *J. Inorg. Nucl. Chem.*, **29**, 2112 (1967).
450. B. N. Ianov-Emin, T. N. Susanina and A. I. Ezhov, *Russ. J. Inorg. Chem.*, **12**, 11 (1967).
451. B. Frlec and H. H. Claassen, *J. Chem. Phys.*, **46**, 4603 (1967).
452. B. Frlec, B. S. Brčić and J. Slivnik, *Inorg. Chem.*, **5**, 542 (1966).
453. B. Frlec and H. H. Hyman, ibid. **6**, 2233 (1967).
454. A. Rosenzweig and D. T. Cromer, *Acta Cryst.*, **23**, 865 (1967).
455. T. K. Keenan, *Inorg. Nucl. Chem. Letters*, **3**, 391 (1967).
456. T. K. Keenan, *Inorg. Nucl. Chem. Letters*, **3**, 463 (1967).
457. P. N. Yocum, ref. 226, p. 51.
458. H. L. Pinch, *J. Am. Chem. Soc.* **86**, 3167 (1964).

Chapter 3

Chlorides and Oxychlorides

HEXAVALENT

Apart from the hexafluorides (p. 21) uranium hexachloride is the only fully halogenated hexavalent actinide halide known and it is unlikely that further hexachlorides will be characterized since thermodynamic calculations[1] indicate that neptunium and plutonium hexachloride are incapable of existence. Even neptunium pentachloride and plutonium tetrachloride are, as yet, unknown. Uranyl chloride, UO_2Cl_2, and the hydrated plutonium (VI) compound, $PuO_2Cl_2 \cdot 6H_2O$, are known and oxychloro complexes of uranium (VI), neptunium (VI), plutonium (VI) and americium (VI), $M_2^I M^{VI} O_2 Cl_4$ (M^I = a univalent cation), have been prepared. Additional chloro complexes are known for uranium (VI), viz. $M_2^I UO_3 Cl_2$, $K_2 U_2 O_5 Cl_4$, $K_2 U_3 O_8 Cl_4$ and $K_2 U_2 O_5 OCl_2$.

Uranium Hexachloride

Several methods have been reported for the preparation of uranium hexachloride but not all are satisfactory. The methods, some of which are reviewed elsewhere[2-5], include (1) thermal decomposition of uranium pentachloride, (2) chlorine oxidation of uranium metal or a lower chloride, (3) the action of chlorine and carbon tetrachloride on U_3O_8 and (4) the reaction between uranium hexafluoride and boron trichloride. Method (3) is quite unsatisfactory since large amounts of the pentachloride are also formed but the disproportionation of UCl_5, if carried out[2] between 120 and 150° at low pressure with a cold finger insertion to collect the hexachloride, is capable of yielding pure material when the initial product is recycled. The recycling is necessary owing to contamination of the first product with uranium pentachloride which itself possesses an appreciable vapour pressure under the above conditions. A suitable apparatus is illustrated in Figure 3.1.

The recently reported[6] conversion of uranium hexafluoride to the hexachloride using boron trichloride (or $AlCl_3$) appears to provide a promising alternative to the above disproportionation but the reaction between

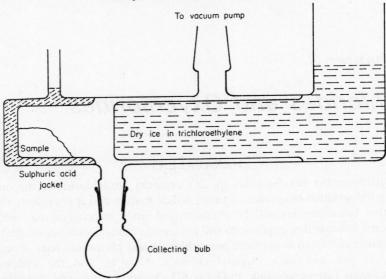

Figure 3.1   Apparatus for the preparation[2] of UCl_6. (*After* O. Johnson, T. Butler and A. S. Newton, U.S. Report TID–5290 (1958) p. 1)

uranium metal[7] and chlorine, which leads to the simultaneous formation of the tetra- and pentachlorides, is not so useful.

Crystal structure. Uranium hexachloride possesses hexagonal symmetry with $a_0 = 10.97$, $c_0 = 6.04$ Å (space group D_{3d}^3–$C\bar{3}m$, $n = 3$) and has a typical molecular structure[8] comprising individual UCl_6 molecules in a three dimensional array. The six chlorine atoms, which form an almost perfect octahedron about each uranium atom, are at a distance of 2.42 Å from the uranium atom and the single bond covalent radius of U^{6+} is calculated to be 1.43 Å. The structure is illustrated in Figure 3.2. The calculated density, 3.56 g cm^{-3} is in fair agreement with direct determinations[9] in tetrahydronaphthalene (3.36 g cm^{-3}).

Properties. Uranium hexachloride is a blackish-green solid (m.p. 177.5°) which may be sublimed at 75–100° (10^{-4} mm Hg). It attacks silver, copper, mercury, aluminium and tantalum to some extent but is stable in contact with molybdenum, gold and stainless steel. There is some disagreement in the reported vapour pressure data[2,10] measured by two independent methods. Thus the results obtained by a transpiration method[2] can be expressed by the relationship $\log p_{mm} = -(3788/T) + 9.52$, whereas measurements by the clicker gauge method give[10] $\log p_{mm} = -(2422/T) + 6.634$. However, the results from the first method were corrected for the evolution of chlorine arising from the decomposition of the hexachloride and are probably the more accurate. No such correction was applied in the clicker gauge determination; the workers here reported negligible thermal decomposition below 130° and it would perhaps be better if the two methods could be applied to samples from the same batch.

Little is known about the chemical behaviour of uranium hexachloride. It hydrolyses in moist air, reacts violently with water to form uranyl chloride and dissolves in carbon tetrachloride and chloroform, but is insoluble in benzene. It reacts with anhydrous liquid hydrogen fluoride at room temperature[11] to form uranium pentafluoride, UF_5, and reduces[6] UF_6 to the tetrafluoride with the evolution of chlorine. The only recorded complex is an incompletely characterized yellow solid formed[12] when dipyridyl is added to uranium hexachloride in carbon tetrachloride.

Hexavalent Oxychlorides

Anhydrous uranyl chloride has been prepared by several methods but only the hexahydrate of plutonyl chloride, $PuO_2Cl_2 \cdot 6H_2O$, has been obtained[13,14] to date and so far neither anhydrous nor hydrated neptunyl chloride has been reported. Chloro complexes of the type $M_2^I NpO_2Cl_4$ have recently been characterized[15,16] and $NpO_2Cl_2 \cdot xH_2O$ should be easily obtained from aqueous solution. The preparation of anhydrous neptunyl chloride may well be much more difficult.

The best way of preparing[2,17-19] pure, anhydrous uranyl chloride is to heat the tetrachloride in oxygen at 300–350°; complete reaction may be

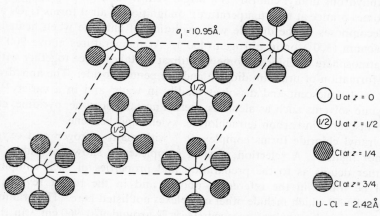

Figure 3.2 The uranium hexachloride structure[8] viewed along the sixfold axis. The U—Cl bonds within the UCl_6 molecules are indicated. (*After* W. H. Zachariasen, *Acta Cryst.*, **1**, 285 (1948))

9

achieved by agitating the bed of uranium tetrachloride. Temperatures in excess of 500° should be avoided because oxides are formed under such conditions. The reaction between uranium dioxide and chlorine, originally due to Péligot[20], is less satisfactory, since incomplete reaction is obtained[21] even at 800° under pressure. Other unsatisfactory procedures which have occasionally been used[22] include the action of carbon tetrachloride, alone or mixed with carbon monoxide or chloroform, on the various uranium oxides and[23] the reaction between uranium trioxide and chlorine at 300°.

In the presence of moisture, hydrogen chloride converts[17] the trioxide to the monohydrate $UO_2Cl_2 \cdot H_2O$ but under anhydrous conditions reaction is very slow. The monohydrate also results when hydrated uranyl chloride ($UO_2Cl_2 \cdot xH_2O$, $x \geqslant 3$) is refluxed with thionyl chloride[24-26] and both it and the trihydrate can be isolated from aqueous solution. The dehydration of these compounds is difficult but can be achieved by[22] heating them in a stream of dry hydrogen chloride at 300°; others[27,28] have found it best to follow this treatment by heating the product in chlorine and hydrogen chloride at 400°.

Properties. X-ray powder diffraction patterns of uranyl chloride powder (unsublimed) have been interpreted on the basis of an ortho-rhombic unit cell ($n = 4$) with[2] $a_0 = 8.71$, $b_0 = 8.39$ and $c_0 = 5.72$ Å, but the powder patterns of crystals obtained by condensing uranyl chloride vapour contain many reflections in addition to those indexed in the orthorhombic symmetry; additional information is obviously required to clarify these observations.

Anhydrous uranyl chloride is a bright yellow solid (m.p.[27] 578°) which becomes orange at high temperature; on ignition in air it forms U_3O_8 and it decomposes to the dioxide with evolution of chlorine when heated in a vacuum (300°) or in nitrogen (400°). At temperatures above 900° in an atmosphere of chlorine some volatilization takes place together with[29] the formation of uranium dioxide and the pentachloride. The anhydrous salt is deliquescent and dissolves readily in water and in a variety[22] of organic solvents such as alcohols, methyl acetate, acetone, pyridine, etc.; it is insoluble in carbon tetrachloride, xylene and benzene.

Uranyl chloride forms complexes[5,22] with many nitrogen and oxygen donor ligands. A selection of such complexes is shown in Table 3.1; further details as to the preparation and properties of such complexes will be found in the references quoted and in the recent reviews by Bagnall[5,375] which include other complexes not listed here. The uranium–oxygen stretching vibration occurs[12,30-32] around 920–960 cm^{-1} in these complexes and infrared studies indicate that in the amide, urea and phosphine oxide complexes coordination occurs via the oxygen atom.

TABLE 3.1

A Selection of the Complexes Formed by Uranyl Chloride

Formula	Ligand (L) and reference
$UO_2Cl_2 \cdot L$	Methyl cyanide[a], 1:10 phenanthroline[b], NNN^1N^1-tetramethyl, α,α-dimethylmalonide[c], NNN^1N^1-tetramethyl-3,-dimethylglutaramide[c].
$UO_2Cl_2 \cdot 1.5L$	NNN^1N^1-tetramethylmalonamide[c], NNN^1N^1-tetramethylglutaramide[c].
$UO_2Cl_2 \cdot 2L$	Aniline[d], pyridine[e], 1:10 phenanthroline[b], p-nitroso dimethylaniline[f], trialkyl or triarylphosphine oxides[g,h,i,j], NN-dimethylacetamide[a], 4-methoxypyridine-N-oxide[k], ethanol[l], acetic anhydride[m], acet-p-phenetidine[n], nitrosyl chloride[o].
$UO_2Cl_2 \cdot 3L$	Acet-p-phenetidine[n], 4-methylpyridine-N-oxide[k], NN-dimethylformamide[p].
$UO_2Cl_2 \cdot 4L$	Urea[q], 4-chloro-pyridine-N-oxide[k].

[a] K. W. Bagnall, D. Brown and P. J. Jones, *J. Chem. Soc. (A)*, **1966**, 1763.

[b] V. P. Markov and V. V. Tsapkin, *Zh. Neorgan. Khim.*, **4**, 2261 (1959).

[c] K. W. Bagnall, D. Brown and P. J. Jones, *J. Chem. Soc. (A)*, **1966**, 741.

[d] A. R. Leeds, *J. Am. Chem. Soc.*, **3**, 145 (1881).

[e] R. Răscanu, *Ann. Sci. Univ. Jassy*, **16**, 461 (1931), according to Katz and Rabinowitch, reference 23.

[f] R. Răscanu, *Ann. Sci. Univ. Jassy*, **16**, 54, 465 (1931); **17**, 133 (1932), according to Katz and Rabinowitch, reference 23.

[g] P. Gans, Thesis, London University (1964).

[h] J. P. Day, Thesis, Oxford University (1965).

[i] P. Gans and B. C. Smith, *J. Chem. Soc.*, **1964**, 4172.

[j] J. P. Day and L. M. Venanzi, *J. Chem. Soc. (A)*, **1966**, 1363.

[k] P. V. Balakrishnan, S. K. Patil and H. V. Venkatasetty, *J. Inorg. Nucl. Chem.*, **28**, 537 (1966).

[l] D. C. Bradley, A. K. Chatterjee and A. K. Chatterjee, *J. Inorg. Nucl. Chem.*, **12**, 71 (1959).

[m] A. Chrétin and G. Oechsel, *Compt. Rend.*, **206**, 254 (1938).

[n] R. Răscanu, *Ann. Sci. Univ. Jassy*, **16**, 480 (1931), according to Katz and Rabinowitch, reference 23.

[o] C. C. Addison and N. Hodge, *J. Chem. Soc.*, **1961**, 2490.

[p] M. Lamisse, R. Heimburger and R. Rohmer, *Compt. Rend.*, **258**, 2078 (1964).

[q] V. P. Markov and I. V. Tsapkin, *Zh. Neorgan. Khim.*, **7**, 2045 (1962).

Hexavalent Oxychloro Complexes

Anhydrous uranyl, neptunyl and plutonyl tetrachloro complexes, $M_2^I M^{VI} O_2 Cl_4$, have been prepared in a variety of ways, several hydrated uranyl complexes, $M_2^I UO_2 Cl_4 \cdot 2H_2O$, are also known and $Cs_2AmO_2Cl_4$ was recently[33] reported. The anhydrous complexes $(NEt_4)_2MO_2Cl_4$ (M = U, Np and Pu), $N(Pr)_4MO_2Cl_4$ (M = U and Pu), $(NHEt_3)_2MO_2Cl_4$ (M = U and Pu) and $Cs_2MO_2Cl_4$ (M = U and Np) are prepared[15,16] by

drying the precipitates obtained by mixing the component halides in hydrochloric acid (6 to 12 M) solution. The tetramethyl- and tetraethylammonium uranium (VI) and plutonium (VI) salts have also been prepared[34] by evaporating 4M hydrochloric acid solutions of the appropriate
halides. Additional anhydrous uranyl complexes which have been isolated
from aqueous hydrochloric acid[15,35–38] solutions include those where
M^I = trimethylammonium, tripropylammonium, diethylammonium, pyridinium, quinolinium, β-lutidinium, xanthylium, 1,10-phenanthrolinium
and 2,2'-dipyridylium. The last and the *o*-phenanthrolinium salt are also
obtained[39] by air oxidation of uranium tetrachloride in dimethylformamide
in the presence of 2,2'-dipyridyl or *o*-phenanthroline whilst the pyridinium
complex[40] precipitates when hydrogen chloride is passed into an ethanolic
solution of a uranyl chloride hydrate. The diphosphonium complexes
$(R_3PH)_2UO_2Cl_4$ (e.g. R = Ph, C_2H_5, and C_3H_7) can be made by hydrogen
peroxide oxidation[30] of the hexachlorouranate (IV) salts or by passing
hydrogen chloride into a methyl cyanide solution of the component
chlorides and adding ether to crystallize[31]. Other complexes prepared
from anhydrous methyl cyanide[31] include $(Ph_3PBz)_2UO_2Cl_4$, $(Ph_3PBu)_2$-
UO_2Cl_4 and $(Ph_4P)_2UO_2Cl_4$.

By treating the americium (V) complex $Cs_3AmO_2Cl_4$ with concentrated
hydrochloric acid a dark red, crystalline solid $Cs_2AmO_2Cl_4$ is obtained[33].
This reaction appears to involve oxidation, rather than disproportionation,
of Am (V) since the supernatant from the reaction contains less than
one third of the original americium present, the quantity which would be
required for disproportionation:

$$3 \text{ Am (V)} \rightarrow 2 \text{ Am (VI)} + \text{Am (III)}$$

The hexavalent chloro complex is slightly soluble in concentrated hydrochloric acid, in which it is rapidly reduced to Am (III) as would be expected
from the known instability of Am (VI) in this medium. The unusual
oxidation is therefore probably due to high lattice energy stabilization
of $Cs_2AmO_2Cl_4$.

Single crystal studies[34] have shown tetramethylammonium dioxytetra-
chlorouranate (VI) and plutonate (VI) to possess tetragonal symmetry,
that the corresponding tetramethylammonium salts are respectively
monoclinic and tetragonal, and[41] that the caesium complex $Cs_2UO_2Cl_4$
is monoclinic. Unit cell parameters are listed in Table 3.2. Crystal data
are lacking for the remaining complexes mentioned. Caesium dioxytetra-
chloroneptunate (VI), $Cs_2NpO_2Cl_4$, which is isostructural with $Cs_2UO_2Cl_4$
shows[16] Curie–Weiss dependence between 300 and 100°K with a Weiss
constant of −216° and a resultant magnetic moment of 2.32 B.M. The

TABLE 3.2

Crystallographic Properties of Some Uranyl and Plutonyl Salts[34,41]

Compound	Colour	Symmetry	Space group	Lattice parameters (Å)		
				a_0	b_0	c_0
$(NMe_4)_2UO_2Cl_4$	Yellow	Tetragonal	C_{4h}^5–$I4/m$	9.12	—	11.77
$(NMe_4)_2PuO_2Cl_4$	Yellow	Tetragonal	C_{4h}^5–$I4/m$	9.20	—	11.90
$(NEt_4)_2UO_2Cl_4$	Yellow	Monoclinic	—	16.30	10.00	12.90
				($\beta = 142°$)		
$(NEt_4)_2PuO_2Cl_4$	Yellow	Tetragonal	C_{4h}^5–$I4/m$	10.0	—	12.90
$Cs_2UO_2Cl_4$	Yellow	Monoclinic	C_{2h}^3–$C2/m$	11.92	7.71	5.83
				($\beta = 99° 40'$)		

neptunium–oxygen and neptunium–chlorine stretching vibrations occur at 919 and 271 cm^{-1} respectively in the infrared spectrum of this compound. The uranium–oxygen stretching vibrations have been observed around 920–930 cm^{-1} in certain of the above complexes (Table B.4, p. 252).

Uranyl complexes containing smaller cations (potassium, rubidium and ammonium) crystallize from aqueous solution as the dihydrates, $M_2^I UO_2Cl_4 \cdot 2H_2O$, e.g. see references 35, 42 and 43. The anhydrous potassium and sodium salts have been prepared by heating the alkali metal halide in uranyl chloride vapour[44] or alternatively $K_2UO_2Cl_4$, which melts at about 290° and forms the dihydrate in moist air, can be prepared[43] by the action of hydrogen chloride on K_2UO_4 at 250° or by fusing potassium chloride and anhydrous uranyl chloride at 280°.

Uranyl salts of the type $M_2^I UO_3Cl_2$ (M^I = K and NH_4) have also been recorded. The potassium salt can be prepared by heating $K_2UO_2Cl_2Br_2$ in oxygen[45] at 250°, by heating K_2UO_4 in hydrogen chloride[43] at about 200° or by heating together stoicheiometric quantities of uranyl chloride monohydrate and potassium hydroxide; $(NH_4)_2UO_3Cl_2$ is formed[45] when uranyl chloride monohydrate is heated in gaseous ammonia. A series of apparently non-stoicheiometric complexes $M_x^I UO_3Cl_x$ (M^I = K, Rb and Cs; $x = \sim 0.9$) have been prepared[46] by heating uranium trioxide with the appropriate alkali metal halide at 900° in a vacuum. These complexes are of monoclinic symmetry and a full structure analysis of the caesium salt has shown that the uranium atom is seven coordinate. Five oxygen and two chlorine atoms form a distorted pentagonal bipyramid around the uranium atom and the bipyramids join up by having edges in common and thereby form infinite two-dimensional strings from which the uranyl oxygens protrude.

Other complex phases which have been prepared[43] by heating potassium diuranate or triuranate (respectively $K_2U_2O_7$ and $K_2U_3O_{10}$) in hydrogen chloride are $K_2U_2O_5Cl_4$ and $K_2U_3O_8Cl_4$ respectively and the complex $K_2U_2O_5OCl_2$ is said to be formed when UO_3 and potassium chloride are heated together in oxygen at 500° and, in a second crystallographic modification, by heating the mixed complex $K_2U_2O_5Cl_2Br_2$ in oxygen at 350°. The analogous bromo complex has also been reported. Obviously further studies on these unusual complexes are desirable.

PENTAVALENT

The pentachlorides of protactinium and uranium are known but neptunium pentachloride has not been characterized although thermodynamic calculations suggest[1] that it ought to exist. The various attempts to prepare this compound have been summarized[16,47]; similarly attempts to prepare hexachloro complexes, of the type M^INpCl_6, analogous to those known for protactinium (v) and uranium (v) have been unsuccessful[16]. Oxychlorides of protactinium (v) and uranium (v), Pa_2OCl_8, $PaOCl_3$, $Pa_2O_3Cl_4$, PaO_2Cl and $UOCl_3$ are known and although similar neptunium (v) compounds have yet to be reported, oxychloro complexes of the types $M_2^INpOCl_5$ and $M_3^INpO_2Cl_4$ are readily obtained from aqueous solution.

Pentachlorides

Protactinium pentachloride was first prepared by von Grosse[48] by reaction of carbonyl chloride with the pentoxide at 550°. More recently[49-52], the pentoxide has been converted to the pentachloride by heating it in a flow of nitrogen saturated with either carbon tetrachloride or a mixture of chlorine and carbon tetrachloride or, on the microgram scale, by heating in carbon tetrachloride vapour[53] in an x-ray capillary. However, the former type of reaction involves rather complicated apparatus, requires rigorously dried nitrogen and results in low yield: in addition serious loss of protactinium pentachloride as a non-condensable smoke has been observed[51]. The freshly precipitated hydroxide reacts to a limited extent with liquid thionyl chloride at room temperature to yield[54] stable concentrated solutions, which on vacuum evaporation deposit the chloro complex $SO(PaCl_6)_2$ (cf. $UCl_5 \cdot SOCl_2$, p. 129); thermal decomposition of this complex gives a partial yield of the pentachloride together with an unidentified black residue. Under similar conditions niobium hydroxide reacts quantitatively with thionyl chloride and the tantalum compound to a lesser extent ($\sim$70%) following which the pentahalides are obtained simply by vacuum evaporation of the excess solvent.

The most satisfactory method[51], since protactinium metal is not available, is to heat the vacuum dried hydroxide (100°) in thionyl chloride vapour at 350° to 500° in a sealed, evacuated Pyrex reaction vessel when the volatile, yellow pentachloride is obtained in better than 95% yield. Alternatively[51] the chlorination can be carried out by heating a mixture of the low-fired pentoxide and carbon with chlorine and carbon tetrachloride at 550° in a sealed tube; the product in this instance, however, is a mixture of the pentachloride and diprotactinium (v) oxyoctachloride, Pa_2OCl_8, from which the former is separated by vacuum sublimation below 200°. At higher temperatures the oxychloride disproportionates in a vacuum (p. 130) to yield more pentachloride, as shown in equation (1).

$$3Pa_2OCl_8 \xrightarrow[10^{-4}\text{mm Hg}]{>250°} 4PaCl_5 + Pa_2O_3Cl_4 \tag{1}$$

Uranium pentachloride occurs as a by-product in several high temperature preparations of the tetrachloride, e.g. uranium metal with chlorine, uranium oxides with carbon tetrachloride vapour and U_3O_8 with a mixture of chlorine and sulphur monochloride, and it was in fact first identified[55] during the preparation of uranium tetrachloride in 1874. However, satisfactory methods for obtaining the pentachloride in good yield and in a pure state are limited, mainly because of its instability towards thermal decomposition and disproportionation. Thus the irreversible reactions (2) and (3)

$$UCl_5 \rightarrow UCl_4 + \tfrac{1}{2}Cl_2 \tag{2}$$

$$2UCl_5 \rightarrow UCl_4 + UCl_6 \tag{3}$$

readily take place below 100° at pressures of 1 atm and 10^{-4} mm Hg respectively.

Probably the most satisfactory procedure is to heat the tetrachloride at 550° in a stream[56-58] of chlorine and quench the pentachloride vapour. The rapid quenching of the vapour and careful control of the chlorine flow are important to prevent contamination of the product with uranium hexachloride. An earlier report describes[59] the conversion of the tetrachloride by liquid chlorine at 350° and 6 atmospheres pressure.

The liquid phase chlorination of uranium oxides with carbon tetrachloride yields, depending on the conditions, the tetrachloride, pentachloride or even the hexachloride. Chlorination of the trioxide or triuranium octoxide, U_3O_8, at 250° with a mixture of carbon tetrachloride and chlorine in a sealed tube is a reasonable way[60] of preparing the

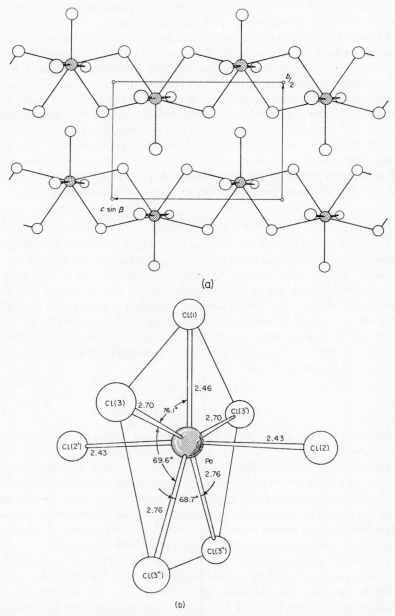

(a)

(b)

Figure 3.3 The structure of protactinium pentachloride.[63] (*a*) Portion of infinite chains in PaCl$_5$. Coordination of Cl around each Pa is pentagonal bipyramidal. (*b*) Bond distances (Å) and angles within an isolated PaCl$_7$ group. (*After* R. P. Dodge, G. S. Smith, Q. Johnson and R. E. Elson, *Acta Cryst.*, **22**, 85 (1967))

pentachloride but the reaction at atmospheric pressure is less satisfactory[61]. For example, the reaction between UO_3 and carbon tetrachloride is extremely slow and for complete conversion it is reported to be essential to add considerable amounts of the pentachloride itself. This autocatalytic action of the pentachloride has been studied in some detail and it is postulated that the first stage proceeds $UO_3 + 2UCl_5 \rightarrow 2UOCl_3 + UOCl_4$ and that the oxychlorides are then chlorinated further by the carbon tetrachloride. The postulated uranium (VI) oxychloride, $UOCl_4$, has yet to be prepared in a pure state and its existence must still be considered speculative. Other references to the detailed investigations of the preparation of uranium pentachloride are to be found in earlier review articles[4,61]. The pentachloride can be separated from uranium tetrachloride by recrystallization from liquid chlorine[62]; it can also be recrystallized from carbon tetrachloride[12].

Crystal structures. Protactinium pentachloride and uranium pentachloride are not isostructural. The former, which possesses monoclinic symmetry[51,63] (Table 3.3), has a structure[63] which comprises (Figure 3.3)

TABLE 3.3

Crystallographic Data for Protactinium and Uranium Pentachloride

Compound	Colour	Lattice dimensions (Å)			Space group	Reference
		a_0	b_0	c_0		
$PaCl_5$	Yellow	8.00	11.42	8.43	$C2/c$ or Cc	51, 63
			($\beta = 106.38°$)			
UCl_5	Brown	7.99	10.69	8.48	$C_{2h}^5 - P2_1/n$	64
			($\beta = 91.5°$)			

infinite chains of non-regular pentagonal bipyramidal $PaCl_7$ groups which share pentagon edges   The Pa–Cl bond distances are 2.43 (2) and 2.46 (1) Å to non-bridging chlorine atoms and 2.70 (2) and 2.76 (2) Å to bridging chlorine atoms.

Uranium pentachloride which is also monoclinic (Table 3.3) possesses[64] a structure (Figure 3.4) based on a cubic close packing of chlorine atoms in which uranium atoms occupy one-fifth of the octahedral holes. Two such octahedra share an edge to form a dimeric U_2Cl_{10} unit. Uranium–chlorine bond distances involving bridging chlorines are 2.67 and 2.70 Å and those involving non-bridging chlorine atoms range from

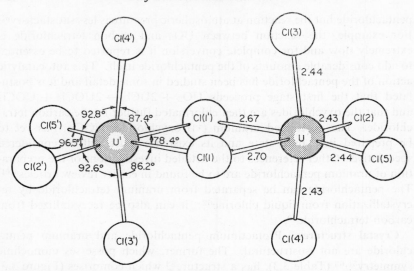

Figure 3.4   Configuration of the U_2Cl_{10} dimer[64] (distances in Å). A primed atom is related to its corresponding unprimed atom by a centre of inversion. (*After* G. S. Smith, Q. Johnson and R. E. Elson, *Acta Cryst.*, **22**, 300 (1967))

2.43 to 2.44 Å. Molecular weight determinations have shown that uranium pentachloride is also dimeric in carbon tetrachloride[65], explaining the observed[66] diamagnetism in this solvent.

Properties. The bright yellow protactinium pentachloride and reddish-brown uranium compound are extremely moisture-sensitive and must be handled in a dry atmosphere. Owing to its ready thermal decomposition there is no measured melting point for the latter compound (estimated as 300°); protactinium pentachloride melts at 301° and can be sublimed above 160° in a vacuum (10^{-4} mm Hg). Vapour pressure data are lacking for both compounds. They are reduced to their respective tetrachlorides by hydrogen[2,53,67] at moderate temperatures and react with oxygen[51,61] to form oxychlorides, protactinium[51] yielding Pa_2OCl_8 or $Pa_2O_3Cl_4$ depending on the conditions employed (p. 130). Uranium pentachloride reacts with chlorine to form the hexachloride, a reaction which proceeds better[68] with a mixture of chlorine and carbon tetrachloride. It reacts with fluorine to give uranium hexafluoride[69] and, like the hexachloride, reacts with anhydrous hydrogen fluoride[62] to form the pentafluoride.

Owing to the ready disproportionation of uranium pentachloride in the presence of moisture any solvents used during complex formation must be perfectly anhydrous. However, solutions in thionyl chloride, are readily obtained by dissolving uranium trioxide in the refluxing solvent[40,70] and vacuum evaporation of such solutions yields the dark red complex $UCl_5 \cdot SOCl_2$ (cf. $SO(PaCl_6)_2$ p. 124) which has also been prepared[71] by the reaction between UO_3 and thionyl chloride in a sealed vessel at 150–200°. Other complexes which can be conveniently prepared direct from the trioxide include[72] the orange-red $UCl_5 \cdot PCl_5$ and an unusual[73] red, crystalline complex, $5UCl_5 \cdot CCl_2 = CClCOCl$, which occurs as a by-product during the conversion of the trioxide to uranium tetrachloride with hexachloropropene (p. 136). Phosphine oxide complexes of both protactinium[74] and uranium pentachloride[75], of the type $MCl_5 \cdot R_3PO$ (R = variously NMe_2, Ph or $PhCH_2$), have also been prepared, either from caesium hexachlorouranate (v) or directly from protactinium pentachloride. However, complexes of the type $MOCl_3 \cdot 2R_3PO$, which are formed[74] by the reaction of niobium and tantalum pentachloride with excess ligand, do not form with the actinide elements.

Protactinium pentachloride is slightly soluble in benzene and carbon tetrachloride and reacts[76] with anhydrous methyl cyanide to form the complex $PaCl_5 \cdot 3CH_3CN$, whereas niobium and tantalum pentachloride form the 1:1 compounds. Like the pentabromide it is converted[77] to protactinium pentaiodide by excess silicon tetraiodide at 150–200°.

The magnetic behaviour of uranium pentachloride (and of $UCl_5 \cdot SOCl_2$[66]) has been discussed in terms of both[66,78] a $6d^1$ and a $5f^1$ configuration for the U (v) ion. The latter appears to be more likely since the absorption spectrum of $UCl_5 \cdot SOCl_2$ in carbon tetrachloride is consistent with a $5f^1$ configuration[79]. However, it must be remembered that the $5f$ and $6d$ energy levels are quite close at uranium and changes in coordination could lead to different electron configurations. Other absorption spectra of uranium (v) chloride solutions have been recorded[70,73,75,80] but the possibility of disproportionation to uranium (IV) and (VI), resulting in the presence of extraneous bands, must always be considered. The ultra-violet absorption spectrum of protactinium pentachloride in methanol and carbon tetrachloride has been recorded[81,82].

Pentavalent Chloro Complexes

Solutions of uranium (v) and protactinium (v) in thionyl chloride are quite stable towards hydrolysis and/or disproportionation and the pale yellow[54] hexachloroprotactinates (v) and yellow to orange[70] hexachlorouranates (v), $M^I M^V Cl_6$ (M^I = variously NMe_4, NEt_4, NH_2Me_2, Ph_4As; M^V = Pa and U) are precipitated from this solvent by the addition of carbon disulphide to equimolar quantities of the component halides. The analogous caesium salts precipitate from a mixture of thionyl chloride and iodine monochloride, the latter being essential to dissolve caesium chloride. The bright yellow tetramethylammonium octachloroprotactinate (v), $(NMe_4)_3PaCl_8$, and the corresponding uranium (v) complex have also been isolated[54,70] from thionyl chloride solution. Hexachloroniobates (v), tantalates (v) and tungstates (v) can be prepared in an identical manner but no octachloro complexes of these elements are known. The metal–chlorine stretching vibration, ν_3, which occurs around 300–310 cm^{-1} for the hexachloro complexes (Table B4) shifts to longer wavelengths (290 cm^{-1}), as would be expected, in the octachloroprotactinate (v) salt. Magnetic susceptibility data have been recorded[70] for the chlorouranates (v) but the results have not been interpreted. The complexes possess low symmetry and no structural information is available. Neptunium (v) hexachloro complexes are unknown but they appear to be capable of transitory existence in thionyl chloride. Thus $(Ph_4As)_2NpOCl_5$ dissolves readily in this solvent to form[16] an intensely red solution which probably contains the $NpCl_6^-$ ion. The spectra of such solutions have been recorded and the decomposition (or reduction) of the neptunium (v) species followed spectrophotometrically. Attempts to obtain NEt_4NpCl_6 by chlorine oxidation of 1 : 1 neptunium tetrachloride-tetraethylammonium chloride mixtures in methyl cyanide have been unsuccessful[384,385] (cf. NEt_4UBr_6 preparation, p. 187). Similar reactions aimed at the preparation of $NpCl_5 \cdot Ph_3PO$ also failed[385].

Pentavalent Oxychlorides

By carefully controlled heating of the initial product from the reaction between protactinium pentoxide–carbon mixtures and chlorine and carbon tetrachloride vapours (p. 125) it is possible[51] to isolate α-diprotactinium (v) oxyoctachloride, α-Pa_2OCl_8, which is analogous to the fluoride Pa_2OF_8 described earlier (p. 47). The procedure involves removal of the pentachloride from the initial product at less than 220° (10^{-4} mm Hg) since above 250° Pa_2OCl_8 disproportionates (4) to yield the pentachloride and diprotactinium (v) trioxytetrachloride, $Pa_2O_3Cl_4$, which is analogous to the recently reported tantalum (v) oxychloride. At higher temperatures, above

$$3Pa_2OCl_8 \xrightarrow[\text{in vacuum}]{>250°} 4PaCl_5 + Pa_2O_3Cl_4 \qquad (4)$$

$$2Pa_2O_3Cl_4 \xrightarrow[\text{in vacuum}]{>520°} PaCl_5 + 3PaO_2Cl \qquad (5)$$

$$4PaCl_5 + O_2 \xrightarrow{350-400°} 2\ \beta\text{-}Pa_2OCl_8 + 2Cl_2 \qquad (6)$$

$$4PaCl_5 + 3O_2 \xrightarrow{500°} 2Pa_2O_3Cl_4 + 6Cl_2 \qquad (7)$$

520°, $Pa_2O_3Cl_4$ disproportionates (5) forming a further white oxychloride, PaO_2Cl, and protactinium pentachloride. A second crystallographic modification of diprotactinium oxyoctachloride, β-Pa_2OCl_8, is obtained[51] when excess protactinium pentachloride and oxygen are heated together (6) in a sealed tube at 350–400°; with more oxygen and at higher temperatures $Pa_2O_3Cl_4$ can be prepared (7) in a similar manner. There is also evidence[51] that an unstable protactinium (v) oxytrichloride, $PaOCl_3$, exists but a satisfactory preparative method has proved elusive. This behaviour contrasts markedly with that of niobium (v) and uranium (v) each of which form stable oxytrichlorides.

Uranium (v) oxytrichloride is, in fact, the only uranium (v) oxychloride which has been prepared. It was first observed[61] as an intermediate in the liquid phase chlorination of uranium trioxide by carbon tetrachloride but is more conveniently prepared[83,84] by heating together stoicheiometric quantities of uranium tetrachloride and uranyl chloride at 370°. It has also been isolated[85] in an essentially pure state by the partial chlorination of a static bed of uranium dioxide by gaseous carbon tetrachloride. Below 300° this reaction yields only a mixture of the dioxide and uranium tetrachloride but between 400 and 500° three definite layers are observed, UCl_4 (top), $UOCl_3$ (middle) and a lower layer which is a mixture of UO_2 and $UOCl_2$. From a detailed study of the gaseous reaction products two possible reactions, (8) and (9), are postulated for the formation of the reddish-brown $UOCl_3$.

$$2UO_2 + 2CCl_4 \rightarrow 2UOCl_3 + 2CO + Cl_2 \tag{8}$$

$$2UO_2 + 3COCl_2 \rightarrow 2UOCl_3 + 2CO_2 + CO \tag{9}$$

There is some spectral evidence for the existence[84] of the dioxymonochloride, UO_2Cl, in fused lithium chloride/potassium chloride solutions of uranyl chloride and it is also possibly formed by[86] the electrolytic reduction of uranyl chloride in molten salts. The analogous plutonium (v) chloride is possibly formed by[87] oxidation of plutonium (III) or (IV) with a 2:1 mixture of chlorine and oxygen in fused chloride melts. Neither of the compounds has been isolated and anhydrous neptunium (v) oxychlorides are also presently unknown although it has been reported that[88,89] evaporation of an aqueous hydrochloric acid solution containing neptunium (v) yields hydrated oxychlorides.

Properties. The oxychlorides are moisture-sensitive, non-volatile solids. The thermal decomposition of the white protactinium (v) compounds has been described above; uranium (v) oxytrichloride decomposes[85] above 700° to give a mixture of the tetrachloride, pentachloride, dioxide and a crystalline phase of composition $U_2O_3Cl_3$. It is soluble, with decomposition, in acetone and water, and is insoluble in benzene and carbon tetrachloride; diprotactinium (v) oxyoctachloride is soluble in anhydrous methyl cyanide and nitromethane and hydrolyses in acetone and water. X-ray powder diffraction data are available[51] for the protactinium (v) oxychlorides but their structures are unknown. The metal–chlorine stretching vibrations occur close to those of the pentachloride (Tables B1 and B2) and the metal–oxygen vibrations are observed[51] below 600 cm^{-1} suggesting oxygen-bridged structures.

The complex $UOCl_3 \cdot EtOH$ is formed[40] by the action of anhydrous ethanol on the thionyl chloride complex $UCl_5 \cdot SOCl_2$ but complexes of protactinium (v) oxychlorides have not been investigated.

Pentavalent Oxychloride Complexes

Although the simple anhydrous neptunium (v) and americium (v) oxychlorides have not been recorded complexes of the type $M_2^I NpOCl_5$ and $M_3^I M^V O_2 Cl_4$ (M^V = Np and Am) are easily prepared[16,33]. Thus the addition of concentrated hydrochloric acid containing caesium or tetraphenylarsonium chloride to neptunium (v) hydroxide results in the formation of the pale yellow complexes Cs_2NpOCl_5 and $(Ph_4As)_2NpOCl_5$ respectively whereas, when the hydroxide is dissolved in the minimum of dilute hydrochloric acid containing caesium chloride, the salt $Cs_3NpO_2Cl_4$ can be precipitated by the addition of acetone. A green americium (v) salt, isostructural with $Cs_3NpO_2Cl_4$, is precipitated by ethanol from a solution of Am (v) hydroxide and caesium chloride in 6M HCl and also by treating $CsAmO_2CO_3$ with concentrated hydrochloric acid saturated with caesium chloride. Obviously the AmO_2^+ ion is more stable in concentrated hydrochloric acid under the above conditions than is the NpO_2^+ ion. When this Am (v) chloro complex is treated with concentrated hydrochloric acid alone a dark red solid, $Cs_2AmO_2Cl_4$ is obtained. The possible mechanism of this oxidation has already been discussed (p. 122). The preparation of analogous plutonium (v) oxychloro salts has not yet been studied. That the above complexes contain the discrete NpO^{3+}, NpO_2^+ and AmO_2^+ ions respectively is shown by the positions of the metal-oxygen stretching vibrations, 921 cm^{-1} (NpO^{3+}) and approximately 800 cm^{-1} (MO_2^+). The metal–chlorine vibrations occur below 300 cm^{-1} (Table B4). Magnetic susceptibility data have been recorded[16] for the neptunium (v) complexes.

Oxychloro complexes of uranium (v) cannot be prepared from aqueous hydrochloric acid solution owing to the rapid disproportionation of uranium (v) in aqueous solutions (apart from hydrofluoric acid, p. 39) but it is claimed[40] that the pyridinium salt $(pyH)_2UOCl_5$ precipitates from ethanol when pyridine and hydrogen chloride are added to a solution of the thionyl chloride complex $UCl_5 \cdot SOCl_2$. The product, however, has not been unambiguously identified since it may be a mixture of $(pyH)_2UCl_6$ and $(pyH)_2UO_2Cl_4$; magnetic susceptibility or x-ray powder diffraction studies would be the most profitable investigations to resolve this question since the infrared stretching vibration of UO_2^{2+} and UO^{3+} are likely to be virtually identical (cf. reference 16, NpO_2^{2+} and NpO^{3+}, Table B4).

Recent preliminary studies[90], however, have shown that caesium hexachlorouranate (V), $CsUCl_6$, reacts with antimony trioxide below 200° to yield the oxytetrachloro salt $CsUOCl_4$. This method is analogous to the one recently used for the preparation of the first tantalum (V) oxychloro complexes and will undoubtedly be applicable to the preparation of similar protactinium (V) compounds.

TETRAVALENT

Simple lanthanide tetrachlorides are unknown but hexachloro complexes of cerium (IV) and praseodymium (IV) have been characterized[91-93,380]. In the actinide series thorium, protactinium, uranium and neptunium tetrachloride are solids at room temperature but plutonium tetrachloride exists only in the vapour state in the presence of excess chlorine. It seems unlikely, in view of the decreasing stability of the higher valence states of the actinides with increasing atomic number, that tetrachlorides of the elements beyond plutonium can exist, at least as the simple compounds. Thermodynamic calculations[94] first indicated that $PuCl_4$ could exist in the gaseous state in equilibrium with chlorine and $PuCl_3$ but the dissociation pressure of chlorine over solid $PuCl_4$ was estimated at 10^7 atmospheres. Following the report[95] that plutonium trichloride was appreciably more volatile in the presence of chlorine the absorption spectrum of gaseous $PuCl_4$ was recorded[381] at 928°. All attempts to prepare solid plutonium tetrachloride have so far been unsuccessful.

Hexachloro complexes of the actinide elements are easily prepared and some pentachloro- and octachlorothorates (IV) have been recorded. The oxydichlorides, $MOCl_2$, of thorium (IV), protactinium (IV), uranium (IV) and neptunium (IV) have been prepared and although thorium (IV) oxychloro complexes have been reported their existence remains doubtful.

Tetrachlorides

Thorium and uranium tetrachloride have been known for many years and numerous methods are available for their preparation. They were first prepared by Berzelius[97] and Peligot[98], respectively, each of whom

TABLE 3.4

Some High Temperature Reactions for the Preparation of
Thorium and Uranium Tetrachloride

Reaction	References[aa] Th	U
Metal + chlorine[bb]	a, c, d, e	m, n
Metal + hydrogen chloride[cc]	f	y
Hydride + chlorine	b	o
Hydride + hydrogen chloride[cc]	b	o
Dioxide + carbon + chlorine	i, j, z	m, p, q
Oxide + sulphur monochloride + chlorine	c, f, g	q, r
Oxide + carbonyl chloride	h	s, t
Oxide + carbon tetrachloride (vapour)	k, l	q, w
Oxide + carbon tetrachloride (liquid)	—	v, x
Carbide + chlorine	d	u

[aa] Other references will be found in the review articles mentioned in the text.
[bb] Difficult to control with uranium.
[cc] Yields UCl₃ which can be carefully chlorinated with chlorine at 250°.

[a] L. F. Nilsen, *Chem. Ber.*, **9**, 1142 (1876); *Chem. Ber.*, **15**, 2537 (1882); *Chem. Ber.*, **16**, 153 (1883); *Compt. Rend.*, **95**, 727 (1882).
[b] H. Lipkind and A. S. Newton, U.S. Report TID-5223, p. 398 (1957).
[c] H. von Wartenberg, *Z. Electrochem.*, **15**, 866 (1909).
[d] H. Moissan and A. Étard, *Ann. Chim. Phys.*, **12**, 427 (1897); *Compt. Rend.*, **140**, 1510 (1905).
[e] A. G. W. Fowles and F. H. Pollard, *J. Chem. Soc.* **1953**, 4128.
[f] G. Krüss and L. F. Nilsen, *Chem. Ber.*, **20**, 1665 (1887).
[g] F. Bourion, *Compt. Rend.*, **148**, 170 (1909); *Ann. Chim. Phys.*, **21**, 49 (1910).
[h] C. Baskerville, *J. Am. Chem. Soc.*, **23**, 762 (1901); E. Chauvenet, *Compt. Rend.*, **147**, 1046 (1908); *Ann. Chim. Phys.*, **23**, 425 (1911).
[i] J. J. Berzelius, *Pogg. Ann.*, **16**, 385 (1829).
[j] J. M. Matthews, *J. Am. Chem. Soc.*, **20**, 815 (1898).
[k] C. Matignon and M. Delépine, *Compt. Rend.*, **132**, 36 (1901); *Ann. Chim. Phys.*, **10**, 130 (1907).
[l] W. Fischer, R. Gewehr and H. Wingchen, *Z. Anorg. Chem.*, **242**, 161 (1939).
[m] E. Peligot, *Ann. Chim. Phys.*, **5**, 5 (1842); *Ann. Chem.*, **43**, 225 (1842).
[n] L. T. Reynolds and G. Wilkinson, *J. Inorg. Nucl. Chem.*, **2**, 246 (1956).
[o] O. Johnson, T. Butler and A. S. Newton, U.S. Report TID-5290, p. 1 (1958).
[p] O. Hönigschmidt and W. E. Schilz, *Z. Anorg. Chem.*, **170**, 145 (1928).
[q] A. Colani, *Ann. Chim. Phys.*, **12**, 66 (1907).
[r] R. W. Moore, *Trans. Am. Electrochem. Soc.*, **43**, 317 (1923).
[s] E. Chauvenet, *Compt. Rend.*, **152**, 87 and 1250 (1911).
[t] A. Rosenheim and M. Kelmy, *Z. Anorg. Chem.*, **206**, 31 (1932).
[u] H. Moissan, *Bull. Soc. Chim. France*, **17**, 266 (1897); J. Aloy, *Bull. Soc. Chim. France*, **21**, 264 (1899).
[v] A. Micheal and A. Murphy, *Am. J. Chem.*, **44**, 365 (1910).

treated the appropriate dioxide and carbon with chlorine at high temperature; in the case of uranium, oxidation to higher chlorides, mainly the pentachloride, also takes place. Some general preparative methods which are available are listed in Table 3.4; selected references are also listed and further references to these and other methods discussed below are to be found in earlier review articles[99-104].

With the present availability of thorium and uranium metal the role of oxide chlorination becomes less important than it was several years ago, particularly if only gram amounts are required for research purposes. However, where kilogram amounts are required the lower cost of the oxides may become important. The most convenient preparations of thorium tetrachloride are those involving direct combination of the elements or the reaction between the hydride and chlorine or hydrogen chloride. Whereas these reactions are easily carried out, the corresponding reactions involving uranium or uranium hydride and chlorine are difficult to control and lead to the formation of mixtures of the tetra-, penta- and hexachloride whilst hydrogen chloride converts the hydride to uranium trichloride[2]. This difficulty is overcome by the use of a 10 % chlorine–90 % helium mixture which is reported[2] to result in a smooth conversion of the hydride to uranium tetrachloride; uranium trichloride and chlorine also react to yield pure tetrachloride at 250°. In all reactions involving chlorine and uranium compounds the production of higher uranium chlorides is inevitable unless careful temperature control is exercised. The use, in the above reactions, of thorium hydride which is itself prepared from the metal, assists the rate and completion of the reactions since at the temperature of chlorination decomposition to finely divided, highly reactive metal takes place.

Although heating thorium tetrachloride hydrates alone or in hydrogen chloride results in oxychloride formation, the dehydration can be successfully achieved by refluxing with thionyl chloride[105,106] or by heating with pyridine hydrochloride[107]. In a recent[108] survey of the methods available for the preparation of anhydrous thorium tetrachloride the reaction between freshly decomposed oxalate and a carbon monoxide–chlorine mixture at 350–530° followed by heating of the product in a carbon tetrachloride–chlorine mixture at 675° was recommended. Thorium

Continuation of TABLE 3.4

^{w} I. V. Budayev and A. N. Vol'sky, *Proc. U.N. Intern. Conf. Peaceful Uses At. Energy, 2nd Geneva*, **28**, 316 (1958).
^{x} E. R. Harrison, Report AERE GP/R, 2409 (1958).
^{y} Reference 3, p. 474.
^{z} M. V. Smirnov and L. E. Ivanovskii, *Zh. Neorgan. Khim.*, **2**, 238 (1957).

10

tetrachloride can be purified[109] by heating with ammonium chloride followed by sublimation through thorium metal turnings.

The reactions between carbon tetrachloride and uranium oxides, particularly the dioxide, have been examined in some detail and there is a wealth of literature pertaining to this topic (see, for example, references 3, 5, 13 and 103). As would be expected use of the higher oxides U_3O_8 and UO_3 results in increased contamination with uranium pentachloride and although this may be converted to the tetrachloride by decomposition at 100–250° in an inert gas flow, such dechlorinating procedures make the preparation tedious and time consuming in addition to resulting in losses owing to the high volatility of the pentachloride. As a result of a recent investigation[110] into the vapour phase chlorination of the dioxide by carbon tetrachloride, which proceeds[85] by way of $UOCl_3$, conditions have been established and an apparatus designed suitable for the rapid production of kilogram quantities of uranium tetrachloride. The reaction is carried out at 500–650° and is followed by vacuum sublimation of the tetrachloride at 700°. Mixtures of carbon tetrachloride and chloroform have also been used[111] to chlorinate UO_2 at 400–500°. The condition of the dioxide used in any chlorination reaction is of prime importance and a highly active form of UO_2 can be prepared by hydrogen or methane reduction of U_3O_8 or UO_3 above 500°. Active thorium dioxide is best prepared by decomposition of the oxalate at about 450°.

Probably the most convenient method of preparing uranium tetrachloride is the liquid phase chlorination[112,113] of UO_3 or U_3O_8 with hexachloropropene, $Cl_2C = CClCCl_3$ (b.p. 210°). Provided a five-fold excess of the chlorinating agent is employed the reaction is virtually quantitative at atmospheric pressure; the product is washed with carbon tetrachloride to remove organic by-products (mainly trichloroacrylyl chloride, $CCl_2 = CClCOCl$) before purification by sublimation. As mentioned earlier (p. 129) the chlorination of uranium trioxide proceeds via the soluble pentachloride and the complex $5UCl_5 \cdot CCl_2 = CClCOCl$ has been isolated as an intermediate reaction product. Although hexachloropropene has been successfully employed to convert UO_3, U_3O_8, $UO_4 \cdot 2H_2O$, UO_2Cl_2, UO_2SO_4 and $UO_2(NO_3)_2$ to the tetrachloride, the reaction with uranium dioxide appears to be unsatisfactory. Other organic chlorinating reagents have been investigated, but with less success[102].

Other less satisfactory reactions which are reported to yield uranium tetrachloride include[102] the reaction of chlorine with the nitride or sulphide, reaction of phosphorus pentachloride with the dioxide, the action of thionyl chloride vapour on the various oxides and the reaction of sulphur monochloride[114] with the trioxide. An examination of the interaction of inactive uranium dioxide and various inorganic chlorides failed to furnish a satisfactory chlorinating agent but aluminium or boron trichloride convert[115] uranium tetrafluoride to the tetrachloride at 250–500° in a sealed tube, and recently boron trichloride has been used to convert uranium dioxide to the tetrachloride[116]. The reaction between thorium tetraiodide and carbon tetrachloride, yields[117] thorium tetrachloride but it is merely of academic interest owing to the necessity of initially preparing the tetraiodide.

The tetrachlorides of protactinium and neptunium have been little investigated in contrast to those of thorium and uranium. Protactinium tetrachloride was first prepared[118] on the microgram scale by reducing the pentachloride with hydrogen at 800°, a reaction which has since been found[67] to proceed smoothly on the 50 milligram scale in a sealed vessel at 400°; this reduction, however, is better done[67] with excess aluminium. Anhydrous neptunium tetrachloride, also first prepared[119] and identified in microgram amounts, is obtained[119,120] by the action of carbon tetrachloride vapour on the oxalate or dioxide above 500°; the latter reaction has also been employed[53] for the preparation of protactinium tetrachloride but protactinium dioxide is a relatively inert material since it is only conveniently obtained by hydrogen reduction of the pentoxide, Pa_2O_5, at 1500°. Although hexachloropropene converts the higher oxides of uranium to the tetrachloride (p. 136) it reacts with $NpO_3 \cdot H_2O$ and Np_2O_5 to yield a mixture of neptunium tetrachloride and trichloride[16] from which the former can be separated by vacuum sublimation above 600°.

The isolation of protactinium tetrachloride from aqueous solution is impracticable owing to the ready aerial oxidation of this valence state. No attempt has been made to dehydrate the product obtained by vacuum evaporation of neptunium (IV) in hydrochloric acid solution but thionyl chloride may well prove useful for such work.

The existence of gaseous plutonium tetrachloride has recently been confirmed[381] by measurement of the absorption spectra of gaseous plutonium trichloride and a mixture of the trichloride and chlorine at 928°. The thermodynamics of the reaction,

$$PuCl_3(l) + \tfrac{1}{2}Cl_2(g) \rightleftharpoons PuCl_4(g)$$

have been studied spectroscopically and the molar free energy shown to be represented by the equation,

$$\Delta F = 23{,}000 - 14.1T \,(\text{cal})$$

Crystal structures. The above four tetrachlorides are isostructural[67,118,119,121] (Table 3.5) possessing body-centred tetragonal symmetry; an interpretation[121] of x-ray powder data indicates that there are two sets of 4 chlorine atoms at distances of 2.46 and 3.11 Å respectively from the thorium atom and 2.41 and 3.09 Å respectively from the uranium atom. A projection of the structure is illustrated in Figure 3.5. Experimental

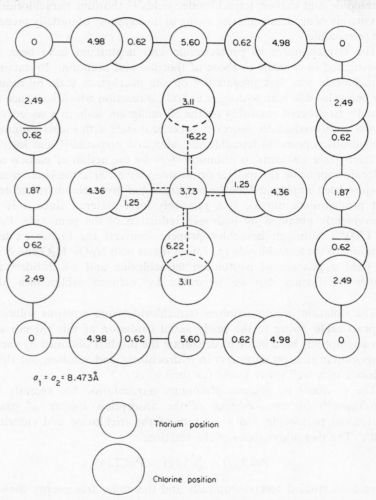

$a_1 = a_2 = 8.473 \text{Å}$

Thorium position

Chlorine position

Figure 3.5   Projection of the structure of thorium tetrachloride on the 001 face (distances in the *c* direction are in Å). The coordination is indicated by connecting lines for one thorium. (*After* R. L. Mooney, *Acta Cryst.*, **2**, 189 (1949))

determinations of the densities of uranium and thorium tetrachloride are in good agreement with those calculated from x-ray data (4.866 and 4.60 g cm^{-3} respectively).

TABLE 3.5

Crystallographic Data[a] for the Actinide Tetrachlorides[67,118,119,121]

Compound	Colour	Lattice parameters (Å)		Density (g cm^{-3})
		a_0	c_0	
ThCl$_4$	White	8.473	7.468	4.60
PaCl$_4$	Greenish-yellow	8.377	7.482	4.72
UCl$_4$	Green	8.296	7.487	4.87
NpCl$_4$	Red-brown	8.250	7.460	4.92

[a] The tetrachlorides are tetragonal, space group D_{4h}^{19}–$I4/amd$ with $n = 4$.

Properties. The most reliable vapour pressure data for thorium[122] and uranium[102] tetrachloride are represented by the equations in Table 3.6; a recent redetermination[123] of the properties of solid uranium tetrachloride is in agreement with the data listed. Similar measurements are lacking for protactinium and neptunium tetrachloride (m.p. 538°) which are volatile in a vacuum above 450° and 500° respectively. The high values of Trouton's constant for thorium tetrachloride (31) and uranium tetrachloride (31 or 36) suggest that these compounds are highly associated in the liquid phase; the vapour density (13.31) indicates no association of uranium tetrachloride in the vapour state.

Although the chemical properties of thorium and uranium tetrachloride have been extensively studied there is scope for further

TABLE 3.6

Vapour Pressure Data for Thorium and Uranium Tetrachloride[102,122]

Compound	m.p.	$\log p_{mm} = -A/T + B$	
		A	B
ThCl$_4$ (solid)	770	12,910	14.30
ThCl$_4$ (liquid)		7,987	9.57
UCl$_4$ (solid)	590	10,427	13.30
UCl$_4$ (liquid)		7,205	9.65

investigations into the chemical behaviour of protactinium and neptunium tetrachloride. The tetrachlorides are all moisture-sensitive and must be handled in a dry atmosphere; thorium tetrachloride forms hydrates at room temperature, protactinium tetrachloride oxidizes to protactinium (V) and the deliquescent uranium compound is slowly oxidized to uranium (VI). At high temperature steam converts thorium tetrachloride to the dioxide and uranium tetrachloride to triuranium octoxide, U_3O_8. Oxygen converts uranium tetrachloride to uranyl chloride, UO_2Cl_2 (p. 119), at 300° but above 500° uranium oxides are formed; an earlier report[124] states that U_3O_8 is formed at 250° but this may have been due to the presence of water vapour.

Chlorine reacts with uranium tetrachloride at moderate temperatures to yield the pentachloride or hexachloride and will presumably convert protactinium tetrachloride to the pentachloride but it fails to oxidize neptunium tetrachloride. Fluorine converts thorium tetrachloride to the tetrafluoride at room temperature and oxidizes the uranium compound to the hexafluoride (p. 21): the behaviour of protactinium and neptunium tetrachloride with fluorine has not been investigated but the products would probably be the respective penta- and hexafluorides. Hydrogen fluoride converts both thorium and uranium tetrachloride to the corresponding tetrafluoride and hydrogen bromide converts uranium tetrachloride to either the tetrabromide[125] (above 350°) or the tribromide (300–350°) whilst hydrogen iodide reduces it to the trichloride at 350–420° and forms uranium (IV) iodochlorides at higher temperatures.

Both uranium and neptunium tetrachloride are reduced to the trichloride by hydrogen at moderate temperatures (p. 150) but thorium and protactinium tetrachloride are unreactive at 800°. Metallic zinc reduces uranium tetrachloride to the trichloride and the reduction of thorium and uranium tetrachloride to the elements by sodium, potassium, calcium, magnesium and aluminium has been described[102,103] but these reactions are inferior to analogous reductions of the tetrafluorides (p. 58) for producing the metal. Neptunium tetrachloride is reduced[126] to the trichloride by ammonia gas at 350°. The interaction of the tetrachlorides with their respective dioxides is described later (p. 146).

The positions of the metal–chlorine stretching vibrations of the solid tetrachlorides have been recorded (Table B1). The magnetic susceptibility of uranium tetrachloride has been measured; Curie–Weiss dependence is observed between 77 and 550°K and the calculated magnetic moment is[127] 3.29 B.M. for a Weiss constant of −62° (or[128] 3.14 B.M. with a Weiss constant = −39°). This value is lower than that predicted by Russell–Saunders coupling (3.58 B.M.) or j–j coupling (3.83 B.M.) and

Dawson[127], who predicts a $5f^2$ configuration for the U^{4+} ion, suggests that exchange effects and crystal field splitting are responsible for the low value.

Thorium and uranium tetrachloride are insoluble in liquid chlorine, carbon disulphide, carbon tetrachloride, benzene and toluene and all the tetrachlorides dissolve in, and form complexes with, the more polar organic solvents such as acetone and the lower alcohols. Thorium and uranium tetrachloride form numerous complexes mainly with oxygen donor ligands containing the C=O, P=O and S=O groups and to a lesser extent with various nitrogen donor molecules, but the only known complex with a monodentate sulphur donor is that with 1,2 dimethylthio-ethane, $UCl_4 \cdot 2L$; complexes with simple ligands containing phosphorus as the donor atom are unknown. These complexes have recently been exhaustively reviewed by Bagnall[5,375] and a brief selection is listed in Table 3.7. Similar complexes of protactinium tetrachloride have scarcely been studied[76,129,376,377] and only $PaCl_4 \cdot 4CH_3CN$, $PaCl_4 \cdot xDMA$ ($x = 3$ and 2.5), $PaCl_4 \cdot xDMSO$ ($x = 5$ and 3) and $PaCl_4 \cdot 2HMPA$ are known. Reported complexes of neptunium tetrachloride and plutonium tetrachloride are limited to the DMSO complexes, $MCl_4 \cdot 5DMSO$ and $MCl_4 \cdot 3DMSO$[376], and to the stable N,N-dimethylacetamide and acetamide compounds[130], $MCl_4 \cdot 2.5DMA$ and $MCl_4 \cdot 6A$ respectively, analogous to the uranium tetrachloride complexes. One of the most interesting facets of the chemistry of these complexes is the volatility[131] of the hexamethyl-phosphoramide complexes $MCl_4 \cdot 2HMPA$ (M = Th and U). Infrared studies indicate that coordination occurs via the oxygen atom in the various amide, phosphine oxide and sulphoxide complexes, for example, see references 12, 31, 32, 67, 129–131, and some metal–chlorine stretching vibrations have been observed below 300 cm^{-1}.

Thorium tetrachloride also reacts[132–135] with various organic acids and aldehydes with replacement of 1, 2, 3, or all 4 chlorine atoms; alkoxides of the type $U(OR)_4$ are formed[136] when uranium tetrachloride is treated with the appropriate lithium alkoxide.

Tetravalent Chloro Complexes

Cerium, praseodymium and the actinide elements thorium to plutonium inclusive form anhydrous tetravalent hexachloro complexes of the type $M_2^I M^{IV} Cl_6$. Hydrated hexachlorothorates (IV) (M^I = Cs, Rb and NH_4) and pentachlorothorates (IV) (M^I = Li, Na and K) are also known; the ammonium salt $(NH_4)_2 ThCl_6 \cdot 10H_2O$ decomposes to the pentachloro complex $NH_4 ThCl_5$ on heating and the caesium and rubidium salts can be dehydrated at 150° in hydrogen chloride but the hydrated pentachloro complexes decompose on heating. Evidence for the existence of the

TABLE 3.7

A Comparison of the Complexes Formed by Thorium and
Uranium Tetrachloride with Certain Donor Ligands[aa]

Ligand (L)	ThCl₄	UCl₄
C=O Complexes		
Acetamide	—	$UCl_4 \cdot 6L^a$
NN-Dimethylacetamide	$ThCl_4 \cdot 4L^b$	$UCl_4 \cdot 2.5L^a$
Dimethylformamide	$ThCl_4 \cdot 4L^c$	$UCl_4 \cdot 2.5L^d$
		$UCl_4 \cdot 3L^e$
N-Methylacetamide	—	$UCl_4 \cdot 4L^b$
NNN′N′-Tetramethylglutaramide	$ThCl_4 \cdot 1.5L^f$	$UCl_4 \cdot 1.5L^f$
NNN′N′-Tetramethyl-3,3-dimethylglutaramide	$ThCl_4 \cdot 1.5L^f$	$UCl_4 \cdot 1.5L^f$
NNN′N′-Tetramethylmalonamide	$ThCl_4 \cdot 2L^f$	$UCl_4 \cdot 1.5L^f$
NNN′N′-Tetramethyl-α,α-dimethylmalonamide	$ThCl_4 \cdot 1.5L^f$	$UCl_4 \cdot L^f$
P=O complexes		
Hexamethylphosphoramide	$ThCl_4 \cdot 2L^g$	$UCl_4 \cdot 2L^g$
Trialkyl or triarylphosphine oxides, R₃PO	$ThCl_4 \cdot 2L^h$	$UCl_4 \cdot 2L^{d,i,j,k}$
Phosphorus oxytrichloride	—	$UCl_4 \cdot 4L^l$
Diphosphineoxides, (R₂PO)₂CH₂	$ThCl_4 \cdot L^m$	$UCl_4 \cdot L^{d,i,m}$
Octamethylpyrophosphoramide	$ThCl_4 \cdot 1.5L^m$	$UCl_4 \cdot 1.5L^m$
S=O complexes		
Dimethylsulphoxide	$ThCl_4 \cdot 5L^g$;	$UCl_4 \cdot 7L^z$;
	$ThCl_4 \cdot 3L^z$	$UCl_4 \cdot 5L^z$;
		$UCl_4 \cdot 3L^{g,z}$
Disulphoxides (R₂SO)₂CH₂	$ThCl_4 \cdot 2L^n$	$UCl_4 \cdot 2L^n$
N complexes		
Methyl cyanide	$ThCl_4 \cdot 4L^o$	$UCl_4 \cdot 4L^o$
Ammonia	$ThCl_4 \cdot 6L^p$	$UCl_4 \cdot 12L$
		$\rightarrow UCl_4 \cdot 2L^{q,r}$
Aliphatic amines	$ThCl_4 \cdot 4L^s$	$UCl_4 \cdot L^q$;
		$UCl_4 \cdot 2L$
Hydrazine	—	$UCl_4 \cdot 6L^q$
Miscellaneous oxygen complexes		
Primary alcohols	$ThCl_4 \cdot 4L^t$	$UCl_4 \cdot 4L^u$
Nitrosyl chloride	$ThCl_4 \cdot 2L^v$	$UCl_4 \cdot 2L^w$
Tetrahydrofuran	$ThCl_4 \cdot 3L^x$	$UCl_4 \cdot 3L^x$
Dioxan	—	$UCl_4 \cdot 3L^d$
1,2-Dimethoxyethane	—	$UCl_4 \cdot 2L^y$

[aa] These and other complexes are discussed in detail in other reviews[5,375].

[a] K. W. Bagnall, A. M. Deane, T. L. Markin, P. S. Robinson and M. A. A. Stewart, *J. Chem. Soc*, **1961**, 1611.

[b] K. W. Bagnall, D. Brown, P. J. Jones and P. S. Robinson, *J. Chem. Soc.*, **1964**, 2531.

pentachlorouranate (IV) ion in the fused UCl_4–KCl–$CuCl_2$ mixture has been reported[137] and the anionic species[138] $ThCl_5^-$, $ThCl_6^{2-}$ and $ThCl_7^{3-}$ are formed in fused mixtures of thorium tetrachloride with sodium, potassium, caesium and cerous chlorides.

The anhydrous thorium (IV) complexes, $M_2^IThCl_6$, (M^I = Li, Na, K, Rb and Cs) have been prepared by heating together the appropriate amounts of the component halides[139]. The analogous uranium (IV) salts (M^I = Li, Na and K), and those of the type $M^{II}UCl_6$ (M^{II} = Ca, Sr and Ba), are reported[140,141] to be formed by passing uranium tetrachloride vapour over the alkali or alkaline earth metal halide. Rubidium and caesium octachlorothorates (IV) have also been reported[139]; they were identified by measurement of their heats of solution which are different from those of the corresponding hexachloro salts and the lithium, sodium and potassium hexachlorothorates (IV). Octachlorothorates (IV) are not formed by the smaller, alkali metal cations. Further work on the rubidium and caesium salts would be of value.

Although it has been stated that[142] hexachlorouranates (IV) cannot be prepared from aqueous solution and that[143] they are oxidized on

Continuation of TABLE 3.7.

c T. Moeller and D. S. Smith, Report AFOSR-TN-58-559 (1958).

d P. Gans, Thesis, London University (1964).

e M. Lamisse, R. Heimburger and R. Rohmer, *Compt. Rend.*, **258**, 2078 (1964).

f K. W. Bagnall, D. Brown and P. J. Jones, *J. Chem. Soc. (A)*, **1966**, 741.

g K. W. Bagnall, D. Brown, P. J. Jones and J. G. H. du Preez, *J. Chem. Soc. (A)*, **1966**, 737.

h B. W. Fitzsimmons, P. Gans, B. C. Smith and M. A. Wassef, *Chem. Ind. London*, **1965**, 1698.

i P. Gans and B. C. Smith, *J. Chem. Soc.*, **1964**, 4172.

j P. Day, Thesis, Oxford University (1965).

k P. Day and L. M. Venanzi, *J. Chem. Soc. (A)*, **1966**, 197.

l R. E. Panzer and J. F. Suttle, *J. Inorg. Nucl. Chem.*, **15**, 67 (1960).

m J. G. H. du Preez, unpublished observations.

n K. W. Bagnall, J. G. H. du Preez and A. J. Basson, to be published.

o K. W. Bagnall, D. Brown and P. J. Jones, *J. Chem. Soc. (A)*, **1966**, 1763.

p G. W. A. Fowles and F. H. Pollard, *J. Chem. Soc.*, **1953**, 4128.

q I. Kalnins and G. Gibson, *J. Inorg. Nucl. Chem.*, **7**, 55 (1958).

r H. J. Berthold and H. Krecht, *Angew. Chem.*, **77**, 453 (1965).

s J. M. Matthews, *J. Am. Chem. Soc.*, **20**, 815 (1898).

t D. C. Bradley, A. A. Saad and W. Wardlaw, *J. Chem. Soc.*, **1954**, 2002.

u D. C. Bradley, K. N. Kapoor and B. C. Smith, *J. Inorg. Nucl. Chem.*, **24**, 863 (1962).

v R. Perrot and C. Devin, *Compt. Rend.*, **246**, 772 (1958).

w C. C. Addison and N. Hodge, *J. Chem. Soc.*, **1961**, 2490.

x S. Herzog, K. Gustav, E. Krueger, H. Oberender and R. Schuster, *Z. Chem.*, **3**, 428 (1963).

y H. C. E. Mannerskantz, G. W. Parshall and G. Wilkinson, *J. Chem. Soc*, **1963**, 3163.

z K. W. Bagnall, D. Brown, D. G. Holah and F. Lux, *J. Chem. Soc. (A)*, **1968**, 465.

exposure to air, subsequent research has shown that these statements are incorrect. Thus the hexachloro complexes, $M_2^I M^{IV} Cl_6$, of uranium, neptunium, plutonium and cerium precipitate from hydrochloric acid solutions[34,93,130,144–147] containing a univalent cation of sufficient size (e.g. $M^I = Cs$, NMe_4, NEt_4 and Ph_4As) and such uranium (IV) salts, which are non-hygroscopic, are perfectly stable[130,146] in the atmosphere. High yields of the uranium (IV), neptunium (IV) and plutonium (IV) complexes are obtained by saturating an aqueous hydrochloric acid[130] or ethanolic hydrochloric acid solution with hydrogen chloride and it is unnecessary to vacuum evaporate the solutions as recommended by some authors[34,148]. In addition, the complexes are easily obtained anhydrous by washing the precipitate with alcohol and vacuum drying it at room temperature[130], a simpler expedient than[146,148] prolonged drying over sodium hydroxide, magnesium perchlorate or sulphuric acid. Caesium hexachloroprotactinate (IV) has been isolated from aqueous hydrochloric acid[151] but reactions involving smaller univalent cations were less successful.

The preparation of anhydrous thorium (IV) hexachloro complexes from aqueous solution is not recommended since they are difficult to dehydrate completely without partial hydrolysis and even those containing large cations are hygroscopic[148]. However, $(NEt_4)_2 ThCl_6$ and $(NMe_4)_2 ThCl_6$ can be made by reacting hydrated thorium tetrachloride with the appropriate tetralkylammonium chloride in thionyl chloride[149,150], a method also employed for the preparation[149] of $(NEt_4)_2 UCl_6$, or by reacting the appropriate halide[150] with anhydrous thorium tetrachloride in methyl cyanide. Numerous[31] hexachlorouranates (IV) (e.g. $M^I = Ph_3PH$, Ph_3PBz, Et_3PH) and some[151] tetralkylammonium hexachloroprotactinates (IV) ($M^I = NMe_4$ and NEt_4) have also been prepared from anhydrous methyl cyanide and it is probable that the alkali metal salts of thorium (IV) and uranium (IV) could be prepared by using a mixture of methyl cyanide or thionyl chloride and iodine monochloride as in the preparation of hexachlorouranates (V) (p. 129). Protactinium (IV) chloro complexes are, however, oxidized[151] by thionyl chloride. Anhydrous pyridinium and quinolinium hexachlorothorates (IV) have been precipitated[152,153] from alcoholic hydrochloric acid solutions. The analogous, and other, cerium (IV) complexes have been prepared in a similar manner[91,92] and various hexachlorouranates (IV) (e.g. $M^I = Et_3PH$, Pr_3PH, Ph_2PH_2 and Et_4P) have been prepared[12,30,39] from ethanolic solution.

The preparation of rubidium hexachloroplutonate (IV), $Rb_2 PuCl_6$, by chlorination of a plutonium dioxide–rubidium chloride mixture with carbon tetrachloride at 750° has been reported[154] and the formation of the sodium, potassium, rubidium and caesium salts occurs[155] when a mixture of the alkali metal chloride and plutonium trichloride is heated in chlorine at about 50° above the melting point of the alkali metal chloride. The yield and stability of these salts increases with the atomic weight of the alkali metal chloride; oxidation to a stable complex does not take place in the presence of lithium, calcium or barium chloride.

TABLE 3.8

Lattice Parameters for Anhydrous Tetravalent Hexachloro Complexes

Compound	Colour	Symmetry	Space group	Lattice parameters (Å)			Reference
				a_0	b_0	c_0	
Cs$_2$ThCl$_6$	White	Trigonal	D_{3d}^3–$C\bar{3}m$	7.590	—	6.026	157
Cs$_2$PaCl$_6$	Bright yellow	Trigonal	D_{3d}^3–$C\bar{3}m$	7.546	—	6.056	151
Cs$_2$UCl$_6$	Green	Trigonal	D_{3d}^3–$C\bar{3}m$	7.478	—	6.026	157
Cs$_2$NpCl$_6$	Yellow	Trigonal	D_{3d}^3–$C\bar{3}m$	7.460	—	6.030	16
Cs$_2$PuCl$_6$	Pale yellow	Trigonal	D_{3d}^3–$C\bar{3}m$	7.430	—	6.030	156
Cs$_2$CeCl$_6$	Yellow	Trigonal	D_{3d}^3–$C\bar{3}m$	7.476	—	6.039	336
(NMe$_4$)$_2$ThCl$_6$	White	Face centred cubic	O_h^5–$Fm3m$	13.12	—	—	150
(NMe$_4$)$_2$PaCl$_6$	Yellow	Face centred cubic	O_h^5–$Fm3m$	13.08	—	—	151
(NMe$_4$)$_2$UCl$_6$	Green	Face centred cubic	O_h^5–$Fm3m$	13.06	—	—	34
(NMe$_4$)$_2$NpCl$_6$	Yellow	Face centred cubic	O_h^5–$Fm3m$	13.02	—	—	150
(NMe$_4$)$_2$PuCl$_6$	Pale yellow	Face centred cubic	O_h^5–$Fm3m$	12.94	—	—	34
(NMe$_4$)$_2$CeCl$_6$	Yellow	Face centred cubic	O_h^5–$Fm3m$	13.05	—	—	93
β-(NEt$_4$)$_2$ThCl$_6$	White	Ortho-rhombic	D_{2h}^{23}–$Fmmm$	14.26	14.84	13.37	150
(NEt$_4$)$_2$PaCl$_6$	Yellow	Ortho-rhombic	D_{2h}^{23}–$Fmmm$	14.22	14.75	13.35	151
(NEt$_4$)$_2$UCl$_6$	Green	Ortho-rhombic	D_{2h}^{23}–$Fmmm$	14.23	14.73	13.33	34
(NEt$_4$)$_2$NpCl$_6$	Yellow	Ortho-rhombic	D_{2h}^{23}–$Fmmm$	14.20	14.69	13.30	150
(NEt$_4$)$_2$PuCl$_6$	Yellow	Ortho-rhombic	D_{2h}^{23}–$Fmmm$	14.20	14.70	13.30	34

Crystal structures and properties. The structure of caesium hexachloroplutonate (IV), Cs_2PuCl_6, was determined[156] from powder data by Zachariasen. It possesses trigonal symmetry with the K_2GeF_6-type structure. There is one molecule per unit cell and the plutonium atom is surrounded by six chlorine atoms, at the corners of an octahedron, each at a distance of 2.62 Å from the plutonium atom. The other known caesium salts are isostructural (Table 3.8) and the Th–Cl and U–Cl bond lengths are[157] 2.81 Å and 2.75 Å respectively. The tetramethylammonium salts are all face-centred cubic, possessing the K_2PtCl_6-type structure, whereas the tetraethylammonium complexes are orthorhombic (Table 3.8). $(NEt)_2UCl_6$ and $(NEt_4)_2PuCl_6$ undergo[34] a reversible phase change at 94° and 97° respectively to become face-centred cubic and $(NEt_4)_2ThCl_6$ exists[150] in two stable forms. The α-form, which is of unknown symmetry, crystallizes from anhydrous methyl cyanide at ice-temperature and is converted to the β-form, which is isostructural with $(NEt_4)_2UCl_6$, at 130° in a vacuum. The phase change can be reversed by cooling the β-form in liquid nitrogen. Similar phase changes are not observed[150,151] for the analogous protactinium (IV), uranium (IV) and neptunium (IV) complexes. Magnetic susceptibility data have been recorded[31,158–160] for certain of the uranium (IV) salts. They show temperature-independent paramagnetism ($\chi_m \approx 2,000 \times 10^{-6}$ c.g.s. units) in agreement with a $5f^2$ electronic configuration and a non-magnetic ground state[158] of $A_1(^3H_4)$ with no permanent thermal population of the excited states (the first excited state $T_1(^3H_4)$ is at 920 cm^{-1}). Cs_2NpCl_6 exhibits Curie–Weiss dependence between 90 and 300°K with a Weiss constant of 71° and $\mu_{eff} = 3.10$ B.M.; Cs_2PuCl_6 also shows[161] Curie–Weiss dependence but deviates below 150°. Paramagnetic resonance and spectral studies[162] on Cs_2PaCl_6 in a matrix of the zirconium (IV) hexachloro salt have shown the electronic configuration of Pa^{4+} in this compound to be $5f^1$. Detailed magnetic susceptibility studies on this or other hexachloroprotactinates (IV) have not yet been reported.

The metal–chlorine stretching vibrations of the tetravalent thorium, protactinium, uranium and neptunium complexes are observed below 300 cm^{-1} in their infrared spectra (Table B.4).

Tetravalent Oxydichlorides

Thorium (IV), protactinium (IV), uranium (IV) and neptunium (IV) oxydichlorides, $MOCl_2$, have been reported. $NpOCl_2$ was first observed[119] when a sample of the tetrachloride was heated in a sealed capillary at 450°; presumably it was formed by interaction of traces of water vapour with the tetrachloride. The light yellow, crystalline solid, which rather surprisingly was stated to be volatile above 550°, was identified by x-ray powder diffraction analysis which showed it to be isostructural with the uranium (IV) analogue. ThOCl$_2$ is obtained when the hydrated tetrachloride is heated in air or hydrogen chloride but one of the most satisfactory methods of preparing this compound, and UOCl$_2$, is by[163–165] heating the dioxide with excess tetrachloride and finally removing the excess tetrachloride by vacuum sublimation. An alternative reaction[166], recently used for the first preparation of PaOCl$_2$[67], is to heat together stoicheiometric amounts of the tetrachloride and antimony trioxide *in vacuo*. This method possesses the advantage that it does not require the preparation of a reactive form of the appropriate dioxide, a requirement which, at present, cannot be fulfilled for PaO$_2$.

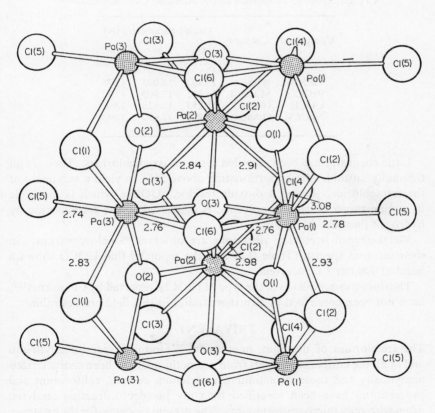

Figure 3.6 The structure of PaOCl$_2$.[378] (*After* R. P. Dodge, G. S. Smith, Q. Johnson and R. E. Elson, *Acta Cryst.*, **B24**, 304 (1968))

The actinide (IV) oxydichlorides possess orthorhombic symmetry, space group *Pbam*. The structure of $PaOCl_2$ has been reported only recently[378]. It consists of infinite polymeric chains which extend along the short *c* axis and which are cross linked in the *ab* plane by bridging chlorine atoms. The repeating unit of structure along the chain is the aggregate $Pa_3O_3Cl_6$. Protactinium atoms are 7-, 8- and 9-coordinate with oxygen atoms 3- and 4-coordinate. The arrangement is shown in Figure 3.6 which also shows certain bond distances. Pa–O and Pa–Cl distances are in the ranges 2.19–2.38 and 2.76–3.08 Å respectively. The unit cell dimensions for the four isostructural actinide oxydichlorides are given in Table 3.9.

TABLE 3.9

Crystallographic Properties of the Actinide Oxydichlorides[166]

Compound	Colour	Lattice parameters (Å)		
		a_0	b_0	c_0
$ThOCl_2$	White	15.494	18.095	4.078
$PaOCl_2$	Mustard	15.332	17.903	4.012
$UOCl_2$	Green	15.255	17.828	3.992
$NpOCl_2$	Orange	15.209	17.670	3.948

Little chemistry has been recorded for these oxydichlorides. They are all thermally unstable, disproportionating above 550° to yield a sublimate of the tetrachloride, leaving a dioxide residue. $UOCl_2$, which is insoluble in a wide range of organic solvents, is converted to the tetrafluoride by hydrogen fluoride.

Metal-oxygen stretching vibrations are observed[166] below 600 cm^{-1} in their infrared spectra (Table B.2) and the report[167] that $UOCl_2$ shows a band at 730 cm^{-1} is incorrect.

Thorium (IV) complexes of the type $MThOCl_3$, reported by Chauvenet[139], have not been confirmed and further studies in this field are desirable.

TRIVALENT

The trichlorides of uranium, neptunium, plutonium, scandium, yttrium and of all the lanthanides, apart from promethium, have been characterized analytically and those of actinium, americium, curium, californium and promethium have been identified by x-ray powder diffraction analysis, often with only microgram amounts. The present evidence for the existence of thorium trichloride is unreliable[168,169] but the compound is probably capable of existence; the $ThCl_4$-Th system is currently being studied[170]. Numerous chlorinating agents have been employed, particularly for the lanthanide elements where oxidation or reduction is limited, but satisfactory methods for uranium and neptunium trichlorides, which are easily oxidized, require reducing conditions, a condition which is not necessary for the preparation of the transneptunium trichlorides. In the following discussion references will mainly be given to general preparative procedures rather than to the preparation of a particular compound since, particularly with the lanthanides, the methods are often of general application.

A few trivalent chloro complexes such as $(Ph_3PH)_3M^{III}Cl_6$ (M^{III} = Ce, Pr, Nd, Sm, Gd, Dy, Ho, Er, Tm and Yb) have been prepared from solution, and other complexes obtained as hydrates, e.g. $Cs_3PuCl_6 \cdot 2H_2O$. Several complex halides of the types $M^IM_3^{III}Cl_{10}$, $M_2^IM^{III}Cl_5$ and $M_3^IM^{III}Cl_6$ have been identified by phase studies of the lanthanide (III) and plutonium (III) chloride systems. Oxychlorides of the general type MOCl have been characterized for the majority of the trivalent lanthanides and actinides.

Trichlorides

The oxides or hydrated trichlorides have almost invariably been employed as starting materials for the preparation of the trichlorides although the reaction between the appropriate metal or metal hydride and either chlorine or hydrogen chloride is undoubtedly the most convenient method, for example, see references 2, 96, 171–176, provided the metal is available and suitable reaction vessels, e.g. molybdenum, are employed.

Rare earth oxides have been converted directly to the trichlorides by high temperature reactions with carbon tetrachloride vapour[177–179], carbon tetrachloride–chlorine mixtures[180–182], sulphur monochloride[183,184], sulphur monochloride–chlorine mixtures[181,182,185–187], hydrogen chloride[188,189], carbonyl chloride[190], phosphorus pentachloride[188,191], ammonium chloride[192–195] or thionyl chloride[71,196] and also by mixing them with carbon and heating the mixture[197–199] in a stream of chlorine. Several of these methods have been evaluated for the preparation of plutonium trichloride, for example see reference 96, and actinium trichloride has been prepared[200] by heating the hydroxide or oxalate in carbon tetrachloride vapour.

The lanthanide trichloride hydrates have frequently been used as a source of the pure anhydrous trichlorides, dehydration being achieved by heating them in dry hydrogen chloride, for example see references 201–206, chlorine[204], carbonyl chloride[208] or, less satisfactorily, in air (p. 160), by refluxing with thionyl chloride[105] or by mixing them with excess ammonium chloride and heating the mixture in air[209], or better, in a vacuum[199,210]. Plutonium and americium trichloride, the latter first prepared[211] by heating the dioxide with carbon tetrachloride vapour, have also been obtained by dehydration of their respective hydrates by[212] heating them in hydrogen chloride followed by vacuum sublimation of the product. Americium and curium trichloride are conveniently prepared by vacuum dehydration of the product obtained on evaporation of an aqueous solution containing ammonium chloride[213], a reaction which should also be applicable to the preparation of the higher actinide trichlorides. Actinium trichloride is obtained[214] by heating the hydroxide with ammonium chloride at 250° in a vacuum.

Of the above methods for preparing lanthanide trichlorides, apart from those involving direct union of the elements or the action of hydrogen chloride on the metal, the dehydration of the chloride hydrates by heating them in hydrogen chloride, or better with ammonium chloride under reduced pressure, and the direct conversion of the oxides by heating with ammonium chloride appear to be the best. The dehydration of the hydrates with hydrogen chloride is usually carried[203] out between 80° and 400°; when large amounts are involved it is essential to remove most of the water at low temperature to minimize oxychloride formation. Lower contamination with oxychlorides can be achieved by employing ammonium chloride as the dehydrating agent and a successful technique for preparing yttrium trichloride in 45 lb batches has been described[215].

The recently reported[216] conversion of lanthanum and erbium oxalates to their respective trichlorides using a chlorine–carbon tetrachloride mixture at 400–450° appears to be another promising method.

Other, less satisfactory, procedures involving treatment of the trivalent lanthanide benzoate in ether with dry hydrogen chloride[217] or chlorination of the sulphide[218] or carbide[219] at high temperature, have been reported.

Although many of the methods used for preparing lanthanide trichlorides will yield impure materials unless carefully controlled, the fact that the trichlorides can be purified by distillation does mean that the less tedious methods, e.g. oxide–carbon mixtures heated in chlorine or chlorine-carbon tetrachloride vapours, may be employed to obtain a crude product prior to vacuum distillation. This aspect is, of course, of greater importance when large quantities are required. Experimentally-determined temperatures necessary to yield a trichloride vapour pressure of up to 4 mm Hg are listed in Table 3.10 where the values for a 2 mm pressure are compared with estimated values.

Uranium[2,175,220] and neptunium[119] tetrachloride are reduced to the trichlorides in hydrogen above 500° but protactinium tetrachloride is unaffected at 800°. The reduction of uranium tetrachloride is achieved

TABLE 3.10

Vapour Pressures, Heats of Vaporization and Some
Physical Constants of Rare Earth Chlorides[242]

Compound	b.p. (°C)[203]	Temperature (°C) for vapour pressure (mm Hg)					ΔH_v (Kcal/mole)
		4 mm	2 mm	1 mm	0.1 mm	2 mm Est.[203]	
YCl_3	1510	1050	975	909	735	950	30.9
$LaCl_3$	1750	1027	997	969	886	1100	78.9
$CeCl_3$	1730	1195	1125	1065	888	1090	40.8
$PrCl_3$	1710	1144	1085	1031	878	1080	44.8
$NdCl_3$	1690	1166	1106	1048	892	1060	44.4
$SmCl_2$	2030	1400	1229	1087	764	1310	19.9
$EuCl_3$	decomp.	998	930	869	703	940	30.9
$GdCl_3$	1580	1048	995	947	808	980	44.0
$TbCl_3$	1550	1121	1068	1010	854	960	42.0
$DyCl_3$	1530	979	939	899	779	950	48.2
$HoCl_3$	1510	986	953	919	827	950	62.7
$ErCl_3$	1500	1155	1076	1000	809	950	32.9
$TmCl_3$	1490	951	925	899	824	940	77.5
$YbCl_3$	decomp.	1091	1039	994	(856)[a]	940	47.7
$LuCl_3$	1480	998	959	926	(819)[a]	950	57.2

[a] Extrapolated value.

more readily at a hydrogen pressure of 7 atmospheres and it is recommended[175] that the temperature should be maintained below 575° in the initial stages to prevent the formation of liquid uranium tetrachloride. Whereas reduction of uranium tetrachloride by ammonia gas yields a product contaminated with uranium nitrides[175] it is reported that even at 1000° neptunium trichloride can be obtained[126] from the tetrachloride in a reasonable state of purity. Uranium trichloride is best prepared[2,175] directly from the metal or hydride by the action of hydrogen chloride at 250–300°, the adsorbed chloride being removed by heating the product at 150° in a vacuum; reduction of the tetrachloride by metallic zinc also appears[221] attractive. An interesting purification[222] of uranium trichloride, by distillation, as UCl_3I, in a stream of iodine vapour, obviates the high temperature necessary for direct sublimation of the trichloride; on cooling, the trichloroiodide decomposes to uranium trichloride and iodine.

The preparation of plutonium trichloride has been extensively investigated[96,223–228]; it is conveniently obtained from the metal by reaction with chlorine[96] at 450° or by conversion to the hydride[96,226] and subsequent treatment with hydrogen chloride at 400°. Other attractive methods for large-scale preparations are the conversion of the trivalent oxalate[225,228] by reaction with hexachloropropene or[227] with phosgene or hydrogen chloride at 140–500°; plutonium (III) carbonate can be employed in place of the oxalate in the latter reactions. As mentioned earlier (p. 137), neptunium (V) hydroxide reacts under similar conditions[16] with hexachloropropene to yield a mixture of neptunium tetra- and trichloride but it is not apparent at which stage the reduction to Np (III) occurs.

Californium[229] and berkelium trichloride have been prepared in only microgram amounts by the action of hydrogen chloride on the oxides, a reaction which, with AmO_2, yields americium trichloride[211,230]

Crystal structures. Three structure types are exhibited by the lanthanide trichlorides. Yttrium trichloride is isostructural with one of these whereas scandium trichloride[231] possesses the different rhombohedral $FeCl_3$-type structure as do $ScBr_3$ (p. 198) and ScI_3 (p. 218). Bommer and Hohmann[232] first reported the trichlorides of lanthanum to gadolinium inclusive to be isostructural, those of terbium and dysprosium to possess a different structure and those of holmium to lutetium to belong to a third structure class together with a second crystallographic modification of $DyCl_3$. Zachariasen[233–235] has shown that the trichlorides La–Nd and those of Ac–Am possess the hexagonal[235] uranium trichloride-type (or yttrium hydroxide) structure (Figure 3.7), space group $C_{6h}^2-C6_3/m$ in which each metal atom is bonded to nine chlorine atoms, the bonding being pre-

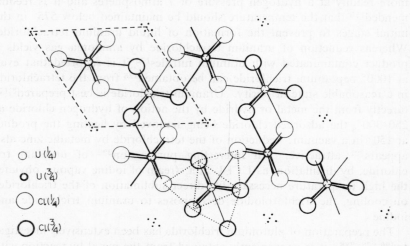

O U (¼)

O U (¾)

◯ Cl (¼)

◯ Cl (¾)

Figure 3.7 Projection of the crystal structure of UCl_3. (*After* A. F. Wells, *Structural Inorganic Chemistry*, Clarendon Press, Oxford, 1962)

dominantly ionic in character; for uranium trichloride there are three chlorine atoms at 2.95 Å from the uranium atom and a further six at 2.96 Å. The lattice dimensions of the above trichlorides are listed in Tables 3.11 and 3.12; data are also available for the remaining isostructural trichlorides in each series[213,236,237].

The trichloride of yttrium and those of dysprosium (β-form) to lutetium inclusive[189] crystallize with the monoclinic aluminium chloride-type structure, space group C_{3h}^2–$C2/m$, which is a distorted sodium chloride structure in which two-thirds of the metal atoms are omitted. The structure of YCl_3 is shown in Figure 3.8; the three sets of two chlorine atoms are at distances of 2.58, 2.63, and 2.69 Å respectively from the yttrium atom.

More recently the third structure type has been identified[238] as the orthorhombic plutonium tribromide structure (p. 198). Terbium tri-

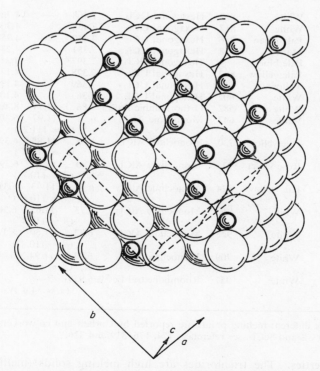

Figure 3.8   The idealized yttrium trichloride structure[189], showing relation to sodium chloride structure. The broken lines outline two of the monoclinic unit cells; several atoms are missing from one of them. (*After* D. H. Templeton and G. F. Carter, *J. Phys. Chem.*, **58**, 940 (1954))

chloride, and presumably α-DyCl$_3$ possess this structure; the eight chlorine atoms to which the terbium atom is bonded are arranged at the following distances, 2 at 2.70 Å, 4 at 2.79 Å and 2 at 2.95 Å. These changes in structure with decreasing cation size, also exhibited by the tribromides and triiodides (pp. 197 and 222), are consistent with the change in ratio of the cation:anion radius.

TABLE 3.11

Crystallographic Properties of the Lanthanide Trichlorides

Compound	Colour	^{a}m.p.[239] (°C)	Symmetry and structure-type	Lattice parameters (Å)			Refer-ence
				a_0	b_0	c_0	
LaCl$_3$	White	862	Hexagonal UCl$_3$	7.468	—	4.366	235
CeCl$_3$	White	817	Hexagonal UCl$_3$	7.436	—	4.304	235
PrCl$_3$	Pale green	786	Hexagonal UCl$_3$	7.410	—	4.250	235
NdCl$_3$	Mauve	758	Hexagonal UCl$_3$	7.381	—	4.231	235
PmCl$_3$	Pale blue	—	Hexagonal UCl$_3$	7.397	—	4.211	236
SmCl$_3$	Pale yellow	682	Hexagonal UCl$_3$	7.378	—	4.171	237
EuCl$_3$	Yellow	d.	Hexagonal UCl$_3$	7.369	—	4.133	237
GdCl$_3$	White	602	Hexagonal UCl$_3$	7.363	—	4.105	237
TbCl$_3$	White	582	Orthorhombic PuBr$_3$	3.86	11.71	8.48	238
DyCl$_3$	White	647	Monoclinic AlCl$_3$	6.91	11.97	6.40	189
				(β = 111.2°)			
HoCl$_3$	Pale yellow	720	Monoclinic AlCl$_3$	6.85	11.85	6.39	189
				(β = 110.8°)			
ErCl$_3$	Rose-violet	776	Monoclinic AlCl$_3$	6.80	11.79	6.39	189
				(β = 110.7°)			
TmCl$_3$	Pale yellow	824	Monoclinic AlCl$_3$	6.75	11.73	6.39	189
				(β = 110.6°)			
YbCl$_3$	White	865	Monoclinic AlCl$_3$	6.73	11.65	6.38	189
				(β = 110.4°)			
LuCl$_3$	White	925	Monoclinic AlCl$_3$	6.32	11.60	6.39	189
				(β = 110.4°)			
YCl$_3$	White	709	Monoclinic AlCl$_3$	6.92	11.94	6.44	189
				(β = 111.0°)			
ScCl$_3$	White	957	Rhombohedral FeCl$_3$	6.979	—	—	231
				(α = 54.4°)			

^{a} Slightly different melting points are reported by Corbett and co-workers for GdCl$_3$, DyCl$_3$, YCl$_3$ and ScCl$_3$, see references 174, 176, 319 and 326.

Properties. The trichlorides are high melting solids (melting points 500–957°, Tables 3.11 and 3.12) which are volatile at high temperature in a vacuum. A few vapour pressure studies have been reported[175,181,241–246] and some results are listed in Table 3.13. There are discrepancies in these data and in the heats of vaporization and the boiling points quoted for the lanthanide trichlorides by different authors. Additional results would help to establish the most reliable values.

TABLE 3.12

Crystallographic Properties of the Actinide Trichlorides[a]

Compound	Colour	m.p.[240] (°C)	Lattice parameters (Å)		Density (g cm^{-3})	Refer- ence
			a_0	c_0		
AcCl$_3$	White	—	7.62	4.55	4.81	235
UCl$_3$	Red	835	7.442	4.320	5.51	235
NpCl$_3$	Green	~800	7.405	4.273	5.58	235
PuCl$_3$	Emerald-green	760	7.380	4.238	5.70	235
AmCl$_3$	Pink	—	7.390	4.234	5.78	213
CmCl$_3$	White	500	7.380	4.186	5.81	372
CfCl$_3$	Green	—	7.383	4.090	5.88	373

[a] The trichlorides all possess the hexagonal UCl$_3$-type structure, space group C_{6h}^2–$C6_3/m$, $n = 2$.

TABLE 3.13

Vapour Pressure Data for the Trichlorides

Compound	$\log p_{mm} = -A/T + B$		Reference
	A	B	
ScCl$_3$ (solid)	14,223	14.370	245
YCl$_3$ (solid)	17,470	42.169 − 9.061 log T	246
LaCl$_3$	15,796	11.828	241
	18,392	41.983 − 9.061 log T	243
CeCl$_3$	15,544	12.035	241
	18,153	42.011 − 9.061 log T	243
PrCl$_3$	15,439	12.121	241
	17,946	41.981 − 9.061 log T	243
NdCl$_3$	15,145	12.014	241
	17,691	41.841 − 9.061 log T	243
ErCl$_3$	16,624	41.671 − 9.061 log T	243
UCl$_3$	12,000	10.000	175
PuCl$_3$ (solid)	15,910	12.726	244
PuCl$_3$ (liquid)	12,587	9.428	244

The rare earth trichlorides are hygroscopic and hydrates of the types $MCl_3 \cdot 7H_2O$ (M = La, Ce, and Pr) and $MCl_3 \cdot 6H_2O$ (M = Sc, Y, and Nd–Lu) are easily prepared from aqueous hydrochloric acid solutions. The only actinide trichloride hydrates reported to date are those of plutonium (III) and americium (III) which crystallize with six molecules of water, the former loses three molecules of water at room temperature in a vacuum. It is likely that uranium and neptunium trichloride will form stable solid hydrates despite the ready oxidation of the trivalent state of these elements in water (cf. p. 199). Crystallographic data are available, for example, see references 96, 247–251 and 374, for many of the hydrates. Their conversion to anhydrous trichlorides has already been discussed (p. 149) and the conditions under which they are decomposed to the oxychlorides are discussed later (p. 159).

Hydrogen bromide and hydrogen iodide convert the lanthanide trichlorides to their respective tribromides and triiodides (p. 195 and 219) and americium and curium trichloride undergo metathesis with ammonium halides (pp. 196 and 219). Hydrogen fluoride oxidizes uranium trichloride to the tetrafluoride at high temperature and oxygen converts it to uranyl chloride, whereas chlorine oxidation at 250° results in the formation of uranium tetrachloride (p. 135) but yields a mixture of the last with higher uranium chlorides at higher temperatures. Uranium trichloride reacts with bromine and iodine to form mixed uranium (IV) halides (p. 255).

Only samarium, europium and ytterbium trichlorides are reduced to stable divalent chlorides by hydrogen. The reported reduction of uranium trichloride is unreliable but neodymium trichloride is reduced to the dichloride by the metal (p. 164). The reduction of yttrium and certain rare earth trichlorides to the elements by metals such as calcium or lithium has been described[199]; uranium trichloride[252] (like the tetrachloride[253]) and plutonium trichloride, see reference 254, for example, have been reduced in a similar manner, but such reactions have little application in metal production.

Scandium, yttrium and the lanthanide trichlorides form complexes with oxygen and nitrogen donor ligands, of which the following are representative: $MCl_3 \cdot x\,NH_3$[255,256,257] (M = Sc, La, Ce, Pr and Nd; x = variously 1 to 20), $ScCl_3 \cdot 2C_6H_5N$[258], $NdCl_3 \cdot 3C_6H_5N$[255], $ScCl_3 \cdot 4NH(C_2H_5)_2$[259], $ScCl_3 \cdot 3CH_3CN$[260], the dioxan[260] complexes, $ScCl_3 \cdot 2L$ and $ScCl_3 \cdot 3L$, and tetrahydrofuran complexes[261] of lanthanum $LaCl_3 \cdot 1.5L$ and of scandium and yttrium, $MCl_3 \cdot 3L$ (M = Sc and Y). Various alcoholates have been reported for lanthanum[262], cerium[262,263], neodymium[264] and yttrium[265]. Complexes of the actinide trichlorides are unknown; in fact the chemistry of this group of compounds, apart from that of uranium and plutonium trichloride, has scarcely been studied. With the exception of the case of neptunium trichloride this is a consequence of the small quantities (usually only of the order of a few milligrams or less) which have been available.

Magnetic susceptibility studies have been reported for uranium trichloride[78,127,266], plutonium trichloride[267] and several of the lanthanide[268-271] compounds. The effective magnetic moments of the lanthanide trichlorides are in agreement with theoretical values, e.g. $GdCl_3$[268], 7.85 B.M. (theoretical value = 7.94 B.M.) and $NdCl_3$[270], 3.78 B.M. (theoretical value 3.64 B.M.). The most reliable results for uranium trichloride[78] were obtained from measurements involving UCl_3–$LaCl_3$ mixtures from which it was shown that the susceptibility at infinite dilution closely fits the relationship $\chi_M = 0.785/T + 0.0028$ over the range 14–300°K. From a comparison of these data with those of the magnetically dilute neodymium (III) salt, $Nd(C_2H_5SO_4)_3 \cdot 9H_2O$, it was predicted that the electronic configuration of uranium trichloride was $5f^3$. This was confirmed[266] by the interpretation of the paramagnetic resonance spectrum of uranium trichloride. The magnetic properties[267] of plutonium trichloride are similar to those of samarium (III) salts ($4f^5$) and it is probable that the electronic configuration of plutonium (III) in the trichloride is $5f^5$.

Low temperature magnetic susceptibility studies with single crystals of gadolinium trichloride[337] have shown that this compound becomes ferromagnetically ordered at 2.2°K. Anomalous behaviour has also been reported[338] below 4°K for cerium, praseodymium and gadolinium trichloride crystals. For the last exchange interaction between nearest neighbours leads to ferromagnetic ordering at 1.745°K whilst at a lower temperature (1.035°K) an antiferromagnetic interaction between next nearest neighbours becomes important.

Paramagnetic resonance absorption has been reported[339-343] for many of the trivalent lanthanide and actinide ions in either a lanthanum or yttrium trichloride matrix.

Group theoretical analysis shows that lanthanum trichloride should have 3 infrared active, 6 Raman active and 5 non-active vibrational modes. The three infrared active modes have been observed[272] at 138, 165 and 210 cm^{-1} respectively and five of the Raman active vibrations are reported[273] at 106.7, 177.3, 185.2, 208.6 and 215.5 cm^{-1} respectively.

There are numerous recent reports, particularly by Dieke and his colleagues, dealing with the absorption and fluorescence spectra of trivalent lanthanide and actinide ions in a lanthanum trichloride host lattice (see, for example, references 344–371). The various data will not be discussed here and it suffices to say that in many instances the results are described in terms of energy level assignments for the trivalent ions.

Trivalent Chloro Complexes

There have been relatively few investigations of the preparation of chloro complexes of the trivalent lanthanide and actinide elements. Hydrated lanthanide (III) complexes can be prepared from aqueous or alcoholic solution; thus the following lanthanide complexes have been reported, $K_2CeCl_5 \cdot xH_2O$ (which can be dehydrated[274] at 145°), $CsCeCl_4 \cdot 7H_2O$[274], $Cs_3M^{III}Cl_6 \cdot 5H_2O$ (M^{III} [275,276] = La, Ce, Pr, Nd and Sm), dimethylammonium chloride salts[277] of the type $MCl_3 \cdot xNH_2Me_2Cl \cdot yH_2O$ (M = La, Ce, Pr and Nd; x = 1, 2, 4, 6 or 7; y = 0 or 2), and the tetraphenylphosphonium complexes[278] $Ph_4PM^{III}Cl_4 \cdot 8H_2O$ (M^{III} = La, Ce and Nd).

TABLE 3.14

Examples[aa] of Trivalent Chloro Complexes Characterized by Phase Studies

Type	Element (M)	Reference
KM_3Cl_{10}	Y, La, Ce	a, c
KM_2Cl_7	Sm	d, j
RbM_2Cl_7	Pu, Sm	f, j
CsM_2Cl_7	Pu	f
$KMCl_4$	La	a
Na_2MCl_5	Eu, Sm	i, j
K_2MCl_5	Ce, Pr, Nd, Sm, Pu	a, b, d, e, j
Rb_2MCl_5	Pu, Sm	f, j
Cs_2MCl_5	La, Ce, Pr	g
$K_3M_2Cl_9$	Ce, Pr, Nd	a
$Cs_3M_2Cl_9$	Sc	h
Na_3MCl_6	Y, Eu, Ho, Er	c, i
K_3MCl_6	Y, Ce, Pr, Nd, Sm, Yb, Pu	a, b, c, d, e, j
Rb_3MCl_6	Pu, Sm	f, j
Cs_3MCl_6	Sc, La, Ce, Pr, Pu	f, g, h
Sr_3MCl_9	Pu	f
Ba_3MCl_9	Pu	f

[aa] This list is not exhaustive.

[a] A. K. Baev and G. I. Novikov, *Zh. Neorgan. Khim.*, **6**, 2610 (1961).

[b] I. S. Morozov, V. I. Ionov and S. G. Korshunov, *Zh. Neorgan. Khim.*, **4**, 1457 (1959).

[c] B. G. Korshunov and D. V. Drobot, *Zh. Neorgan. Khim.*, **9**, 222 (1964).

[d] G. I. Novikov, O. G. Polyachenok and S. A. Frid, *Zh. Neorgan. Khim.*, **9**, 472 (1964).

[e] Yu-Lin Sung and G. I. Novikov, *Zh. Neorgan. Khim.*, **8**, 700 (1963).

[f] J. A. Leary, U.S. Report LA-2661 (1962).

[g] In-Chzhu Sun and I. S. Morozov, *Zh. Neorgan. Khim.*, **3**, 1914 (1958).

[h] R. Gut and D. M. Gruen, *J. Inorg. Nucl. Chem.*, **21**, 259 (1961).

[i] B. G. Korshunov and D. V. Drobot, *Russ. J. Inorg. Chem.*, **10**, 1156 (1965).

[j] B. G. Korshunov, D. V. Drobot, V. V. Bukhtiyarov and Z. N. Shevtsova, *Russ. J. Inorg. Chem.*, **9**, 773 (1964).

Anhydrous hexachloro complexes, $(NHMe_3)_3M^{III}Cl_6$ (M^{III} = La, Pr, Nd but not Ce) are said[279] to crystallize slowly from ethanol and a series of triphenylphosphonium[280] complexes $(Ph_3PH)_3M^{III}Cl_6$ (M^{III} = Ce, Pr, Nd, Sm, Gd, Dy, Ho, Er, Tm and Yb) have been isolated from alcohol saturated with hydrogen chloride.

The hydrated plutonium complex[281], $Cs_3PuCl_6 \cdot 2H_2O$, which loses its water of crystallization above 155° in a vacuum, deposits slowly from 4M hydrochloric acid solution containing excess caesium chloride. Americium (III) yields[282] the tetrachloro salt $CsAmCl_4 \cdot 2H_2O$ from 11M hydrochloric acid solution but in the presence of sodium chloride it is the mixed salt, $Cs_2NaAmCl_6$, which crystallizes. This compound possesses body-centred cubic symmetry with $a_0 = 10.86$ Å. The anhydrous hexachloro complexes Cs_3AmCl_6 and $(Ph_3PH)_3AmCl_6$ have been isolated[282,283] from alcohol–hydrogen chloride mixtures.

There are several reports concerning the identification of complex chlorides observed in $M^ICl–MCl_3$ (M^I = univalent cation) phase studies. A selection of the complexes identified in this manner is listed in Table 3.14. Similar studies[284–288] failed to characterize any complex halides in the systems, NaCl with UCl_3, $PuCl_3$, $LaCl_3$, $CeCl_3$, $PrCl_3$ and $NdCl_3$, $LiCl–PuCl_3$, $CaCl_2$ with $PuCl_3$ and $NdCl_3$, $BaCl_2–UCl_3$ and $MgCl_2–PuCl_3$. The properties of the known complexes have not been investigated.

Trivalent Oxychlorides

Oxychlorides of the type MOCl have been characterized for all the trivalent lanthanides and for actinium, uranium, plutonium, americium, berkelium and californium (Table 3.15). The absence of the thorium (III) and protactinium (III) analogues is not surprising in view of the instability of this valence state of these elements but this aspect of neptunium chemistry merely appears to have been neglected.

The various lanthanide oxychlorides have been obtained[289,290] by heating the oxide in chlorine, by heating the trichloride or its hydrate in oxygen and, in a few instances (La, Nd and Sm), by thermally decomposing the oxalate–chlorides $M(\overline{OX})Cl$. It is also reported[334,335] that thermal decomposition of the perchlorate hydrates yields oxychlorides. The most satisfactory method, which has been employed for the preparation of the oxychlorides of yttrium and of the elements lanthanum to lutetium inclusive (excepting promethium), is to heat[291] the oxide in a mixture of hydrogen chloride and water vapour. The majority have also been identified during the controlled thermal decomposition of the trichloride hydrates in air[292–295], a procedure in which $CeCl_3 \cdot 6H_2O$ decomposes to the dioxide and for which it is reported[292] that $ScCl_3 \cdot 6H_2O$ yields the sesquioxide Sc_2O_3. However, others[296] appear to have isolated ScOCl by this technique. The oxychlorides are obtained in this manner, in general, at lower temperatures[292,293] as the basicity decreases and whereas phases approximating to pure $LaCl_3$, $PrCl_3$, $NdCl_3$, $SmCl_3$ and $GdCl_3$ are observed as intermediates during the thermal decomposition, the trichlorides of the elements europium to lutetium inclusive are not formed (cf. bromides, p. 201). The oxychlorides increase in thermal stability from lanthanum to lutetium.

Actinium oxychloride, AcOCl, is obtained[200] by heating the trichloride in ammonia vapour at 900° and it has also been observed to form on ignition of the hydroxide, precipitated from 0.1 M hydrochloric acid, in air at 1000°, possibly as a result of the coprecipitation of ammonium chloride. Uranium oxychloride, which has only recently been identified[297] in the residue from the sublimation of uranium trichloride contaminated with the dioxide, is a dark red, crystalline solid and, like the lanthanide oxychlorides, it is stable to water, but dissolves in dilute acids. The blue-green plutonium analogue is formed[96] by heating the trichloride or plutonium (IV) hydroxide at 650° in a mixture of hydrogen and hydrogen chloride vapour. The americium (III) analogue, first prepared accidentally[298] when the dioxide was heated in hydrogen, and also berkelium and californium oxychloride, are prepared[229,299,382,383] by heating the sesquioxide in a mixture of hydrogen chloride and water vapour at 450–500°.

TABLE 3.15

Crystallographic Properties of Trivalent Lanthanide
and Actinide Oxychlorides[a]

Compound	Lattice parameters (Å)		Reference	Compound	Lattice parameters (Å)		Reference
	a_0	c_0			a_0	c_0	
YOCl	3.903	6.597	291	TbOCl	3.927	6.645	291
LaOCl	4.119	6.883	291	DyOCl	3.911	6.620	291
CeOCl	4.080	6.831	291	HoOCl	3.893	6.602	291
PrOCl	4.051	6.810	291	ErOCl	3.880	6.580	291
NdOCl	4.018	6.782	291	AcOCl	4.240	7.070	300
PmOCl	4.020	6.740	236	UOCl	4.000	6.850	297
SmOCl	3.982	6.721	291	PuOCl	4.004	6.779	300
EuOCl	3.965	6.695	291	AmOCl	4.000	6.780	298
GdOCl	3.950	6.672	291	BkOCl	3.966	6.710	382
				CfOCl	3.956	6.662	383

[a] All the oxychlorides are tetragonal, space group D_{4h}^7–P4/nmm with two molecules per unit cell.

A standard preparative technique which could probably be applied to the preparation of these oxychlorides is the interaction of the trichlorides with a reactive oxide such as Sb_2O_3 or Bi_2O_3 as described for the preparation of certain of the higher valence oxyhalides (e.g. pp. 147 and 187).

Crystal structures and properties. The actinide oxychlorides and those of yttrium and the lanthanides apart from TmOCl, YbOCl and LuOCl possess the[291,297,298,300,382,383] lead chlorofluoride-type structure. Erbium oxychloride is dimorphic but the structure of its second crystal form and of thulium, ytterbium and lutetium oxychloride is unknown. Available unit cell data are listed in Table 3.15 and Table 3.16 summarizes the known metal–oxygen and metal–chlorine bond distances.

The vapour phase hydrolysis of several lanthanide and actinide trichlorides has been investigated[299,301–304] and certain thermodynamic

TABLE 3.16

Bond Distances in the Trivalent Oxychlorides (Å)

Compound	M–4Cl	M–1Cl	M–4O	Reference
LaOCl	3.18	3.14	2.39	290
SmOCl	3.11	3.09	2.30	291
HoOCl	3.05	3.04	2.25	291
PuOCl	3.11	3.08	2.35	96
AmOCl	3.08	3.08	2.34	298
BkOCl	3.07	3.05	2.23	382

TABLE 3.17

Crystallographic Properties[a] of the Trivalent Lanthanide Thiohalides[379]

Compound	Colour	Lattice parameters (Å)		
		a_0	b_0	c_0
LaSCl	White	7.04	13.69	6.83
CeSCl	White	7.04	13.46	6.76
LaSBr	White	7.19	13.99	7.02
CeSBr	Yellow-green	7.12	13.82	6.94
LaSI	Green	7.38	14.57	7.10
CeSI	Yellow	7.33	14.35	7.05

[a] All possess orthorhombic symmetry, space group D_{2h}^{15}–Pcab.

functions calculated. It has, however, been pointed out by Koch and Cunningham[302] that any values for the heats of formation of the oxychlorides calculated using the data of Bommer and Hohmann for the trichlorides (listed in The National Bureau of Standards, Bulletin No. 500, 1952) will be in error since more recent studies, e.g.[305] have yielded values for the heats of formation which are up to 10 kcal/mole lower than those reported previously. (See Table A.1, p. 238.)

A few trivalent lanthanide thiohalides have recently been reported[379], including the chlorides LaSCl and CeSCl. They are easily prepared by heating together the component elements at 500° in a sealed tube. These thiochlorides, which are white, possess orthorhombic symmetry, space group $D_{2h}^{15}-Pcab$. Unit cell dimensions for these compounds and the isostructural thiobromides and thioiodides are listed in Table 3.17.

DIVALENT

Dichlorides of samarium, europium and ytterbium are well known and more recently those of neodymium and dysprosium have been characterized. The existence of scandium dichloride is doubtful and some preliminary work has been done on the preparation of thulium dichloride[306]. Details of M–MCl$_3$ phase systems have been reported for scandium, yttrium and all the lanthanides excepting promethium, terbium, holmium, thulium and lutetium; phases intermediate between the tri- and dichlorides have been recorded in certain instances. Some of the earlier studies have been reviewed by Asprey and Cunningham[307] and by Novikov and Polyachenok[308].

Dichlorides

Samarium[309], europium[184] and ytterbium[186] dichloride have been known for many years and each was, in fact, the first divalent compound to be prepared for the particular element. Hydrogen, see, for example, references 181, 184, 186, 187, 191, 196, 202, 310–313, or ammonia gas, for example, see references 309 and 314 will reduce the trivalent chlorides of the above elements at high temperature but the trichlorides of the remaining lanthanides and those of the actinides are stable under similar conditions. There is an unsubstantiated report[315] that hydrogen reduces cerium trichloride but this is probably erroneous since no evidence of reduction is observed[316,317] in the Ce–CeCl$_3$ system. There is, however, evidence[171] that divalent cerium may exist in solid solutions of cerium in the phase NdCl$_{2.37}$. Europium and ytterbium dichloride can be prepared[181,318] by reduction of the trichlorides with zinc but this procedure does not yield pure samarium dichloride.

During recent years interest has centred on the M–MCl$_3$ phase diagrams and the electrical conductivity and saturated vapour pressures of such systems. Accounts are available for the systems Sc–ScCl$_3$[174,180,318], Y–YCl$_3$[246,318,319], La–LaCl$_3$[320–322], Ce–CeCl$_3$[316,317], Pr–PrCl$_3$[172,182,321–324], Nd–NdCl$_3$[171,172,182,321,325], Sm–SmCl$_3$[246,318], Gd–GdCl$_3$[176], Dy–DyCl$_3$[326,327], Ho–HoCl$_3$[327], and Er–ErCl$_3$[319]. Similar studies on the actinide systems are limited to those for U–UCl$_3$[328], and Pu–PuCl$_3$[329].

Such studies are complicated by the corrosive properties of the melts at the high temperatures employed and it is necessary to use molybdenum or tantalum containers. The phenomenon of solution of the metal in the molten trichloride is observed for the systems involving scandium, yttrium, lanthanum, cerium, holmium, erbium, uranium and plutonium with their respective trichlorides and although the properties of the lanthanide element systems indicate solubility as the M^{2+} ion there is no

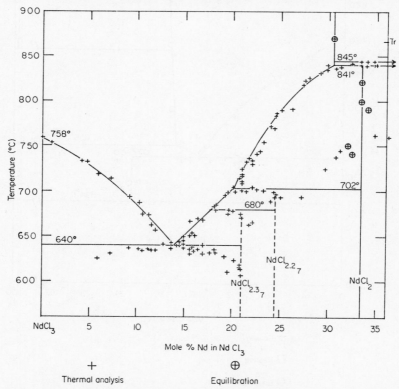

Figure 3.9 The system NdCl$_3$–NdCl$_2$.[171] (*After* L. F. Druding and J. D. Corbett, *J. Am. Chem. Soc.*, **83**, 2462 (1961))

evidence for the formation of stable intermediate halide phases in any of these systems. The information on holmium, however, is limited to electrical conductivity data and a complete phase study has yet to be reported.

In the Nd–NdCl₃ system the green dichloride, $NdCl_2$, is formed[171,172] (Figure 3.9) and the intermediate phases $NdCl_{2.37}$ and $NdCl_{2.27}$ have been observed. Others[182] have confirmed the existence of this dichloride for which the measured[325] heat of formation is -163.2 kcal/mole, but report the existence of the intermediate phases $NdCl_{2.33}$, $NdCl_{2.27}$ and $NdCl_{2.25}$.

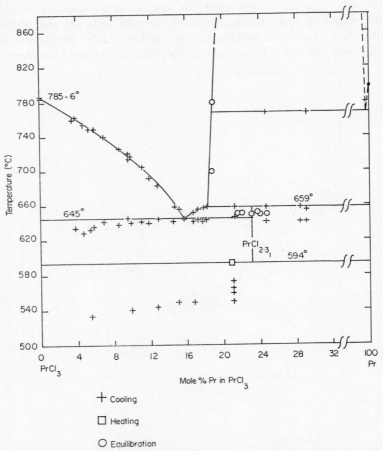

Figure 3.10   The system PrCl₃–Pr.[324]   (*After* L. F. Druding, J. D. Corbett and B. N. Ramsey, *Inorg. Chem.*, **2**, 869 (1963))

The Pr–PrCl$_3$ system (Figure 3.10) is[324] intermediate to the solution behaviour found with lanthanum and cerium and the formation of the above dichloride by neodymium. Thus the slightly stable, reduced phase PrCl$_{2.31}$ has been reported and has been shown to be stabilized by the addition of only 3 mole % of neodymium. The x-ray powder diffraction pattern of the stabilized phase is the same as that of GdCl$_{2.37}$. This behaviour may be compared with that of the analogous bromide system (p. 206) in which the more stable PrBr$_{2.38}$ is formed but whereas this latter compound exhibits a moderate conductivity suggestive of semiconduction,

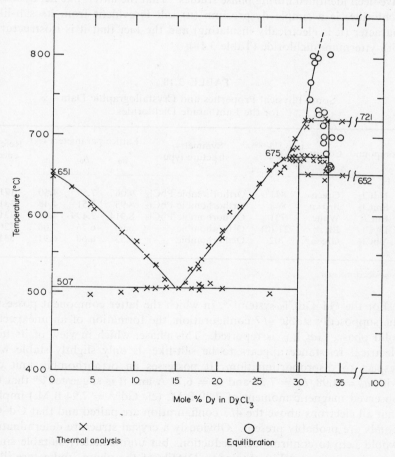

Figure 3.11   The system DyCl$_3$–Dy.[326]   (*After* J. D. Corbett and B. C. McCollum, *Inorg. Chem.*, **5**, 938 (1966))

the stabilized $Pr(Nd)Cl_{2.31}$ has megohm resistance typical of an ionic salt (cf. the metallic behaviour of the diiodide, PrI_2, p. 232). Vapour pressure studies[323] of $PrCl_3$ above $Pr + PrCl_3$ solutions support the existence of divalent praseodymium in the liquid state but the thermal analysis data reported in the same publication are not in accordance with the above results of Druding and co-workers[324].

Dworkin and co-workers[327], in the course of a conductivity study, found the metal reaction limit in the $Dy–DyCl_3$ system to be about 28%. Subsequently[326] the two intermediate phases $DyCl_{2.1}$ and $DyCl_2$ (Figure 3.11) have been identified during phase studies. That the latter, like neodymium dichloride, is a simple dysprosium (II) salt is evident from its salt-like character (it is electrically insulating) and the fact that it is isostructural with ytterbium dichloride (Table 3.18).

TABLE 3.18

Some Physical Properties and Crystallographic Data
for the Lanthanide Dichlorides

Compound	Colour	m.p.[172,239] (°C)	Symmetry/ structure type	Lattice parameters (Å)			Reference
				a_0	b_0	c_0	
$NdCl_2$	Green	841	Orthorhombic $PbCl_2$	9.06	7.59	4.50	171
$SmCl_2$	Brown	848	Orthorhombic $PbCl_2$	8.95	7.51	4.48	331
$EuCl_2$	White	731	Orthorhombic $PbCl_2$	8.914	7.499	4.493	331
$DyCl_2$	Black	721 (d)	Orthorhombic —	6.69	6.76	7.06	326
$YbCl_2$	Green	702	Orthorhombic —	6.53	6.68	6.91	331

d = decomposes.

For the $Gd–GdCl_3$ system[176], in which the latter component possesses the supposedly stable $4f^7$ configuration, the formation of an unexpected solid phase, $GdCl_{1.6}$, is reported. This phase, which in view of its high electrical resistance appears to be salt-like, is only slightly stable with respect to disproportionation. It possesses an orthorhombic unit cell with $a_0 = 8.98$, $b_0 = 7.22$ and $c_0 = 6.72$ Å and it is suggested[330] that the observed magnetic moment of 7.89 B.M. (cf. $Gd^{3+} = 7.94$ B.M.) implies that all electrons above the $4f^7$ configuration are paired and that Gd–Gd bonds are probably present. Obviously a crystal structure determination would help to confirm such deductions but unfortunately suitable single crystals are not readily obtained. Details of the phase studies are illustrated in Figure 3.12.

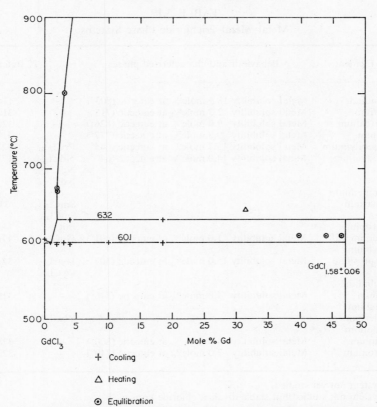

+ Cooling

△ Heating

⊙ Equilibration

Figure 3.12   The salt-rich portion of the GdCl₃–Gd phase system[176]. Metal is in equilibrium with the GdCl₃-rich solution above 632° and with the GdCl₁.₅₈ at lower temperatures. (*After* J. E. Mee and J. D. Corbett, *Inorg. Chem.*, **4**, 88 (1965))

Some information on the metal–metal trichloride phase studies is shown in Table 3.19.   The reduction characteristics of the molten trichlorides for the elements lanthanum to europium exhibit a relatively simple trend: the sole reduction product is apparently the dipositive salt which increases in stability with increasing atomic number. However, at gadolinium the reduced state shows a sharp decrease in stability and, furthermore, the available data show an extremely irregular trend in the heavier lanthanide elements. A general comparison of such reduction reactions has been made in terms of the composition of the salt-rich phase in equilibrium with solid metal above the melting point of any intermediate phase; that is, the apparent metal 'solubility' which leads to the presumed

TABLE 3.19

Metal–Metal Trichloride Phase Systems

Element	Behaviour and characterized phases		Reference
Scandium	Metal solubility 18.5 mole% at eutectic (803°)	—	174
Yttrium	Metal solubility 2.3 mole% at eutectic (715°)	—	319
Lanthanum	Metal solubility 9.0 mole% at eutectic (826°)	—	320
Cerium	Metal solubility 9.3 mole% at eutectic (777°)	—	317
Praseodymium	Metal solubility 15.7 mole% at eutectic (645°);	$PrCl_{2.31}$	324
Neodymium	Metal solubility 14.0 mole% at eutectic (640°);	$NdCl_{2.37}$	171
		$NdCl_{2.27}$	
		$NdCl_{2.0}$	
[a]Promethium		—	
Samarium		$SmCl_{2.2}$,	318
		$SmCl_{2.0}$	
[b]Europium		$EuCl_2$	—
Gadolinium	Metal solubility 1.0 mole% at eutectic (601°);	$GdCl_{1.6}$	176
[a]Terbium			
Dysprosium	Metal solubility 15.0 mole% at eutectic (507°);	$DyCl_{2.1}$,	326
		$DyCl_{2.0}$	
[a]Holmium			
Erbium	Metal solubility 4.8 mole% at eutectic (746°)	—	319
[a]Thulium		—	
[b]Ytterbium		$YbCl_2$	—
[a]Lutetium		—	
Uranium	Metal solubility 4.0 mole% at eutectic (800°)	—	328
Plutonium	Metal solubility 7.0 mole% at eutectic (740°)	—	329

[a] System not yet studied.
[b] System not studied but stable divalent chloride known.

solution of MCl_3 and MCl_2 in equilibrium with the metal M. A plot of the limit of reduction for the chlorides of the scandium family and of the lanthanide elements[319], shown in Figure 3.13, illustrates the irregularities mentioned above. Corbett and co-workers[319] have discussed these data in detail and have shown that changes in sublimation energies of the metals appear to be primarily responsible for the observed trends (apart from the small reduction of Y (III) relative to La (III) and Sc (III)) with relatively small effects arising from the minor changes in the first two ionization steps and from other terms in the complete Born–Haber cycle. The resulting correlation is shown in Figure 3.14 where log K_{app} for the disproportionation of MCl_2 (*l*) is plotted as a function of $\Delta F^0_{subl\,1073} + I_1 + I_2$ for M. K_{app} is evidently very small for disproportionation of $SmCl_2$, $EuCl_2$ and $YbCl_2$ which melt congruently. On the basis of these considerations Corbett and colleagues[319] predict that the melt interactions

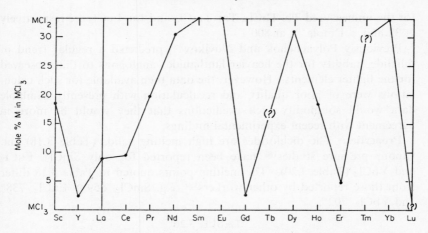

Figure 3.13 Composition of molten chloride melts[319] of scandium, yttrium and lanthanide metals in equilibrium with excess solid metal, expressed as apparent mole percentage dissolved metal ($MCl_2 = 33.3$ per cent M in MCl_3). (*After* J. D. Corbett, D. L. Pollard and J. E. Mee, *Inorg. Chem.*, **5**, 761 (1966))

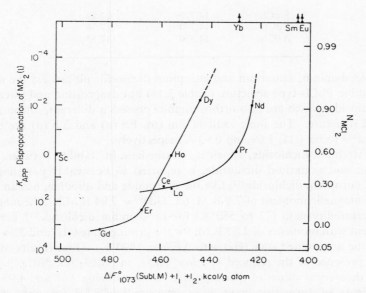

Figure 3.14 Log K_{app} at 800° for the disproportionation $3MCl_2$ (soln) $= 2MCl_3$ (soln) $+ M$ (s) as a function of $I_1 + I_2 + \Delta F^0_{subl}$ for the metal.[319] (*After* J. D. Corbett, D. L. Pollard and J. E. Mee, *Inorg. Chem.*, **5**, 761 (1966))

for the unmeasured Tb–TbCl$_3$, Tm–TmCl$_3$ and Lu–LuCl$_3$ are respectively 7, 32 and $\leqslant 1$ mole % at 800°.

Previously Polyachenok and Novikov[325] predicted a regular trend of dihalide stability for the heavier lanthanides analogous to that observed for the lighter elements. However, the data then available for such calculations were of poor quality and recalculation with presently available data would so modify such predictions that they would be more in agreement with recent experimental findings.

Properties. The dichlorides are high melting solids (Table 3.18) and vapour pressure studies[181] have been reported for only SmCl$_2$, EuCl$_2$ and YbCl$_2$ (Table 3.20). The melting points quoted in Table 3.18 differ from those reported by other workers[181] e.g. SmCl$_2$ 859°; EuCl$_2$ 738° and YbCl$_2$ 702°.

TABLE 3.20

Vapour Pressure Data for the Lanthanide Dichlorides[181]

| Compound | $\log p_{mm} = -(A/T) + B - 2.8 \log T$ | |
	A	B
SmCl$_2$	14,770	17.89
EuCl$_2$	14,770	17.36
YbCl$_2$	14,770	17.53

Neodymium, samarium and europium dichloride all possess the orthorhombic PbCl$_2$-type structure (Table 3.18) but dysprosium and ytterbium dichloride, which are also orthorhombic, possess a different, uncharacterized structure. The ionic radii of Sm (II), Eu (II) and Yb (II) are calculated[332] to be 1.11, 1.09 and 0.93 Å respectively.

Ytterbium dichloride, which is intermediate in stability between europium and samarium dichloride, is reported to be weakly paramagnetic and europium dichloride[311], like the dibromide and diiodide, has an effective magnetic moment of 7.9 B.M. (cf. Gd^{3+} = 7.94 B.M.). The magnetic susceptibility data (77 to 550°K) for neodymium dichloride[270] are consistent with moments of 2.87 B.M. for the ground state ($^5I_4$) and 5.36 B.M. for the first excited state ($^5I_5$) with $\Delta E/k = 1400$°K. These results confirm the presence of the reduced $4f^4$ ion, Nd^{2+}, in NdCl$_2$ (cf. NdI$_2$, p. 232), the theoretical values of the respective moments being 2.68 and 4.93 B.M.

Heats of formation have been measured[196,325,333] for only NdCl$_2$, SmCl$_2$ and YbCl$_2$ (Table A.3). The calculated values[325] for ScCl$_2$, YCl$_2$ and other lanthanide dichlorides are in error as discussed above.

REFERENCES

1. L. Brewer, L. Bromley, P. W. Giles and N. Logfren in 'The Transuranium Elements' (G. T. Seaborg, J. J. Katz and W. M. Manning, Eds.), *Nat. Nucl. Energy Ser.*, *Div. IV*, McGraw-Hill, New York, **14B**, 874, 1116 (1949).
2. O. Johnson, T. Butler and R. S. Newton, U.S. Report TID-5290, p. 1 (1958).
3. J. J. Katz and E. Rabinowitch, 'The Chemistry of Uranium', *Nat. Nucl. Energy Ser.*, *Div. VIII*, Vol. 5. McGraw-Hill, New York, 1951, p. 497.
4. O. Johnson, T. Butler, J. Powell and R. Nottorf, U.S. Report CC-1974 (1944).
5. K. W. Bagnall in *Halogen Chemistry* (V. Gutman, Ed.), Vol. 3, Academic Press, London, 1967, p. 303.
6. T. A. O'Donnell, D. F. Stewart and P. Wilson, *Inorg. Chem.*, **5**, 438 (1966).
7. J. C. Warf, U.S. Report TID-5290, p. 29 (1958).
8. W. H. Zachariasen, *Acta Cryst.*, **1**, 285 (1948).
9. P. H. Davidson and J. A. Holmes, according to reference 3, p. 500.
10. D. Altman, D. Lipkin and S. Wiessman, Report RL.-4.6.22 (1943), according to reference 3, p. 500.
11. C. A. Kraus, U.S. Reports A-1090; A-1091 (1943).
12. P. Gans, Thesis, London University (1964).
13. 'The Actinide Elements' (G. T. Seaborg and J. J. Katz, Eds.), *Nat. Nucl. Energy Ser.*, *Div. IV*, McGraw-Hill, New York, **14A**, 387 (1954).
14. I. F. Alenchikova, L. L. Zaitseva, L. V. Lipis and V. V. Fomin, *Russ. J. Inorg. Chem.*, **4**, 435 (1959).
15. J. L. Ryan, *Inorg. Chem.*, **2**, 348 (1963).
16. K. W. Bagnall and B. J. Laidler, *J. Chem. Soc. (A)*, **1966**, 516.
17. E. Greenberg and E. F. Westrum, *J. Am. Chem. Soc.*, **78**, 4526 (1956).
18. S. A. Shchukarev, I. V. Vasil'kova, N. S. Martynova and G. G. Mal'chev, *Zh. Neorgan. Khim*, **3**, 2467 (1958).
19. J. A. Leary and J. F. Suttle, *Inorg. Syn.*, **5**, 148 (1957).
20. E. Péligot, *Ann. Chem.*, **43**, 278 (1842).
21. J. Prigent, *Compt. Rend.*, **247**, 1737 (1958).
22. Reference 3, page 577.
23. C. Jangg, W. Ochsenfeld and A. Burker, *Atompraxis*, **7**, 332 (1961).
24. J. D. Hefley, D. M. Mathews and E. S. Amis, *Inorg. Syn.*, **7**, 146 (1963).
25. J. D. Hefley, D. M. Mathews and E. S. Amis, *J. Inorg. Nucl. Chem.*, **12**, 84 (1959).
26. K. W. Bagnall, D. Brown and P. J. Jones, *J. Chem. Soc. (A)*, **1966**, 1763.
27. L. Ochs and F. Strassman, *Z. Naturforsch.*, **7B**, 637 (1952).
28. D. C. Bradley, A. K. Chatterjee and A. K. Chatterjee, *J. Inorg. Nucl. Chem.*, **12**, 71 (1959).
29. O. Johnson, U.S. Report CC-1781 (1944), according to reference 3, p. 577.
30. P. Gans and B. C. Smith, *J. Chem. Soc.*, **1964**, 4172.
31. J. P. Day, Thesis, Oxford University (1965).
32. K. W. Bagnall, D. Brown and P. J. Jones, *J. Chem. Soc. (A)*, **1966**, 741.
33. K. W. Bagnall, B. J. Laidler and M. A. A. Stewart, *Chem. Commun.*, **1967**, 24.
34. E. Staritzky and J. Singer, *Acta Cryst.*, **5**, 536 (1952).
35. E. Rimbach, *Ber.*, **37**, 461 (1904).

36. H. Loebel, Dissertation, Berlin (1907), according to reference 38.
37. V. P. Markov and V. V. Tsapkin, *Russ. J. Inorg. Chem.*, **6**, 1052 (1961).
38. V. P. Markov and V. V. Tsapkin, *Russ. J. Inorg. Chem.*, **4**, 1030 (1959).
39. P. Gans and B. C. Smith, *J. Chem. Soc.*, **1964**, 4177.
40. D. C. Bradley, B. B. Chakravarti and A. K. Chatterjee, *J. Inorg. Nucl. Chem.*, **3**, 367 (1957).
41. D. Hall, A. D. Rae and T. N. Waters, *Acta Cryst.*, **20**, 160 (1966).
42. E. Péligot, *Ann. Chim. Phys.*, **44**, 387 (1842).
43. J. Lucas, *Rev. Chim. Minérale*, **1**, 479 (1964).
44. J. Aloy, *Bull. Soc. chim. France*, **25**, 153 (1901).
45. J. Prigent and J. Lucas, *Compt. Rend.*, **253**, 474 (1961).
46. J. G. Allpress and A. D. Wadsley, *Acta Cryst.*, **17**, 41 (1964).
47. G. Gibson, D. Gruen and J. J. Katz, *J. Am. Chem. Soc.*, **74**, 2103 (1952).
48. A. V. Grosse, *J. Am. Chem. Soc.*, **56**, 2200 (1934).
49. J. Flegenheimer in *Physico Chimie du Protactinium*, Publication No. 154, Centre National Recherche Scientifique, Paris, **1966**, p. 135.
50. D. Brown and R. G. Wilkins, *J. Chem. Soc.*, **1961**, 3804.
51. D. Brown and P. J. Jones, *J. Chem. Soc. (A)*, **1966**, 874.
52. P. Conte, J. Vernois, D. Robertson and R. Muxart, *Bull. Soc. Chim. France*, **1964**, 2115.
53. P. A. Sellers, S. Fried, R. E. Elson and W. H. Zachariasen, *J. Am. Chem. Soc.*, **76**, 5935 (1954).
54. K. W. Bagnall and D. Brown, *J. Chem. Soc.*, **1964**, 2031.
55. H. Roscoe, *Ber.*, **7**, 1131 (1874).
56. A. D. Webb, U.S. Report RL-4.6.102 (1943), according to reference 3.
57. H. P. Kyle, U.S. Report RL-4.6.50 (1944), according to reference 3.
58. A. D. Webb and H. P. Kyle, U.S. Report RL-4.6.51 (1943), according to reference 3.
59. O. Ruff and A. Heinzlemann, *Z. Anorg. Chem.*, **72**, 65 (1911).
60. A. Micheal and A. Murphy, *Am. Chem. J.*, **44**, 365 (1910).
61. Reference 3, p. 488.
62. A. V. Grosse, U.S. Report TID-5290, 300 (1958).
63. R. P. Dodge, G. S. Smith, Q. Johnson and R. E. Elson, U.S. Report, UCRL-14581 (1966); *Acta Cryst.*, **22**, 85 (1967).
64. G. S. Smith, Q. Johnson and R. E. Elson, *Acta Cryst.*, **22**, 300 (1967).
65. H. L. Goren, R. S. Lowrie and J. V. Hubbard, U.S. Report CD-5.350.8 (1946), according to reference 3.
66. W. Rüdorff and W. Menzer, *Z. Anorg. Chem.*, **292**, 197 (1957).
67. D. Brown and P. J. Jones, *Chem. Commun.*, **1966**, 280; *J. Chem. Soc. (A)*, **1967**, 719.
68. C. A. Kraus, U.S. Report A-726 (1943), according to reference 3.
69. Reference 3, p. 495.
70. K. W. Bagnall, D. Brown and J. G. H. du Preez, *J. Chem. Soc.*, **1964**, 2603.
71. H. Hecht, G. Jander and H. Schlappman, *Z. Anorg. Chem.*, **254**, 255 (1947).
72. R. E. Panzer and J. F. Suttle, *J. Inorg. Nucl. Chem.*, **20**, 229 (1961).
73. R. E. Panzer and J. F. Suttle, *J. Inorg. Nucl. Chem.*, **13**, 244 (1960).
74. D. Brown, J. F. Easey and J. G. H. du Preez, *J. Chem. Soc. (A)*, **1966**, 258.
75. K. W. Bagnall, D. Brown and J. G. H. du Preez, *J. Chem. Soc.*, **1965**, 5217.

76. D. Brown and P. J. Jones, unpublished observations.
77. D. Brown, J. F. Easey and P. J. Jones, *J. Chem. Soc. (A)*, **1967**, 1698.
78. P. Handler and C. A. Hutchinson, *J. Chem. Phys.*, **25**, 1210 (1956).
79. D. G. Karracher, *Inorg. Chem.*, **3**, 1618 (1964).
80. R. Rohmer, R. Freymann, R. Freymann, A. Chevet and P. Hamon, *Bull. Soc. Chim. France*, **1952**, 598, 603.
81. A. M. Pissot, R. Muxart and C. F. de Miranda, *Bull. Soc. Chim. France*, **1966**, 1757.
82. R. Muxart, reference 49, p. 140.
83. S. A. Shchukarev, I. V. Vasil'kova, N. S. Martynova and Yu. G. Mal'tsev, *Zh. Neorgan. Khim.*, **3**, 2647 (1958).
84. M. D. Adams, D. A. Wenz and R. K. Steunenberg, *J. Phys. Chem.*, **67**, 1939 (1963).
85. I. V. Budayev and A. N. Vol'sky, *Proc. U.N. Intern. Conf. Peaceful Uses At. Energy, 2nd Geneva*, **28**, 316 (1958).
86. R. S. Wilks, *J. Nucl. Mater.*, **7**, 157 (1962).
87. J. L. Swanson, *J. Phys. Chem.*, **68**, 438 (1964).
88. T. J. La Chapelle, U.S. Report UCRL-336 (1949).
89. T. J. La Chapelle, U.S. Pat. 3,149,908 (1964), according to *Chem. Abs.*, **61**, 15700g.
90. K. W. Bagnall, D. Brown and J. F. Easey, unpublished observations.
91. F. di Stefana, *Ann. Chim. Applicata*, **12**, 130 (1919).
92. O. Stelling, *Z. Physik. Chem.*, **24B**, 282, 292 (1934).
93. D. T. Cromer and R. J. Kline, *J. Am. Chem. Soc.*, **76**, 5282 (1954).
94. Reference 1, p. 861.
95. R. Benz, *J. Inorg. Nucl. Chem.*, **24**, 1191 (1962).
96. B. M. Abrahams, B. B. Brody, N. R. Davidson, F. Hagemann, I. Kark, J. J. Katz and M. J. Wolf, reference 1, p. 740.
97. J. J. Berzelius, *Pogg. Ann.*, **16**, 385 (1829).
98. E. Péligot, *Ann. Chim. Phys.*, **5**, 5 (1842).
99. J. Flahaut in *Nouveau Traite de Chimie Minérale* (P. Pascal, Ed.), Vol. IX, Masson et Cie, Paris, 1963, p. 1061.
100. L. I. Katzin, reference 13, p. 66.
101. E. L. Wagner, U.S. Report C.D.-0.350.9 (1946), according to reference 3, p. 463.
102. Reference 3, p. 461.
103. M. Oxley in *Nouveau Traite de Chimie Minérale* (P. Pascal, Ed.), Vol. XV, Masson et Cie, Paris, 1961, p. 135.
104. D. Bradley, British Report AERE C/R 2215 (1957).
105. J. H. Freeman and M. L. Smith, *J. Inorg. Nucl. Chem.*, **7**, 224 (1958).
106. D. C. Bradley, M. A. Saad and M. Wardlaw, *J. Chem. Soc.*, **1954**, 2002.
107. R. Didchenko, *Trans. AIME*, **215**, 401 (1959).
108. O. C. Dean and A. M. Chandler, *Nuc. Sci. Eng.*, **2**, 57 (1957).
109. R. L. Skaggs and D. Peterson, U.S. Report IS-1059 (1958).
110. E. R. Harrison, Report AERE GP/R 2409 (1958).
111. S. Rosenfeld, British Pat. 841,681 (1960).
112. J. A. Hermann and J. F. Suttle, *Inorg. Syn.*, **5**, 145 (1957).
113. Reference 3, p. 468.
114. E. Uhlemann and W. Fischback, *Z. Chem.*, **3**, 431 (1963).

115. Reference 3, p. 472.
116. K. W. Bagnall and P. S. Robinson, unpublished observations.
117. G. W. Watt and S. C. Malhotra, *J. Inorg. Nucl. Chem.*, **14**, 184 (1960).
118. R. Elson, S. Fried, P. A. Sellers and W. H. Zachariasen, *J. Am. Chem. Soc.*, **72**, 5791 (1950).
119. S. Fried and N. Davidson, *J. Am. Chem. Soc.*, **70**, 3539 (1948); reference 13, p. 1072.
120. J. G. Malm, unpublished observations.
121. R. L. Mooney, *Acta Cryst.*, **2**, 189 (1949).
122. W. Fischer, R. Gewehr and H. Wingchen, *Z. Anorg. Chem.*, **242**, 161 (1939).
123. S. A. Shchukarev, I. V. Vasìlkova, A. I. Efimov and V. P. Kirdyashev, *Zh Neorgan. Khim.*, **1**, 2272 (1956).
124. Reference 3, p. 486.
125. Reference 103, p. 155.
126. I. Sheft and S. Fried, *J. Am. Chem. Soc.*, **75**, 1236 (1953).
127. J. K. Dawson, *J. Chem. Soc.*, **1951**, 429.
128. R. Stoenner and M. Elliott, *J. Chem. Phys.*, **19**, 950 (1951).
129. D. Brown, unpublished observations.
130. K. W. Bagnall, A. M. Deane, T. L. Markin, P. S. Robinson and M. A. A. Stewart, *J. Chem. Soc.*, **1961**, 1611.
131. K. W. Bagnall, D. Brown, P. J. Jones and J. G. H. du Preez, *J. Chem. Soc.* (*A*), **1966**, 737.
132. R. N. Kapoor, K. C. Pande and R. C. Mehrotra, *J. Indian Chem. Soc.*, **35**, 157 (1958).
133. K. L. Jaura, H. S. Banga and R. L. Kaushik, *J. Indian Chem. Soc.*, **39**, 531 (1962).
134. J. Prasad and S. Kumar, *J. Indian Chem. Soc.*, **39**, 444 (1962).
135. K. L. Jaura and P. S. Bajura, *J. Sci. Ind. Res.* (*India*), **20B**, 391 (1961).
136. D. C. Bradley, R. N. Kapoor and B. C. Smith, *J. Inorg. Nucl. Chem.*, **24**, 863 (1963).
137. M. Taube, *Symp. Power Reactor Exp., Vienna, Austria*, No. 1, 353 (1962).
138. V. I. Ionov, B. G. Korshunov, V. V. Kokorev and I. S. Morozov, *Izv. Vysshikh Uchebn. Zavedenii, Tsvetn. Met.*, **3**, 102 (1960), according to *Chem. Abs.*, **55**, 5101e.
139. E. Chauvenet, *Ann. Chim. Phys.*, **23**, 425 (1911); *Compt. Rend.*, **148**, 1519 (1909).
140. H. Moissan, *Compt. Rend.*, **122**, 1089 (1896).
141. J. Aloy, *Bull. Soc. Chim. France*, **21**, 264 (1899); *Ann. Chim. Phys.*, **24**, 412 (1901).
142. Reference 3, p. 487.
143. G. H. Dieke and A. B. F. Duncan, *The Spectroscopic Properties of Uranium Compounds*, McGraw-Hill, New York, 1949.
144. H. H. Anderson, reference 13, p. 793.
145. K. W. Bagnall, D. Brown, P. J. Jones and J. G. H. du Preez, *J. Chem. Soc.* **1965**, 350.
146. J. L. Ryan, *J. Chem. Phys.*, **65**, 1856 (1961).
147. F. J. Miner, R. P. de Grazio and J. T. Byrne, *Anal. Chem.*, **35**, 1219 (1963).
148. J. R. Ferraro, *J. Inorg. Nucl. Chem.*, **4**, 283 (1957).

149. D. M. Adams, J. Chatt, J. M. Davidson and J. Gerratt, *J. Chem. Soc.*, **1963**, 2189.
150. D. Brown, *J. Chem. Soc. (A)*, **1966,** 766.
151. D. Brown and P. J. Jones, *J. Chem. Soc. (A)*, **1967,** 243.
152. A. Rosenheim and J. Schilling, *Ber.*, **33,** 977 (1900).
153. A. Rosenheim, V. Samter and J. Davidsohn, *Z. Anorg. Chem.*, **35,** 424 (1903).
154. V. V. Fomin, N. A. Dimitrieva and V. E. Reznikova, *Zh. Neorgan. Khim.*, **3,** 1999 (1958).
155. R. Benz and R. M. Douglass, *J. Inorg. Nucl. Chem.*, **23,** 134 (1961).
156. W. H. Zachariasen, *Acta Cryst.*, **1,** 268 (1948).
157. S. Siegel, *Acta Cryst.*, **9,** 827 (1956).
158. G. A. Candela, C. A. Hutchinson and W. B. Lewis, *J. Chem. Phys.*, **30,** 246 (1959); C. A. Hutchinson and G. A. Candela, *J. Chem. Phys.*, **27,** 707 (1957).
159. K. W. Bagnall, D. Brown and R. Colton, *J. Chem. Soc.*, **1964,** 2527.
160. J. P. Day and L. Venanzi, *J. Chem. Soc. (A)*, **1966,** 197.
161. W. B. Lewis and N. Elliott, *J. Chem. Phys.*, **27,** 904 (1957).
162. J. D. Axe, H. J. Stapleton and R. Kyi, *J. Chem. Phys.*, **32,** 1216 (1960); J. D. Axe, H. J. Stapleton and C. D. Jefferies, *Phys. Rev.*, **121,** 1630 (1960).
163. C. A. Kraus, U.S. Report CC-342 (1942); CC-1717 (1944).
164. M. V. Smirnov and L. E. Ivanovskii, *Zh. Neorgan. Khim.*, **1,** 1843 (1956).
165. Kung-Fan Yen, Shao-Chung Li and G. I. Novikov, *Zh. Neorgan. Khim.*, **8,** 89 (1963).
166. K. W. Bagnall, D. Brown and J. F. Easey, *J. Chem. Soc. (A)*, **1968,** 288.
167. J. Selbin and M. Schöber, *J. Inorg. Nucl. Chem.*, **28,** 817 (1966).
168. G. Jantsch, J. Homayr and F. Zemek, *Monatsh.*, **85,** 526 (1954).
169. E. Hayek, Th. Rehner and A. Frank, *Monatsh.*, **82,** 575 (1951).
170. P. Chiotti, U.S. Reports IS-1200 (1965); IS-1500 (1966).
171. L. F. Druding and J. D. Corbett, *J. Am. Chem. Soc.*, **83,** 2462 (1961).
172. L. F. Druding and J. D. Corbett, *J. Am. Chem. Soc.*, **81,** 5512 (1959).
173. C. V. Banks, O. N. Carlson, A. H. Daane, V. A. Fassel, R. W. Fisher, E. H. Olsen, J. E. Powell and F. H. Spedding, U.S. Report IS-1 (1959).
174. J. D. Corbett and B. N. Ramsey, *Inorg. Chem.*, **4,** 260 (1965).
175. Reference 3, p. 450.
176. J. E. Mee and J. D. Corbett, *Inorg. Chem.*, **4,** 88 (1965).
177. M. Meyer, *Ber.*, **20,** 681 (1887).
178. J. F. Miller, S. E. Miller and R. C. Himes, *J. Am. Chem. Soc.*, **81,** 4449 (1959).
179. C. T. Stubblefield and L. Eyring, *J. Am. Chem. Soc.*, **77,** 3004 (1955).
180. O. G. Polyachenok and G. I. Novikov, *Zh. Neorgan. Khim.*, **8,** 2819 (1963).
181. O. G. Polyachenok and G. I. Novikov, *Zh. Neorgan. Khim.*, **8,** 2631 (1963).
182. G. I. Novikov and O. G. Polyachenok, *Zh. Neorgan. Khim.*, **8,** 1053 (1963).
183. F. Bourion, *Compt. Rend.*, **148,** 170 (1909); *Ann. Chim. Phys.*, **20,** 547 (1910).
184. G. Urbain and F. Bourion, *Compt. Rend.*, **153,** 1155 (1911).
185. C. Matignon and F. Bourion, *Compt. Rend.*, **138,** 131 (1904); *Ann. Chim. Phys.*, **15,** 127 (1905).
186. W. Klemm and W. Schüth, *Z. Anorg. Chem.*, **184,** 352 (1929).

187. W. Klemm and J. Rockstroh, *Z. Anorg. Chem.*, **176**, 181 (1928).

188. C Matignon, *Ann. Chim. Phys.*, **8**, 364 (1906).

189. D. H. Templeton and G. F. Carter, *J. Phys. Chem.*, **58**, 940 (1954).

190. E. Chauvenet, *Compt. Rend.*, **152**, 87 (1911).

191. G. Jantsch, H. Alber and H. Grübitsch, *Monatsh.*, **53–54**, 304 (1929).

192. M. J. Schmidt and L. F. Audrieth, *Trans. Illinois State Acad. Sci.*, **28** (2), 133 (1935), according to *Chem. Abs.*, **30**, 3349[7].

193. R. C. Young and J. L. Hastings, *J. Am. Chem. Soc.*, **59**, 765 (1937).

194. J. B. Reed, B. S. Hopkins and L. F. Audrieth, *J. Am. Chem. Soc.*, **57**, 1159 (1935); *Inorg. Syn.*, **1**, 28 (1939).

195. N. H. Kiesse, *J. Res. Nat. Bur. Std.*, **67A**, 343 (1963).

196. G. R. Machlan, Thesis, Iowa State University (1952).

197. F. R. Hartley and A. W. Wylie, *Nature*, **161**, 241 (1948).

198. R. Gut and D. M. Gruen, *J. Inorg. Nucl. Chem.*, **21**, 259 (1961).

199. T. T. Campbell, F. E. Block, R. E. Mussler and G. B. Robidart, Bureau of Mines Report BM-RI-5880 (1960).

200. S. Fried, F. Hagemann and W. H. Zachariasen, *J. Am. Chem. Soc.*, **72**, 771 (1950).

201. J. H. Kleinheksel and H. C. Kremers, *J. Am. Chem. Soc.*, **50**, 959 (1928).

202. W. Prandtl and H. Kogl, *Z. Anorg. Chem.*, **172**, 265 (1928).

203. F. E. Block and T. T. Campbell in *The Rare Earths* (F. H. Spedding and A. H. Daane, Eds.), Wiley, New York, 1961, Chap. 7, p. 90.

204. G. Jantsch, H. Grübitsch, F. Hoffmann and H. Alber, *Z. Anorg. Chem.*, **185**, 49 (1929).

205. G. Jantsch, N. Skalla and H. Grübitsch, *Z. Anorg. Chem.*, **216**, 75 (1933).

206. E. R. Harrison, *J. Appl. Chem.*, **2**, 601 (1952).

207. G. Jantsch, H. Jawarek, N. Skalla and H. Galowski, *Z. Anorg. Chem.*, **207**, 353 (1932).

208. W. Heap and E. Newbury, British Pat. 130,365 (1918); U.S. Pat. 1,331,257 (1918).

209. R. Hermann, *J. Prakt. Chem.*, **32**, 385 (1861).

210. M. D. Taylor and C. P. Carter, *J. Inorg. Nucl. Chem.*, **24**, 387 (1960).

211. S. Fried, *J. Am. Chem. Soc.*, **73**, 416 (1951).

212. J. Fuger and B. B. Cunningham, U.S. Report UCRL-10722 (1963).

213. L. B. Asprey, T. Keenan and F. H. Kruse, *Inorg. Chem.*, **4**, 985 (1965).

214. J. D. Farr, A. L. Giorgi, M. G. Bowman and R. K. Money, U.S. Report LA-1545 (1953).

215. Reference 203, p. 95.

216. G. I. Novikov and V. D. Talmachaev, *Russ. J. Appl. Chem.*, **38**, 1142 (1965).

217. P. Brauman and S. Takvorian, *Compt. Rend.*, **194**, 1579 (1932).

218. W. Müthmann and L. Stützel, *Ber.*, **32**, 3413 (1899).

219. H. Moissan, *Compt. Rend.*, **122**, 357 (1895).

220. J. F. Suttle, *Inorg. Syn.*, **5**, 145 (1957).

221. H. S. Young, U.S. Report TID-5290, p. 757 (1958).

222. C. H. Barkelew, U.S. Report RL-4.6.906 (1945), according to reference 3, p. 452.

223. F. Hagemann, U.S. Report CK-1327 (1944), according to reference 1, p. 740.

224. M. J. Rassmussen and H. H. Hopkins, *Ind. Eng. Chem.*, **53**, 453 (1961).
225. B. R. Harder, F. Hudswell and K. L. Wilkinson, Report AERE C/R-2445 (1958).
226. J. G. Reavis, K. W. R. Johnson, J. A. Leary, A. N. Morgan, A. E. Ogard and K. A. Walsh, *Extract. Phys. Met. Plutonium Alloys, Symp. San Francisco, Calif.* 1959, p. 89 (1960).
227. D. Boreham, T. H. Freeman, E. W. Hooper, I. L. Jenkins and J. L. Woodhead, *J. Inorg. Nucl. Chem.*, **16**, 154 (1960).
228. E. L. Christensen and L. J. Mullens, U.S. Report LA-1431 (1952).
229. B. B. Cunningham, *Microchem. J., Symp.*, **1961**, 69.
230. A. Broido and B. B. Cunningham, U.S. Report AECD-2918 (1950).
231. W. Klemm and E. Krose, *Z. Anorg. Chem.*, **253**, 218 (1947).
232. H. Bommer and E. Hohmann, *Z. Anorg. Chem.*, **248**, 373 (1941).
233. W. H. Zachariasen, reference 1, p. 1473.
234. W. H. Zachariasen, *Acta Cryst.*, **1**, 265 (1948).
235. W. H. Zachariasen, *J. Chem. Phys.*, **16**, 254 (1948).
236. F. Weigel and V. Scherer, *Radiochim. Acta*, **7**, 40 (1967).
237. *The Properties of the Rare Earth Metals and Compounds*, compiled by J. A. Gibson, J. F. Miller, P. S. Kennedy and G. W. Prengstorff, for The Rare Earth Research Group (1959).
238. J. D. Forrester, A. Zalkin, D. H. Templeton and J. C. Wallman, *Inorg. Chem.*, **3**, 185 (1964).
239. F. H. Spedding and A. H. Daane, *Metallurgical Rev.*, **5**, 297 (1960).
240. J. J. Katz and I. Sheft, in *Advances in Inorganic and Radiochemistry* (H. J. Eméleus and A. G. Sharpe, Eds.), Vol. 2, Wiley, New York, 1960, p. 196.
241. V. E. Shimazaki and K. Niwa, *Z. Anorg. Chem.*, **314**, 21 (1962).
242. J. L. Moriarty, *J. Chem. Eng. Data*, **8**, 422 (1963).
243. O. G. Polyachenok and G. I. Novikov, *Zh. Neorgan. Khim.*, **8**, 1526 (1963).
244. T. E. Phipps, G. W. Sears, R. L. Seifert and O. C. Simpson, *J. Chem. Phys.*, **18**, 718 (1950); reference 1, p. 682.
245. W. Fischer, R. Gewehr and H. Wingchen, *Z. Anorg. Chem.*, **242**, 161 (1939).
246. O. G. Polyachenok and I. G. Novikov, *Zh. Neorgan. Khim.*, **8**, 2818 (1963).
247. A. Pabst, *Am. J. Sci.*, **24**, 426 (1931).
248. K. H. Hellewege and H. G. Kahle, *Z. Physik.*, **129**, 62 (1951).
249. V. I. Ivernova, V. P. Tarasova and M. M. Umanskiï, *Izv. Akad. Nauk SSSR Ser. Fiz.*, **15**, 164 (1951).
250. M. Marezio, H. A. Plettinger and W. H. Zachariasen, *Acta Cryst.*, **14**, 234 (1961).
251. T. Kojima, T. Inoue and T. Ishiyama, *J. Electrochem. Soc., Japan*, **19**, 285 (1951).
252. Reference 3, p. 127.
253. Reference 3, p. 486.
254. W. B. Tolley, U.S. Report HW-30121 (1953).
255. *Nouveau Traite de Chimie Minerale*, P. Pascal (Ed.), Vol. VII, Masson et Cie, Paris, 1959, p. 763.
256. I. V. Tananaev and V. B. Orlovskiï, *Zh. Neorgan. Khim.*, **7**, 2299 (1962); **8**, 1104 (1963).

257. Y. Y. Kharitonov, V. P. Orlovskiï and I. V. Tananaev, *Zh. Neorgan. Khim.*, **8**, 1093 (1963).
258. F. Petru and F. Jost, *Chem. Listy*, **52**, 164 (1958), according to *Chem. Abs.*, **52**, 19641e.
259. I. V. Tananaev and V. B. Orlovskiï, *Zh. Neorgan. Khim.*, **7**, 2022 (1963).
260. H. Funk and B. Koehler, *Z. Anorg. Chem.*, **325**, 67 (1963).
261. S. Herzog, K. Gustav, E. Krueger, H. Oberender and R. Schuster, *Z. Chem.*, **3**, 428 (1963).
262. Z. A. Sheka and E. E. Kriss, *Zh. Neorgan. Khim.*, **4**, 1809 (1962).
263. F. R. Hartley and A. W. Wylie, *J. Chem. Soc.*, **1962**, 679.
264. T. Moeller and D. S. Smith, P.B. Report 136.180 (1959).
265. E. M. Kirmse and G. Zschischang, *Z. Chem.*, **3**, 396 (1963).
266. C. A. Hutchinson, P. M. Llewellyn, E. Wong and P. Dorain, *Phys. Rev.*, **102**, 292 (1956).
267. J. K. Dawson, C. J. Mandelberg and D. Davies, *J. Chem. Soc.*, **1951**, 2047.
268. P. W. Selwood, *J. Am. Chem. Soc.*, **55**, 4869 (1933).
269. H. M. Crosswhite and G. H. Dieke, *J. Chem. Phys.*, **35**, 1535 (1961).
270. R. A. Sallach and J. D. Corbett, *Inorg. Chem.*, **3**, 993 (1964).
271. B. Schneider, *Z. Physik.*, **177**, 179 (1964).
272. J. Murphy, H. H. Caspers and R. A. Buchanan, *J. Chem. Phys.*, **40**, 743 (1964).
273. J. T. Houghen and S. Singh, unpublished observations, according to reference 272.
274. T. Kojima, *J. Electrochem. Soc. Japan*, **21**, 119 (1953).
275. R. J. Meyer, A. Wassjuchnow, N. Drapier and E. Bodländer, *Z. Anorg. Chem.*, **86**, 257 (1914).
276. I. N. Belyaev and Le T'yuk, *Russ. J. Inorg. Chem.*, **10**, 664 (1965).
277. N. N. Sakharov and V. P. Ardeer, *Russ. J. Inorg. Chem.*, **10**, 1104 (1965).
278. N. N. Sakharov, *Dokl. Akad. Nauk SSSR*, **77**, 73 (1951).
279. N. N. Sakharov and S. V. Zemskov, *Dokl. Akad. Nauk SSSR*, **120**, 539 (1958).
280. B. K. Jørgensen and J. Ryan, *J. Phys. Chem.*, **70**, 2845 (1966).
281. R. E. Stevens, *J. Inorg. Nucl. Chem.*, **27**, 1873 (1965).
282. K. W. Bagnall, B. J. Laidler and M. A. A. Stewart, *J. Chem. Soc. (A)*, **1968**, 133.
283. J. L. Ryan, unpublished observations.
284. Reference 3, p. 460.
285. A. K. Baev and G. I. Novikov, *Zh. Neorgan. Khim.*, **6**, 2610 (1961).
286. C. W. Bjorklund, J. G. Reavis, J. A. Leary and K. A. Walsh, *J. Phys. Chem.*, **63**, 1774 (1959).
287. K. W. R. Johnson, M. Kahn and J. A. Leary, *J. Phys. Chem.*, **65**, 2226 (1961).
288. I. S. Morozov, L. N. Shevtsova and L. V. Klyukina, *Zh. Neorgan. Khim.*, **2**, 1639 (1957).
289. L. Mazza, A. Iandelli and E. Botti, *Gazz. Chim. Ital.*, **70**, 57 (1940), according to *Chem. Abs.*, **34**, 5771[5].
290. L. G. Sillen and A. L. Nylander, *Svensk Kem. Tidskr.*, **53**, 367 (1941).
291. D. H. Templeton and C. H. Dauben, *J. Am. Chem. Soc.*, **75**, 6069 (1953).
292. W. W. Wendlandt, *J. Inorg. Nucl. Chem.*, **5**, 118 (1957).

293. W. W. Wendlandt, *J. Inorg. Nucl. Chem.*, **9**, 136 (1959).
294. J. E. Powell and H. R. Burkholder, *J. Inorg. Nucl. Chem.*, **14**, 65 (1960).
295. G. Haesler and F. Matthes, *J. Less-Common Metals*, **9**, 133 (1965); *Z. Chem.*, **3**, 72 (1963).
296. F. Petru and F. Kutek, *Collection Czech. Chem. Commun.*, **25**, 1143 (1960), according to *Chem. Abs.*, **54**, 12856i.
297. S. A. Shchukarov and A. I. Efimov, *Zh. Neorgan. Kkim.*, **2**, 2304 (1957).
298. D. H. Templeton and C. H. Dauben, *J. Am. Chem. Soc.*, **75**, 4560 (1953).
299. C. W. Koch and B. B. Cunningham, *J. Am. Chem. Soc.*, **76**, 1470 (1954).
300. W. H. Zachariasen, *Acta Cryst.*, **2**, 388 (1948).
301. C. W. Koch, A. Broido and B. B. Cunningham, *J. Am. Chem. Soc.*, **74**, 2349 (1952).
302. C. W. Koch and B. B. Cunningham, *J. Am. Chem. Soc.*, **75**, 796 (1953).
303. C. W. Koch and B. B. Cunningham, U.S. Report UCRL-2286 (1953); *J. Am. Chem. Soc.*, **76**, 1471 (1954).
304. F. Weigel and H. Haug, *Ber.*, **94**, 1548 (1961).
305. F. H. Spedding and C. F. Miller, *J. Am. Chem. Soc.*, **74**, 4195 (1952); F. H. Spedding and J. P. Flynn, *J. Am. Chem. Soc.*, **76**, 1477 (1954).
306. A. H. Daane, personal communication, 1967.
307. L. B. Asprey and B. B. Cunningham in, *Progress in Inorganic Chemistry* (F. A. Cotton, Ed.), Vol. 2, Interscience, New York, 1960, p. 267
308. G. I. Novikov and O. G. Polyachenok, *Russ. Chem. Rev.* (English Transl.), **33**, 342 (1964).
309. C. Matignon and E. Cazes, *Compt. Rend.*, **142**, 83, 176 (1906).
310. G. Jantsch, N. Skalla and H. Grubitsch, *Z. Anorg. Chem.*, **216**, 95 (1933).
311. W. Klemm and W. Döll, *Z. Anorg. Chem.*, **241**, 233 (1939).
312. R. A. Cooley and D. M. Yost, *Inorg. Syn.*, **2**, 69 (1946).
313. A. I. Popov and W. W. Wendlandt, *Proc. Iowa Acad. Sci.*, **60**, 300 (1953).
314. G. Jantsch, H. Rüping and W. Kunze, *Z. Anorg. Chem.*, **161**, 210 (1927).
315. S. A. Shchukarov and G. I. Novikov, *Zh. Neorgan. Khim.*, **1**, 362 (1956).
316. D. Cubicciotti, *J. Am. Chem. Soc.*, **71**, 4119 (1949).
317. G. W. Mellors and S. Senderoff, *J. Phys. Chem.*, **63**, 1110 (1959).
318. O. G. Polyachenok and G. I. Novikov, *Zh. Obshch. Khim.*, **33**, 2797 (1963).
319. J. D. Corbett, D. L. Pollard and J. E. Mee, *Inorg. Chem.*, **5**, 761 (1966).
320. F. J. Keneshea and D. Cubicciotti, *J. Chem. Eng. Data*, **6**, 507 (1961).
321. O. G. Polyachenok and G. I. Novikov, *Zh. Neorgan. Khim.*, **8**, 1785 (1963).
322. A. S. Dworkin, H. R. Bronstein and M. A. Bredig, *Discussions Faraday Soc.*, **32**, 188 (1961).
323. G. I. Novikov and O. G. Polyachenok, *Zh. Neorgan. Khim.*, **7**, 1209 (1962).
324. L. F. Druding, J. D. Corbett and B. N. Ramsey, *Inorg. Chem.*, **2**, 869 (1963).
325. O. G. Polyachenok and G. I. Novikov, *Zh. Neorgan. Khim.*, **8**, 1567 (1963).
326. J. D. Corbett and B. C. McCollum, *Inorg. Chem.*, **5**, 938 (1966).
327. A. S. Dworkin, H. R. Bronstein and M. A. Bredig, *J. Phys. Chem.*, **67**, 2715 (1963).
328. D. Cubicciotti, U.S. Report MDDC-1058 (1946).
329. K. W. R. Johnson and J. A. Leary, *J. Inorg. Nucl. Chem.*, **26**, 103 (1964).
330. J. D. Greiner, J. F. Smith and J. D. Corbett, *J. Inorg. Nucl. Chem.*, **28**, 971 (1966).

331. W. Döll and W. Klemm, *Z. Anorg. Chem.*, **241**, 239 (1939).
332. W. H. Zachariasen, reference 13, p. 775.
333. C. T. Stubblefield, J. L. Rutledge and R. Phillips, *J. Phys. Chem.*, **69**, 991 (1965).
334. R. I. Slavkina, G. E. Sorokina and V. V. Serebrennikov, *Tr. Tomsk. Gos. Univ. Ser. Khim.*, **157**, 135 (1960), according to *Chem. Abs.*, **61**, 7927h.
335. M. M. Bel'kova and L. A. Alekseenko, *Zh. Neorgan. Khim.*, **10**, 1374 (1965).
336. T. Kaatz and M. Marcovich, *Acta Cryst.*, **21**, 1011 (1966).
337. W. P. Wolf, M. J. M. Leask, B. W. Mangum and A. F. G. Wyatt, *J. Phys. Soc. Japan Suppl. B-I*, **487** (1961).
338. J. C. Eisenstein, R. H. Hudson and B. W. Mangum, *Phys. Rev.*, **137**, 1886 (1965).
339. C. A. Hutchinson, B. R. Judd and D. F. D. Pope, *Proc. Phys. Soc. (London)*, **70B**, 514 (1957).
340. P. B. Dorain, C. A. Hutchinson and E. Wong, *Phys. Rev.*, **105**, 1037 (1957).
341. C. A. Hutchinson and E. Y. Wong, *J. Chem. Phys.*, **29**, 754 (1958).
342. H. H. Wickman, U.S. Report UCRL-11538 (1964).
343. G. Gaston, M. T. Hutchings, R. Shore and W. P. Wolf, *J. Chem. Phys.*, **41**, 1970 (1964).
344. B. R. Judd, *Proc. Roy. Soc. (London)*, *Ser. A*, **241**, 414 (1957).
345. E. Carlson and G. H. Dieke, *J. Chem. Phys.*, **29**, 229 (1958).
346. F. Varsanyi and G. H. Dieke, *J. Chem. Phys.*, **33**, 1616 (1960).
347. J. B. Gruber, *J. Chem. Phys.*, **35**, 2186 (1961).
348. K. B. Keating and H. G. Drickamer, *J. Chem. Phys.*, **34**, 140 (1961); *J. Chem. Phys.*, **34**, 143 (1961).
349. G. H. Dieke and S. Singh, *J. Chem. Phys.*, **35**, 355 (1961).
350. J. S. Margolis, *J. Chem. Phys.*, **35**, 1367 (1961).
351. J. B. Gruber, U.S. Report UCRL-9203 (1961).
352. J. D. Axe and G. H. Dieke, *J. Chem. Phys.*, **37**, 2364 (1962).
353. F. Varsanyi and G. H. Dieke, *J. Chem. Phys.*, **36**, 385 (1962).
354. R. M. Agrawal, R. K. Asunki, R. C. Naik, D. Ramakrishnan and S. Singh, *Proc. Indian Acad. Sci.*, *Sect. A*, **55**, 325 (1962).
355. I. Richmann and E. Y. Wong, *J. Chem. Phys.*, **37**, 2270 (1962).
356. S. Huefner, *Z. Physik*, **168**, 74 (1962).
357. F. Varsanyi and G. H. Dieke, *J. Chem. Phys.*, **36**, 2951 (1962).
358. J. G. Conway, J. B. Gruber, E. K. Hulet, R. J. Morrow and R. G. Gutmacher, *J. Chem. Phys.*, **36**, 189 (1962).
359. J. S. Margolis, O. Stafsudd and E. Y. Wong, *J. Chem. Phys.*, **38**, 2045 (1963).
360. E. Y. Wong, O. S. Stafsudd and D. R. Johnston, *J. Chem. Phys.*, **39**, 786 (1963).
361. A. Hadni and P. Strimer, *Compt. Rend.*, **257**, 398 (1963).
362. J. C. Eisenstein, *J. Chem. Phys.*, **39**, 2128 (1963).
363. K. S. Thomas, S. Singh and G. H. Dieke, *J. Chem. Phys.*, **38**, 2180 (1963).
364. L. G. DeShazer and G. H. Dieke, *J. Chem. Phys.*, **38**, 2190 (1963).
365. J. T. Houghen and S. Singh, *Phys. Rev. Letters*, **10**, 408 (1963).
366. J. B. Gruber, *J. Inorg. Nucl. Chem.*, **25**, 1093 (1963).
367. H. Lammerman and J. G. Conway, *J. Chem. Phys.*, **38**, 259 (1963).
368. I. Richman, NASA-Doc. N63-16,520 (1963).
369. A. Hadni, *Phys. Rev.*, **136A**, 758 (1964); *Compt. Rend.*, **258**, 5616 (1964).

370. J. T. Houghen and S. Singh, *Proc. Roy. Soc. (London), Ser. A*, **277**, 193 (1964).
371. G. E. Barasch and G. H. Dieke, *J. Chem. Phys.*, **43**, 988 (1965).
372. J. C. Wallman, J. Fuger, I. R. Peterson and J. L. Green, U.S. Report UCRL-16476 (1965).
373. J. L. Green, U.S. Report UCRL-16516 (1965).
374. E. J. Graeber, G. H. Conrad and S. F. Duliere, *Acta Cryst.*, **21**, 1012 (1966).
375. K. W. Bagnall, *Co-ordination Chemistry Review*, **2**, 145, 1967.
376. K. W. Bagnall, D. Brown, D. G. Holah and F. Lux, *J. Chem. Soc. (A)*. **1968**, 465.
377. K. W. Bagnall, D. Brown and F. Lux, to be published.
378. R. P. Dodge, G. S. Smith, R. Johnson and R. E. Elson, *Acta Cryst.* **B24**, 304 (1968).
379. M. C. Dagron, *Compt. Rend.*, **260**, 1422 (1965); **262**, 1575 (1966).
380. Sw. Pajakoff, *Monatsh.*, **94**, 482 (1963).
381. D. M. Gruen and C. W. Dekock, *J. Inorg. Nucl. Chem.*, **29**, 2569 (1967).
382. J. R. Peterson and B. B. Cunningham, *Inorg. Nucl. Chem. Letters* **3**, 579 (1967).
383. J. C. Copeland, U.S. Report UCRL-17718 (1967).
384. J. L. Ryan, personal communication, 1968.
385. D. Brown, unpublished observations.

Chapter 4

Bromides and Oxybromides

HEXAVALENT

Uranyl bromide is the only hexavalent actinide bromide known. A few complexes of the type $M^I_2UO_2Br_4$ (M^I = univalent cation) have been prepared but little is known about their properties. From the limited amount of information given below it will be obvious that there is scope for further research in this field.

Uranyl Bromide

The earlier work on uranyl bromide has been well reviewed and the reader is referred to those articles[1-5] for a comprehensive collection of references to the original literature. The earliest recorded preparative methods are not satisfactory. One involves bromination of the dioxide mixed with carbon, and results in the simultaneous formation of the tetrabromide, and the other dehydration of hydrated uranyl bromide obtained from aqueous hydrobromic acid solution. The most convenient preparation involves[6-9] the action of oxygen on uranium tetrabromide at 150–160°: careful temperature control is essential since excess oxygen converts uranyl bromide to U_3O_8 at about 185°. The oxidation of the tribromide under similar conditions is not recommended owing to the vigorous nature of the reaction. Uranium (v) oxytribromide, $UOBr_3$, is oxidized[10] smoothly by oxygen at 148° to yield pure uranyl bromide but in view of the difficulty associated with the preparation of $UOBr_3$ itself this is not a convenient route. It is also reported[10] that bromine reacts with uranium dioxide at 230° in a sealed tube at 45 atmospheres pressure to yield UO_2Br_2, but this again has little to recommend it.

Anhydrous uranyl bromide is a dark red solid which is appreciably less stable than uranyl chloride, evolving bromine slowly even at room temperature and decomposing completely above 250° in an inert atmosphere. It is deliquescent and forms[11] the yellow trihydrate, $UO_2Br_2 \cdot 3H_2O$, (previously[12] thought to be $UO_2Br_2 \cdot 7H_2O$) on controlled hydration; this

182

TABLE 4.1
Examples of the Complexes Formed by Uranyl Bromide and the Actinide Tetrabromides

Halide	Ligand (L)	Complex	Reference
UO_2Br_2	Ether	$UO_2Br_2 \cdot 2L$	a
	Ammonia	$UO_2Br_2 \cdot xL$ ($x = 2, 3,$ or 4)	a
	Acetic anhydride	$UO_2Br_2 \cdot 2L$	b
	Methyl cyanide	$UO_2Br_2 \cdot L$	c
	NN-Dimethylacetamide	$UO_2Br_2 \cdot 2L$	c
	NN-Dimethylformamide	$UO_2Br_2 \cdot 3L$	d
	Trialkyl or aryl phosphine oxides, R_3PO (R = Me, Et, Ph)	$UO_2Br_2 \cdot 2L$	e, f, g
	N-Methylacetanilide	$UO_2Br_2 \cdot 2L$	h
	Antipyrine	$UO_2Br_2 \cdot 2L$	h
$ThBr_4$	Methyl cyanide	$ThBr_4 \cdot 4L$	c
	Ammonia	$ThBr_4 \cdot xL$ ($x = 8 \rightarrow 20$)	i
	Ethanol	$ThBr_4 \cdot 4L$	j
	Pyridine	$ThBr_4 \cdot 3L$	j
	Benzaldehyde	$ThBr_4 \cdot 4L$	j
	Ethylamine	$ThBr_4 \cdot 4L$	k
	Hexamethylphosphoramide	$ThBr_4 \cdot 3L$ and $ThBr_4 \cdot 2L$	l
	NN-Dimethylacetamide	$ThBr_4 \cdot 5L$	o
		$ThBr_4 \cdot 4L$	c
	Dimethylsulphoxide	$ThBr_4 \cdot 6L$	l
$PaBr_4$	Methyl cyanide	$PaBr_4 \cdot 4L$	m
	NN-Dimethylacetamide	$PaBr_4 \cdot 5L$ and $PaBr_4 \cdot 2.5L$	o
	Hexamethylphosphoramide	$PaBr_4 \cdot 2L$	m
UBr_4	Methyl cyanide	$UBr_4 \cdot 4L$	c
	NN-Dimethylacetamide	$UBr_4 \cdot 5L$,	o
		$UBr_4 \cdot 4L$ and $UBr_4 \cdot 2.5L$	c
	Dimethylsulphoxide	$UBr_4 \cdot 6L$	l
	Hexamethylphosphoramide	$UBr_4 \cdot 2L$	l
	Triphenylphosphine oxide	$UBr_4 \cdot 2L$	n
$NpBr_4$	Methyl cyanide	$NpBr_4 \cdot 4L$	p
	Hexamethylphosphoramide	$NpBr_4 \cdot 2L$	p
	Triphenylphosphine oxide	$NpBr_4 \cdot 2L$	p, q
$PuBr_4$	Hexamethylphosphoramide	$PuBr_4 \cdot 2L$	p, q
	Triphenylphosphine oxide	$PuBr_4 \cdot 2L$	p, q

[a] A. von Unruh, Dissertation, Rostock (1909), according to reference 1, p. 592.
[b] R. C. Paul, S. S. Sandhu and J. S. Bassi, *J. Indian Chem. Soc.*, **38**, 85 (1961).
[c] K. W. Bagnall, D. Brown and P. J. Jones, *J. Chem. Soc. (A)*, **1966**, 1743.
[d] G. Kaufmann, R. Weiss and R. Rohmer, *Bull. Soc. Chim. France*, **1963**, 1140.
[e] P. Gans, Thesis, London (1964).

(Continued overleaf)

compound is soluble in organic solvents such as acetone, ether and alcohols; UO_2Br_2 forms complexes with a variety of oxygen and nitrogen donor ligands including ammonia (Table 4.1). No structural data are available for uranyl bromide; the uranium–oxygen stretching vibration occurs at 948 cm^{-1} in the infrared spectrum.

Hexavalent Oxybromo Complexes

Anhydrous bromo complexes of the type $M_2^I UO_2Br_4$ (M^I = Ph_3PBu, Ph_3PBz, Ph_3PH, Et_3PH and Pr_3PH) have been prepared by reacting the component halides in anhydrous methyl cyanide[13] and the anhydrous caesium salt, $Cs_2UO_2Br_4$, can be obtained[14] from hydrobromic acid solution. The yellow, hydrated complexes, $M_2^I UO_2Br_4 \cdot 2H_2O$ (M^I = NH_4, K and pyH), which are precipitated[12,15,16] from aqueous or alcoholic hydrobromic acid solution, lose water on heating in nitrogen at 120°. The anhydrous potassium salt reacts[17] with oxygen at 350° to form $K_2UO_3Br_2$, which is also obtained by reacting uranyl bromide monohydrate with the stoicheiometric amount of potassium hydroxide. The analogous ammonium salt[17], $(NH_4)_2UO_3Br_2$ and the potassium salt[18], KUO_3Br, have also been recorded.

Mixed halogeno complexes $M_2^I UO_2Cl_2Br_2$, (M^I = Cs, K and NH_4), $Cs_2UO_2ClBr_3$ and $Cs_2UO_2Cl_3Br$ are also formed. For example, the ammonium complex $(NH_4)_2UO_2Cl_2Br_2$ by the action of hydrogen chloride on[16,19] $NH_4UO_3Br_2$, at 150° and[14] the caesium salts by the reaction of varying amounts of CsBr with uranyl chloride in hydrobromic acid solution.

Single crystal studies[20] have shown that $Cs_2UO_2Br_4$ possesses monoclinic symmetry with a_0 = 9.90, b_0 = 9.808, c_0 = 6.39 Å, β = 103.5° and that each unit cell contains two molecules. The U–Br and U–O distances

---

Continuation of TABLE 4.1.

f B. W. Fitzsimmons, P. Gans, B. Hayton and B. C. Smith, *J. Inorg. Nucl. Chem.*, **28**, 915 (1966).

g J. P. Day, Thesis, Oxford (1965).

h Reference 1, p. 593.

i R. C. Young, *J. Am. Chem. Soc.*, **57**, 997 (1935).

j R. C. Young, *J. Am. Chem. Soc.*, **56**, 29 (1934).

k J. M. Matthews, *J. Am. Chem. Soc.*, **20**, 839 (1898).

l K. W. Bagnall, D. Brown, P. J. Jones and J. G. H. du Preez, *J. Chem. Soc. (A)*, **1966**, 737.

m D. Brown and P. J. Jones, *Chem. Commun.*, **1966**, 280.

n J. P. Day and L. M. Venanzi, *J. Chem. Soc. (A)*, **1966**, 197.

o K. W. Bagnall, D. Brown and F. Lux, to be published (1968).

p D. Brown and C. E. F. Rickard, *J. Chem. Soc. (A)*, in press.

q D. Brown, D. G. Holah and C. E. F. Rickard, *Chem. Commun.* **1968**, 651.

are respectively 2.82 and 1.69 Å. The uranium–oxygen stretching vibration in certain of the anhydrous salts occurs around 920 cm^{-1} as expected (Table B.4).

PENTAVALENT

The pentabromides of protactinium and uranium are known and hexabromo complexes can be crystallized from anhydrous methyl cyanide. The oxybromides $MOBr_3$ and MO_2Br (M = Pa and U) have been prepared but no oxybromo-complexes are yet known.

Pentabromides

Protactinium pentabromide was first prepared[21] in milligram amounts by reacting the pentoxide with aluminium bromide but this method is not suitable for the preparation of larger quantities. Gram amounts are best made by the reaction[22] of excess bromine with a mixture of the pentoxide and carbon at 600–700° in a sealed, evacuated silica vessel. The dark red, crystalline pentabromide is readily separated, by sublimation at 250–300°, from the green oxytribromide, $PaOBr_3$, which is also formed to some extent in this reaction.

Owing to the difficulties associated with the conversion of protactinium tetrafluoride to the metal the preparation of the pentabromide by direct union of the elements has not yet been attempted. In view of the oxidation of metallic protactinium to the pentaiodide by iodine (p. 211) this should prove a satisfactory route.

Uranium pentabromide is appreciably less stable than $PaBr_5$ and despite a not inconsiderable amount of work, has only recently been characterized. A reactive form of uranium trioxide can be obtained by heating $UO_4 \cdot xH_2O$ in oxygen at 400°, and this is converted[10,24,25] to the pentabromide by carbonyl bromide or carbon tetrabromide at 108–128°. The best results are obtained by heating the reactants together for two hours at 126° in a sealed vessel. Careful temperature control is essential since at higher temperatures ($\sim$165°) uranium tetrabromide is the major product.

Structures and Properties. Protactinium pentabromide is dimorphic. The β-form has a similar dimeric structure to that of uranium pentachloride and this is illustrated in Figure 4.1. The monoclinic unit cell, space group C_{2h}^5–$P2_1/n$ has $a_0 = 8.48$, $b_0 = 11.205$, $c_0 = 8.95$ Å and $\beta = 91.1°$. Full details of the structure of $\alpha PaBr_5$ are not yet available[116]. Protactinium pentabromide is moisture-sensitive, dissolving readily in water with hydrolysis and in anhydrous solvents such as methyl cyanide and alcohol, forming[23] the complex $PaBr_5 \cdot 3CH_3CN$ with the former. Uranium pentabromide is a brown, moisture-sensitive solid which dissolves in

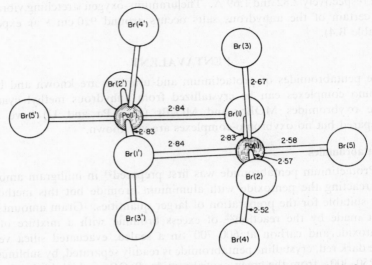

Figure 4.1 The structure of β-protactinium pentabromide.[116] (*After*
D. Brown, T. Petcher and A. J. Smith, *Nature*, **217**, 738 (1968)

alcohol, acetone and ethyl acetate. The absorption spectra reported
for the resulting solutions, however, showed that disproportionation
had occurred[25], presumably because of the presence of water in the
solvents. Recently, stable phosphine oxide complexes of the type
$MBr_5 \cdot L$ (M = Pa and U; L = triphenylphosphine oxide and hexamethyl-
phosphoramide) have been prepared[124]. The uranium pentabromide
complexes are obtained by bromine oxidation of uranium tetrabromide
in oxygen-free methyl cyanide in the presence of the stoicheiometric
amount of ligand. The dark-red solids are stable in an inert atmosphere
and readily soluble in methylene dichloride. Absorption spectra in this
solvent show the characteristic uranium (v) bands.

The chemical and physical properties of protactinium and uranium
pentabromide have scarcely been investigated and further studies will be
of value. Structural data on UBr_5 would be particularly useful since the
corresponding pentachlorides possess different structures (p. 127).

Pentavalent Bromo Complexes

Hexabromoprotactinates (v), $M^I PaBr_6$ (M^I = NMe_4 and NEt_4), can be
prepared[23] from methyl cyanide solutions of the component bromides in
a similar manner to the hexabromoniobates (v) and the hexabromotan-
talates (v). The orange-red hexabromoprotactinates (v) are moisture-

sensitive solids which dissolve in anhydrous methyl cyanide, alcohol and other polar solvents and which are immediately hydrolysed by water. The Pa–Br stretching vibrations occur at about 216 cm^{-1} (Table B.4). Uranium (v) bromo complexes have only recently been prepared[121] by bromine oxidation of tetravalent complexes in non-aqueous solvents, or by reacting the appropriate hexachlorouranate (v) with anhydrous liquid hydrogen bromide. Little is yet known of their chemistry although spectral properties have been recorded[121].

Pentavalent Oxybromides

Protactinium (v) oxytribromide, $PaOBr_3$, a green-yellow solid, is invariably obtained[22] as a by-product of the reaction between the pentoxide, carbon and bromine at 600–700°. It is also formed when stoicheiometric quantities of the pentabromide and either oxygen[22] or antimony trioxide[26] are heated together at 350° in a sealed tube. It disproportionates above 500° in a vacuum to the pentabromide and protactinium (v) dioxymonobromide, PaO_2Br, an off-white, non-volatile solid for which the Pa–O stretching vibrations are observed[26] at 575, 386 and 286 cm^{-1}.

Uranium (v) oxytribromide is prepared from the trioxide by heating it at 110° in a stream of nitrogen[10,24,27] and carbon tetrabromide vapour; the carbonyl bromide so produced is removed by the nitrogen flow thereby preventing the formation of the pentabromide. $UOBr_3$ evolves bromine slowly at room temperature, reacts with carbon tetrabromide at 165° to form uranium tetrabromide and with oxygen at 148° to form uranyl bromide; it decomposes at 300° in dry nitrogen, losing bromine to yield uranium (IV) oxydibromide, $UOBr_2$. It is insoluble in carbon tetrachloride and carbon disulphide and, like the pentabromide, it is moisture-sensitive and disproportionates in solvents such as acetone and alcohol. Its absorption spectrum has been recorded[25] in chloroform and bromoform. The dark brown uranium (v) dioxymonobromide, UO_2Br, is obtained[28] by heating uranium trioxide with hydrogen bromide at 250° or by treating uranyl chloride with hydrogen bromide. At −20° the former reaction yields $UO_2Br_2 \cdot H_2O$ and at room temperature a brown-black solid is obtained which is thought to be $UO_2Br \cdot 2HBr$.

Uranium (v) dioxymonobromide decomposes at 500° in nitrogen to form the dioxide and bromine. Its infrared spectrum is reported to be similar to that of uranyl bromide with bands at 940, 890 and 850 cm^{-1}. These results suggest that some decomposition may have taken place prior to measurement of the infrared spectrum since vibrations associated with the discrete MO_2^+ ion generally occur around 800 cm^{-1} and not in the region of the MO_2^{++} ion vibrations. The reported x-ray powder reflections show that UO_2Br is not isostructural with PaO_2Br but in neither instance have the results been interpreted.

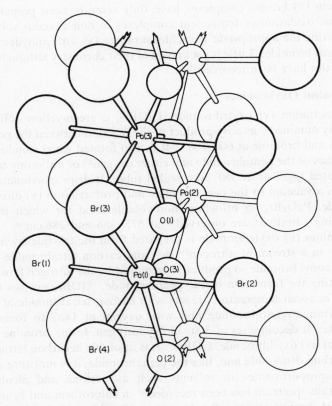

Bond lengths and angles in one pentagonal bipyramid

	(Å)		Degrees
Pa–O(1)	2.14	O(1)–Pa–Br(3)	80.7
Pa–O(2)	2.25	Br(3)–Pa–Br(4)	84.85
Pa–O(3)	2.06	Br(4)–Pa–O(2)	72.4
Pa–Br(1)	2.69	O(2)–Pa–O(3)	60.3
Pa–Br(2)	2.56	O(3)–Pa–O(1)	61.9
Pa–Br(3)	2.76	Br(1)–Pa–Br(2)	174.2
Pa–Br(4)	3.02		
		Pa(1)–O–Pa(2)	121.4
		Pa(2)–O–Pa(3)	116.7
		Pa(3)–O–Pa(1)	120.9

Figure 4.2   The structure of protactinium oxytribromide.[29]   (*After* D. Brown, T. Petcher and A. J. Smith. *Nature* **217**, 738 (1968))

Structures. The only pentavalent oxybromide for which structural data are available is protactinium (v) oxytribromide. A single crystal study[29], using crystals accidentally formed by slow sublimation in the presence of an excess of the pentabromide, has shown it to possess monoclinic symmetry, space group $C2$, with $a_0 = 16.911$, $b_0 = 3.871$, $c_0 = 9.334$ Å and $\beta = 113.67°$. As suggested by the positions of the protactinium–oxygen stretching vibrations (515, 476, 364 and 303 cm^{-1}) the structure comprises chains of protactinium atoms linked by bridging oxygens (Figure 4.2). The oxygens are actually three coordinate, with Pa–O distances ranging between 2.06 and 2.25 Å; a similar situation is found in the structure of $Cs_{0.9}UO_3Cl_{0.9}$ (p. 123). Protactinium–bromine bond lengths range from 3.02 Å for bridging bromine atoms to 2.56 Å for non-bridging bromines.

TETRAVALENT

Only thorium, protactinium, uranium and neptunium form tetrabromides, the last being thermally unstable. Plutonium tetrabromide is unlikely to be obtained since thermodynamic calculations indicate[30] that the equilibrium decomposition pressure of bromine over the tetrabromide would be 10^{18} atmospheres at room temperature. However, hexabromoplutonates (IV), of the type $M_2^I PuBr_6$ and phosphine oxide complexes of the type $PuBr_4 \cdot 2L$ have recently been prepared[31,124]. Analogous tetravalent bromo complexes are known for the elements thorium to uranium inclusive. The oxydibromides, $MOBr_2$, of thorium, protactinium, uranium and neptunium have been reported.

Tetrabromides

The wealth of literature pertaining to the preparation of thorium and uranium tetrabromides has been reviewed in some detail previously[1-5,8,32]. The various reactions investigated are summarized in Table 4.2 and it suffices to say here that for thorium tetrabromide, direct union of the elements[33-35] or the action of hydrogen bromide or bromine on the hydride[36] at 350°, and for uranium tetrabromide, the action of bromine on the tribromide[1-3,8] or direct combination of uranium and bromine[1-3,8,35,37-39] are the most satisfactory methods for large scale preparations. The last reaction may be conveniently carried out either in a gas-flow system or in a sealed, evacuated tube. The recently reported[10] conversion of uranium trioxide to the tetrabromide by carbon tetrabromide at 165° provides the best alternative preparation not involving uranium metal. Thorium and uranium tetrabromides are purified by sublimation at 550° in a vacuum or at higher temperatures in an inert atmosphere containing a small amount of bromine.

TABLE 4.2

Preparative Methods for Thorium and Uranium Tetrabromide[aa]

Reaction	Conditions	Reference
[bb]Metal + bromine	(1) Sealed tube >700°	a, b, c, d, e, f,
	(2) He flow at 650°	g, h, i, j, p
[bb]Hydride + bromine	He flow at 650°	a, b, c, g, p
[bb]Oxide admixed with carbon + bromine	N_2 flow at 700–900°	a, b, c, k, l, p
[bb]Carbide + bromine	N_2 flow at 700–900°	a, c, i, m, p, o
Thorium hydride + hydrogen bromide	HBr flow at 350°	g, p
Thoria + sulphur monochloride + hydrogen bromide	HBr flow at 135°	n, p
Uranium tribromide + bromine	N_2 flow at 300°	a, b, c, o
Uranium dioxide + carbon tetrabromide	N_2 flow at 175°	q
Uranium trioxide + carbon tetrabromide }	N_2 flow at 165°	r
$UOBr_3$ + carbon tetrabromide }		
$UOBr_2$ + heat	800–1000°	c
$U_3O_2S_4$ + bromine	N_2 flow at 400°	a, c, o, s
UO_2 }	CO_2 flow at 700–900°	c, h, t
U_3O_8 } + carbon + bromine		
Uranium nitride + bromine	N_2 flow at >600°	a, c, o
Uranium sulphide + bromine	N_2 flow at >200°	a, c, o

[aa] Other references will be found in review articles a, b, c, k, o, p.
[bb] Both thorium and uranium.

[a] J. J. Katz and E. Rabinowitch, The Chemistry of Uranium, *Nat. Nucl. Energy Ser. Div. VIII*, Vol. 5, McGraw-Hill, New York, 1951; p. 521.
[b] The Actinide Elements (G. T. Seaborg and J. J. Katz, Eds.), *Nat. Nucl. Energy Ser., Div. IV*, McGraw-Hill, New York, **14A**, 153 (1954).
[c] O. Johnson, T. Butler, J. Powell and R. Notorff, USAEC Report CC-1974 (1955).
[d] L. F. Nilsen, *Ber.*, **9**, 1142 (1876); **15**, 2537 (1882); **16**, 153 (1883).
[e] H. Moissan and A. Étard, *Compt. Rend.*, **122**, 573 (1896); *Ann. Chim. Phys.*, **12**, 427 (1897).
[f] K. W. Bagnall, D. Brown and P. J. Jones, *J. Chem. Soc. (A)*, **1966**, 1743.
[g] H. Lipkind and A. S. Newton, Report TID-5223, p. 398 (1952).
[h] C. Zimmerman, *Ann.*, **216**, 2 (1882).
[i] N. W. Gregory, Report TID-5290, p. 498 (1952).
[j] J. D. Corbett, R. H. Clark and T. F. Munday, *J. Inorg. Nucl. Chem.*, **25**, 1287 (1963).
[k] R. C. Young and H. G. Futcher, *Inorg. Syn.*, **1**, 51 (1939).
[l] R. C. Young, *J. Am. Chem. Soc.*, **56**, 29 (1934).
[m] H. Moissan and A. Martinson, *Compt. Rend.*, **140**, 1513 (1905).
[n] F. Burion, *Compt. Rend.*, **145**, 245 (1907).
[o] F. H. Spedding, A. S. Newton, R. Notorff, J. Powell and V. Calkins, Report TID-5290, p. 91 (1958).
[p] Reference b, p. 84.
[q] R. M. Douglass and E. Staritsky, *Anal. Chem.*, **29**, 459 (1957).
[r] J. Prigent, *Ann. Chim. (Paris)*, **5**, 65 (1960).
[s] J. Prigent, *Compt. Rend.*, **247**, 1737 (1958).
[t] R. Rohmer, R. Freymann, R. Freymann, A. Clevet and P. Harman, *Bull. Soc. Chim. France*, **1952**, 598.

In contrast to the above, protactinium and neptunium tetrabromide have been virtually neglected. Protactinium tetrabromide is obtained by hydrogen or aluminium reduction[40] of the pentabromide at 400–450°, the latter being preferred for quantities in excess of 50 mg. The reaction is conveniently carried out in a sealed, evacuated Pyrex vessel following which the product is purified by vacuum sublimation above 500°. Although direct union of the elements has not yet been investigated this would undoubtedly lead to the formation of the pentabromide (cf. PaI_5, p. 211). Neptunium tetrabromide is formed[41] when the dioxide is heated with excess aluminium bromide at 350° or better[117], by heating the metal with excess bromine at 600°. It sublimes at 500° in a vacuum with extensive decomposition to the tribromide. Thermodynamic calculations suggest[42] that neptunium tetrabromide should be unstable above 430°.

Structures and Properties. The known melting points of the tetrabromides and their crystallographic properties are summarized in Table 4.3. Thorium tetrabromide exists in two crystal forms; β-$ThBr_4$, the high-

TABLE 4.3

The Actinide Tetrabromides

Compound	Colour	m.p.	Symmetry	Space group	Lattice parameters (Å)			Reference
					a_0	b_0	c_0	
β-$ThBr_4$	White	679	Tetragonal	D_{4h}^{19}–$I4/amd$	8.945	—	7.930	43
α-$ThBr_4$	White	—	Orthorhombic	—	13.610	12.050	7.821	45
$PaBr_4$	Red	—	Tetragonal	D_{4h}^{19}–$I4/amd$	8.824	—	7.957	40
UBr_4	Brown	519	Monoclinic	$2/c$-/–	10.92	8.69	7.05 ($\beta = 93.15°$)	44
$NpBr_4$	Reddish-brown	464	Monoclinic	$2/c$-/–	10.89	8.74	7.05 ($\beta = 94.19°$)	117

temperature form, and protactinium tetrabromide[40,43] possess the same structure as uranium tetrachloride (p. 138), unlike uranium[44] and neptunium[41] tetrabromide which are monoclinic. The low-temperature form of thorium tetrabromide, α-$ThBr_4$, is obtained[45] when β-$ThBr_4$ is heated at 330° for several hours. It has tentatively been assigned orthorhombic symmetry (Table 4.3) on the basis of x-ray powder results. β-$ThBr_4$ slowly converts to the α-form at room temperature. Calculated densities for β-$ThBr_4$ and UBr_4 are 5.69 and 5.55 g cm^{-3} respectively, but measurements of the density of the latter have yielded values from 4.84 to 5.35 g cm^{-3}. Vapour pressure data for thorium[32] and uranium tetrabromide[1,2] can be represented by the equations (1)–(3) shown below. The calculated boiling point of uranium tetrabromide, from transpiration vapour pressure measurements, is 761° at 1 atmosphere, in good agreement with the observed value of 765°. Similar studies are lacking for $PaBr_4$ and $NpBr_4$.

$$\log Pmm_{(ThBr_4)} = \frac{-9,630}{T} + 11.73 \tag{1}$$

$$\log Pmm_{(UBr_4 \text{ solid})} = \frac{-10,900}{T} + 14.56 \tag{2}$$

$$\log Pmm_{(UBr_4 \text{ vapour})} = \frac{-7060}{T} + 9.71 \tag{3}$$

Thorium and uranium tetrabromide are deliquescent; the latter is oxidized to uranyl bromide in oxygen at 150–160°, reduced to the tribromide in hydrogen above 470° and reacts with chlorine to form the tetrachloride. It is reported to be reduced to the metal by calcium and magnesium but attempts to reduce it with potassium in liquid ammonia have been[46] unsuccessful. The magnetic susceptibility of uranium tetrabromide shows Curie–Weiss dependence between 72 and 569°K and the calculated magnetic moment, with a Weiss constant of −35°, is 3.12 B.M. Electron diffraction of the vapour shows that the uranium tetrabromide molecule is a distorted tetrahedron of C_{2v} symmetry[48].

Thorium and uranium tetrabromide are soluble in solvents such as acetone, alcohol, etc., and they form complexes with ammonia, methyl cyanide and oxygen donor ligands containing the $C = O$, $P = O$ and $S = O$ groups. Such complexes have not been as extensively studied as those formed by the tetrachlorides; a selection of the complexes which have been identified is given in Table 4.1 and further details are to be found in the recent exhaustive reviews by Bagnall[5,125]. Some analogous complexes of neptunium and plutonium tetrabromide have recently been reported[124]; only NN-dimethylacetamide, methyl cyanide and hexamethylphosphoramide complexes (Table 4.1), are known for protactinium (IV). Infrared studies show that coordination of amides[35], phosphine oxides[40,49,50,124] and sulphoxides[49] occurs via the oxygen atom; the metal–bromine stretching vibrations of such complexes occur at 180–190 cm^{-1}, as in the[51] hexabromo complexes.

Tetravalent Bromo Complexes

Hexabromo complexes of thorium (IV), protactinium (IV), uranium (IV), neptunium (IV) and plutonium (IV) are known and mixed chlorobromo complexes have been prepared for uranium (IV) (p. 256). The dark green sodium and potassium hexabromouranates (IV), Na_2UBr_6 and K_2UBr_6, have been made[52] by heating the alkali metal bromide in uranium tetrabromide vapour and the tetramethylammonium complex $(NMe_4)_2UBr_6$ has been isolated[53] from aqueous hydrobromic acid. The white pyridinium complex $(pyH)_2ThBr_6$ is reported[54,55] to be obtained from anhydrous alcoholic hydrobromic acid solution. The tetraethylammonium complexes $(NEt_4)_2UBr_6$, $(NEt_4)_2NpBr_6$ and $(NEt_4)_2PuBr_6$, which are green, yellow and dark red respectively, precipitate from ethanolic hydrobromic acid solution on the addition of acetone[31] and the triphenylphosphonium hexabromouranate (IV) crystallizes[56] from aqueous acetone–hydrobromic acid. However, there is less risk of hydrolysis if non-aqueous solvents are used and recently the hexabromothorates (IV)[51] and hexabromoprotactinates (IV)[57], $(NMe_4)_2M^{IV}Br_6$ and $(NEt_4)_2M^{IV}Br_6$ (M^{IV} = Th and Pa) and several[13,50,51] hexabromouranates (IV), $M_2^IUBr_6$ (M^I = NMe_4, NEt_4, Ph_3PBu, Ph_3PBz, Ph_3PH, Et_3PH and $OctylNMe_3$) have been prepared by crystallization from anhydrous methyl cyanide solutions of the component bromides. Like their chloro-analogues, $(NMe_4)_2ThBr_6$, $(NMe_4)_2PaBr_6$ and $(NMe_4)_2UBr_6$ possess[51,57] face-centred cubic symmetry, space group O_h^5–$Fm3m$ with a_0 = 13.49 Å, 13.40 Å, and 13.37 Å respectively. The metal–bromine stretching vibration, ν_3, occurs[51,57] around 180 cm^{-1} for certain of the uranium (IV) and thorium (IV) complexes (Table B.4). As is the case with uranium (IV) hexachloro complexes (p. 146) the magnetic susceptibilities of the hexabromo salts are almost independent of temperature. This observation is in agreement with the non-magnetic ground state $A_1(^3H_4)$, derived from spectroscopic measurement[53], and assuming no thermal population of excited states the susceptibility is probably due to an unusually large high-frequency term. In this case the latter is primarily due to an interaction between the ground state and the first excited state $T_1(^3H_4)$. Since the energy level separation $A_1(^3H_4)$–$T_1(^3H_4)$ is dependent only on ligand field, magnetic susceptibility changes are inversely proportional to changes in ligand field. Thus at room temperature the susceptibility of UBr_6^{2-} complexes is approximately 10% greater than that of analogous UCl_6^{2-} complexes.

The cerium (IV) complex $(NMe_4)_2CeBr_6$ has been made in methyl cyanide or nitromethane solutions by reacting the corresponding hexachloro complex with hydrogen bromide. The $CeBr_6^{2-}$ species, however, is very unstable at room temperature in these solvents. The reaction between tetravalent actinide hexachloro-complexes and anhydrous liquid hydrogen bromide appears to provide an excellent alternative preparative procedure for the hexabromo complexes[121]. Trivalent lanthanide complexes can also be prepared in an analogous manner.

Tetravalent Oxydibromides

Thorium (IV), protactinium (IV), uranium (IV) and neptunium (IV) oxydibromide have now all been characterized analytically. The first is obtained by boiling an aqueous solution of the tetrabromide and heating[58] the residue at 160°, by heating[59] the hydrated tetrabromide, or by heating the dioxide in a mixture[60] of sulphur monochloride and hydrogen bromide at 125°. Uranium (IV) oxydibromide, $UOBr_2$, was first observed[61] as a pale yellow residue on sublimation of the tetrabromide. It is conveniently prepared[3,8] by heating $U_3O_2S_4$ with bromine at 600° when the volatile tetrabromide distils away leaving the greenish-yellow $UOBr_2$. More recently[10] it has been obtained by decomposition of $UOBr_3$ at 300° in nitrogen. Another satisfactory route[38] is to heat stoicheiometric quantities of the dioxide and tetrabromide at 500°. However the most convenient method for the preparation of the actinide (IV) oxydibromides is[40,62] to react antimony trioxide and the appropriate tetrabromide at 150°. The volatile antimony tribromide distils off following which the residual oxydibromide can be heated to 450° to improve crystallinity. Using this technique $PaOBr_2$ was recently prepared[40] for the first time.

The tetravalent oxydibromides are all non-volatile solids which disproportionate above 500° *in vacuo* according to equation (4).

$$2MOBr_2 \rightarrow MO_2 + MBr_4 \qquad (4)$$

There is some evidence[38] that $UOBr_2$ can be reduced to the trivalent oxybromide, UOBr, but the latter has not been characterized. Metal–oxygen stretching vibrations are found[10,62] below 600 cm^{-1} in their infrared spectra (Table B.2). X-ray powder diffraction data[62] have not yet been interpreted for these isostructural compounds.[62,63]

TRIVALENT

Anhydrous tribromides of scandium, yttrium and of all the lanthanides have been characterized and tribromides are known for the elements actinium through curium excepting thorium and protactinium (Table 4.4). Although it has been claimed that thorium tribromide exists, see, for example, references 64 and 65, there is at present insufficient evidence available to substantiate such claims. Fewer methods are available for the preparation of the lanthanide tribromides than have been used for the corresponding trichlorides, and much less is known of their chemistry. Trivalent oxybromides of the type MOBr have been characterized for many of the elements and bromo-complexes, $(Ph_3PH)_3M^{III}Br_6 \cdot Ph_3PHBr$ (M^{III} = La, Pr, Nd, Sm and Eu) are known.

Tribromides

The general methods of preparation involve the hydrated tribromides, obtainable from aqueous hydrobromic acid, the oxides or the trichlorides. Matignon[66] used the reaction between the anhydrous lanthanide trichlorides and hydrogen bromide at 400–600° but though it has been employed by others[67] and also for the preparation of plutonium tribromide it finds little application owing to the necessity of initially preparing the anhydrous trichloride. The dehydration of the lanthanide tribromide hydrates in a stream of dry hydrogen bromide has been employed with mixed success[68-71], possibly owing to the difficulty of obtaining hydrogen bromide completely free of water and oxygen. This method seems to be most successful with the lighter lanthanides and Harrison[72] recently obtained pure samples of lanthanum, cerium, neodymium and samarium tribromide by first partially dehydrating at relatively low temperature and finally heating the reactants just above the melting points of the anhydrous tribromides. The addition of ammonium bromide to the hydrate prior to dehydration in hydrogen bromide improves the yield and helps to prevent formation of the oxybromides, for example see references 68, 73, and 74. Alternatively, the reaction between the hydrated tribromides and ammonium bromide alone will yield[74,75] pure anhydrous tribromides. Thus lanthanum, samarium, europium and ytterbium tribromides have been prepared by heating the reactants in a vacuum to remove the water and excess ammonium bromide, and yttrium tribromide is obtained[76] at 600° in dry helium.

However, the use of such dehydrating agents is unnecessary since it has recently been shown[122] that by carefully controlled vacuum thermal decomposition of both the lanthanide and actinide tribromide hexahydrates the anhydrous tribromides can be prepared in gram amounts. Temperature control is essential and the method comprises heating the tribromide hexahydrate in a non-static vacuum system at about 70° for 12 hours following which the temperature is gradually raised during a further 12 hours to 170°. Oxybromide contamination is observed only for ytterbium and lutetium and even in these instances is less than 5%.

The oxides are also converted directly to the tribromides by heating them with excess ammonium bromide, for example, see references 77–80, in an inert atmosphere at high temperature. This reaction is undoubtedly superior to the older technique involving bromination of the oxide with a mixture of sulphur monochloride and hydrogen bromide[81,82] which was used by Bourion[81] for the first preparation of several of the lanthanide tribromides. The conversion of lanthanum oxide may alternatively be achieved by heating it in a stream of dry hydrogen bromide and bromine[83] but this reaction is less satisfactory for the other lanthanides and the tribromides of praseodymium, neodymium, samarium and gadolinium are more easily prepared by heating the carbonate[84] in a stream of hydrogen bromide and bromine at approximately 50° below the melting point of the tribromides. Neodymium and gadolinium tribromide have also been obtained, in an impure state, by[83] reacting their respective benzoates with hydrogen bromide in ether. Vacuum sublimation of the lanthanide tribromides is recommended to ensure complete freedom from oxybromide or other impurities.

Of the actinide tribromides, the actinium[85] compound has been prepared on the microgram scale only and has been identified crystallographically. It is obtained[85] by heating the oxide or oxalate at moderate temperatures with aluminium bromide followed by sublimation of the tribromide at higher temperatures. It has been reported that excess aluminium is necessary for the preparation of neptunium tribromide by similar methods to ensure complete reduction of the tetrabromide but above 500° the latter is unstable with respect to bromine and the tribromide[117]. The first identification[86] of americium tribromide was also made with only microgram amounts but both it and curium tribromide have since been prepared[87] in milligram amounts by heating the trichloride with ammonium bromide at 400–450° in an atmosphere of hydrogen.

The preparation of uranium and plutonium tribromide has been investigated in more detail and there are several satisfactory procedures. The best methods of preparing uranium tribromide are the direct union of stoicheiometric amounts of the elements[1-3,88] at 300–500° and the reaction between uranium hydride and dry hydrogen bromide[1-3,8] which proceeds smoothly at 350°. It is also formed by reduction of the tetrabromide by hydrogen[1-3,8,38,88] at 500°, or by uranium metal[38,39] at higher temperatures, the reaction being best carried out in tantalum. Hydrogen reduction of the tetrabromide was used for the first preparations of uranium tribromide prior to the turn of the century and although more recent studies[8,88] suggest that reduction to a lower halide may occur it is likely that traces of oxygen or reaction of the tribromide with the silica apparatus result in partial conversion to an oxybromide or oxide since uranium tribromide is known to react with glass or silica at high temperature and, in addition, although uranium is soluble in its molten tribromide[39] lower halides have not been identified in such mixtures.

Plutonium tribromide has been prepared by several reactions analogous to those described for the trichloride (p. 151). Thus bromination of the 'dioxide' (i.e. gently ignited hydroxide or vacuum dried peroxide) proceeds satisfactorily with hydrogen bromide[89,91] at 600° or sulphur mono-bromide[90,92] at 800°. Alternatively, plutonium (IV) oxalate hexahydrate[91] and the hydrated tribromide[90] react with hydrogen bromide at 500° and 300° respectively although the product from the tribromide hydrate is usually heavily contaminated with oxybromide; the addition of ammonium bromide again improves this reaction. The reaction between plutonium metal and bromine proceeds[89,90] smoothly at 300° and with the ready availability of metallic plutonium this reaction and those involving dehydration[122] of $PuBr_3 \cdot 6H_2O$ or reaction between the hydride and hydrogen bromide[93] at 600° are the most attractive methods for preparing plutonium tribromide. It can be purified by vacuum sublimation above 800°.

Of the remaining tribromides only that of scandium is reported[96] to have been prepared by direct union of the elements and only cerium tribromide[120] appears to have been prepared by reacting the hydride with hydrogen bromide. With increased availability of the metals these routes are obviously to be preferred.

Crystal structures. As the atomic weight increases along each series of elements there is a change of crystal structure in accordance with the

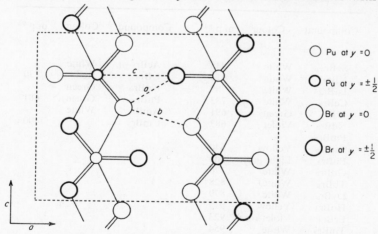

Figure 4.3 The (layer) structure of $PuBr_3$. The planes of the layers are normal to that of the paper. The Br–Br contacts a and b (3.81 and 3.65 Å respectively) prevent the layers from approaching close enough for coordination across c (4.03 Å) and thereby limit the plutonium to eight coordination. (*After* A. F. Wells, *Structural Inorganic Chemistry*, Clarendon Press, Oxford, 1962)

decreasing atomic radii of the elements. Thus[63,97] the tribromides of lanthanum, cerium and praseodymium, like those of actinium, uranium and one form of the neptunium compound are isostructural with uranium trichloride (p. 152). Whereas[87,97], owing to the instability of this structure with the decreasing ratio of cation to anion radius, neodymium, samarium, β-neptunium, plutonium, americium and curium tribromide possess the plutonium tribromide-type structure (Table 4.5). This latter structure (Figure 4.3) is a layer-lattice type with each metal atom bonded to eight halogen atoms with, in PuBr$_3$, two at 3.06 Å and six at 3.08 Å. Although two crystal forms of neptunium tribromide have been observed, details of the specific preparation of each form appear to be lacking and during a recent study[117] only the hexagonal form was observed. A third, as yet unidentified, structure-type has been reported[83] for gadolinium tribromide; the tribromides of gadolinium to lutetium inclusive were recently shown to possess the ferric chloride-type (or BiI$_3$) structure (p. 223) like scandium tribromide[96].

Properties. The tribromides are all fairly high melting solids (Table 4.4) which are volatile at high temperature in a vacuum. The lanthanide tribromides are soluble in water, in which they are only slowly hydrolysed,

TABLE 4.4

The Lanthanide and Actinide Tribromides

Compound	Colour	m.p.[94]	Compound	Colour	m.p.[95]
ScBr$_3$	White	948	AcBr$_3$	White	—
YBr$_3$	White	913	UBr$_3$	Red	730
LaBr$_3$	White	789	NpBr$_3$	Green	—
CeBr$_3$	White	733	PuBr$_3$	Green	681
PrBr$_3$	Green	691	AmBr$_3$	White	—
NdBr$_3$	Violet	682	CmBr$_3$	—	400
PmBr$_3$	—	—			
SmBr$_3$	Yellow	640			
EuBr$_3$	Light grey	d			
GdBr$_3$	White	770			
TbBr$_3$	White	828			
DyBr$_3$	White	879			
HoBr$_3$	Yellow	919			
ErBr$_3$	Violet-rose	923			
TmBr$_3$	White	954			
YbBr$_3$	White	d			
LuBr$_3$	White	1025			

d = decomposes.

and hexahydrates have been characterized (p. 202). Uranium, neptunium, plutonium and americium tribromides also form[122] hexahydrates of which the first is only obtained by controlled hydration in an oxygen-free atmosphere. Like the lanthanide tribromide hexahydrates they possess monoclinic symmetry and unit cell data are now available[122] for both series of compounds. The tribromides of lanthanum, praseodymium, neodymium and samarium react with ethyl benzoate to form insoluble benzoates and like gadolinium tribromide they form methyl cyanide complexes[83] of the type $MBr_3 \cdot xCH_3CN$ (x = variously 3, 4 or 5). A pyridine complex of cerium tribromide[98], $CeBr_3 \cdot 4L$, and tetrahydrofuran adducts[99,123] of certain of the lanthanide tribromides, $MBr_3 \cdot 3L$ (M = La, Ce, Pr), $MBr_3 \cdot 3.5L$ (M = Nd $\rightarrow$ Er inclusive) and $MBr_3 \cdot 4L$ (M = Tm, Yb and Lu) are also known but complexes of the actinide tribromides, apart from the uranium tribromide compound[100], $UBr_3 \cdot 6NH_3$, have not

TABLE 4.5

Crystallographic Properties of the Lanthanide and Actinide Tribromides

Compound	Symmetry	Structure-type	Lattice dimensions (Å)			Density (g cm^{-3})	Reference
			a_0	b_0	c_0		
LaBr$_3$	Hexagonal	UCl$_3$	7.967	—	4.501	5.07	97
CeBr$_3$	Hexagonal	UCl$_3$	7.952	—	4.444	5.18	97
PrBr$_3$	Hexagonal	UCl$_3$	7.930	—	4.390	5.26	97
NdBr$_3$	Orthorhombic	PuBr$_3$	12.65	4.11	9.16	5.35	97
PmBr$_3$	Orthorhombic	PuBr$_3$	12.65	4.08	9.12	5.38	118
SmBr$_3$	Orthorhombic	PuBr$_3$	12.63	4.04	9.07	5.40	97
EuBr$_3$	Orthorhombic	PuBr$_3$	12.71	4.019	9.128	5.40	119
GdBr$_3$	Rhombohedral^a	FeCl$_3$	7.216	—	19.189	4.57	122
TbBr$_3$	Rhombohedral	FeCl$_3$	7.159	—	19.163	4.67	122
DyBr$_3$	Rhombohedral	FeCl$_3$	7.107	—	19.161	4.78	122
HoBr$_3$	Rhombohedral	FeCl$_3$	7.072	—	19.150	4.86	122
ErBr$_3$	Rhombohedral	FeCl$_3$	7.045	—	19.148	4.93	122
TmBr$_3$	Rhombohedral	FeCl$_3$	7.002	—	19.111	5.02	122
YbBr$_3$	Rhombohedral	FeCl$_3$	6.981	—	19.115	5.10	122
LuBr$_3$	Rhombohedral	FeCl$_3$	6.950	—	19.109	5.17	122
YBr$_3$	Rhombohedral	FeCl$_3$	7.072	—	19.150	3.95	122
ScBr$_3$	Rhombohedral	FeCl$_3$	6.643	—	18.765	3.93	96
AcBr$_3$	Hexagonal	UCl$_3$	8.06	—	4.68	5.85	85, 97
UBr$_3$	Hexagonal	UCl$_3$	7.942	—	4.440	6.53	97
α-NpBr$_3$	Hexagonal	UCl$_3$	7.917	—	4.382	6.62	97
β-NpBr$_3$	Orthorhombic	PuBr$_3$	12.65	4.11	9.15	6.62	97
PuBr$_3$	Orthorhombic	PuBr$_3$	12.62	4.09	9.13	6.69	97
AmBr$_3$	Orthorhombic	PuBr$_3$	12.661	4.064	9.144	6.83	87
CmBr$_3$	Orthorhombic	PuBr$_3$	12.660	4.048	9.124	6.87	87

^{a} Hexagonal cell parameters are listed.

14

been reported. Complexes[101] of scandium tribromide with ammonia and several amines have been described. Uranium and plutonium tribromide are oxidized by oxygen at moderate temperatures and bromine converts the former to the tetrabromide at 300° (p. 189). Uranium tribromide, and to a lesser extent plutonium tribromide, attacks glass or quartz at high temperatures and UBr_3 disproportionates above 900°.

Available vapour pressure data[1,89,102,103] are limited to the tribromides of scandium, lanthanum, cerium, praseodymium, neodymium, uranium and plutonium. These data are summarized in Table 4.6. The results for uranium tribromide, although probably the best available, are possibly

TABLE 4.6

Vapour Pressure Data for the Solid Tribromides

Compound	$\log p_{mm} = -(A/T) + B$		Reference
	A	B	
$ScBr_3$	13,785	14.300	103
$LaBr_3$	15,446	12.568	102
$CeBr_3$	14,990	12.334	102
$PrBr_3$	14,916	12.508	102
$NdBr_3$	14,829	12.555	102
UBr_3	15,000	12.500	1
$PuBr_3$	15,281	13.386	89

TABLE 4.7

Estimated Temperatures to give 2 mm Hg Vapour Pressure of the Lanthanide Tribromides[80]

Halide	Estimated temperature (°C)	Boiling point (°C)	Halide	Estimated temperature (°C)	Boiling point (°C)
$LaBr_3$	1010	1580	$TbBr_3$	930	1490
$CeBr_3$	1010	1560	$DyBr_3$	940	1480
$PrBr_3$	990	1550	$HoBr_3$	940	1470
$NdBr_3$	970	1540	$ErBr_3$	940	1460
$PmBr_3$	960	1530	$TmBr_3$	930	1440
$SmBr_3$	d	d	$YbBr_3$	d	d
$EuBr_3$	d	d	$LuBr_3$	920	1410
$GdBr_3$	930	1490	YBr_3	940	1470

d = decomposes.

not accurate owing to corrosion difficulties and those for plutonium tribromide imply a melting point of 655° compared with the observed value, 681°. The estimated temperatures necessary to give 2 mm Hg vapour pressure of the lanthanide tribromides are shown in Table 4.7 (cf. the estimated and measured values for the trichlorides, p. 151).

Dawson[47] has reported magnetic susceptibility data for uranium tribromide over the temperature range 90 to 483°K. His results are very similar to those reported by others for uranium trichloride (p. 157) and it can be confidently predicted that the electronic configuration of the U^{3+} ion is $5f^3$. The effective magnetic moment is 3.29 B.M. ($\theta = 25°$). The magnetic properties of samarium tribromide have also been investigated[71]. Recently the vapour phase spectra of the tribromides of praseodymium, neodymium, erbium and thulium have been reported and discussed in terms of electronic transitions[126].

Obviously there is much scope for research into the chemical and physical properties of the tribromides, particularly those of neptunium, americium and curium which could now be prepared in larger amounts than has previously been possible. It is also probable that the tribromides of thorium and protactinium could be characterized by $M–MBr_4$ reactions or the latter even by heating protactinium hydride with the stoicheiometric amount of hydrogen bromide.

Trivalent Bromo Complexes

Complex bromides of the trivalent lanthanides and actinides have been little investigated. The interaction of tetraphenylphosphonium bromide and lanthanum and cerium tribromide is said[104] to yield complex bromides and only recently Ryan and Jørgensen[105] have prepared the series of complexes $(Ph_3PH)_3M^{III}Br_6 \cdot Ph_3PHBr$ ($M^{III} = La$, Pr, Nd, Sm and Eu) by reacting the tribromides with triphenylphosphonium bromide in ethanol saturated with hydrogen bromide. In solution these complexes give rise to the MBr_6^{3-} ions. The pyridinium complex $(C_5H_5NH)_3NdBr_6$ is obtained in a similar manner and other hexabromo complexes have been prepared[121] by condensing anhydrous hydrogen bromide onto the appropriate hexachloro complex. Spectral studies have been reported but magnetic susceptibility and infrared data are lacking for such complexes.

Trivalent Oxybromides

Compounds of the type MOBr have been characterized for scandium[106], yttrium[107], all the lanthanides[107,108,118], actinium[85] and plutonium[90] (Table 4.8). Yttrium and the lanthanide oxybromides are readily obtained[107] either by heating the appropriate tribromide hydrate at 650–700°

TABLE 4.8

Crystallographic Data for the Oxytribromides[85,107,110,118]

Compound	Lattice parameters (Å)		M–4Br in same layer (Å)	M–Br in next layer (Å)
	a_0	c_0		
LaOBr	4.159	7.392	3.29	3.55
CeOBr	4.138	7.487	3.29	3.59
PrOBr	4.071	7.487	3.25	3.59
NdOBr	4.024	7.597	3.23	3.65
PmOBr	3.980	7.560	—	—
SmOBr	3.950	7.909	3.21	3.80
EuOBr	3.908	7.973	3.19	3.83
GdOBr	3.895	8.116	3.20	3.90
TbOBr	3.891	8.219	3.21	3.95
DyOBr	3.867	8.219	3.19	3.95
HoOBr	3.832	8.241	3.17	3.96
ErOBr	3.821	8.264	3.16	3.97
TmOBr	3.806	8.288	3.16	3.98
YbOBr	3.780	8.362	3.15	4.01
LuOBr	3.770	8.387	3.15	4.03
YOBr	3.838	8.241	3.18	3.96
AcOBr	4.270	7.400	—	—
PuOBr	4.014	7.556	3.21	

in air (La to Sm, excepting Ce) or at 450–500° in a stream of nitrogen saturated with water vapour (Y, Ce, Eu to Lu). Thermogravimetric studies[108] have shown that the tribromide hydrates, $PrBr_3 \cdot 7H_2O$ and $MBr_3 \cdot 6H_2O$ (M = Y, Nd–Lu) decompose on being heated in air by one of two alternative routes, viz.:

$$(1) \quad MBr_3 \cdot xH_2O \ (x = 6 \text{ or } 7) \rightarrow MBr_3 \cdot H_2O \rightarrow MBr_3 \rightarrow \qquad (5)$$

$$MOBr \rightarrow M_2O_3 \ (M = Pr, Nd, Sm \text{ and } Eu)$$

$$(2) \quad MBr_3 \cdot 6H_2O \rightarrow (MBr_3 + MOBr) \rightarrow MOBr \rightarrow M_2O_3 \qquad (6)$$

$$(M = Gd–Lu)$$

The stability of the tribromide phase decreases with increasing atomic number of the lanthanide element and increasing contamination by the oxybromide is observed. $LaBr_3 \cdot 7H_2O$ was found to decompose first to the tribromide following which LaOBr and La_2O_3 were successively formed. During these thermogravimetric studies no evidence was found for the existence of CeOBr; the hydrated salt appeared to decompose through cerium tribromide to the dioxide.

Actinium oxybromide, AcOBr, has been prepared[85] only on the microgram scale, and identified crystallographically. It is obtained by heating the tribromide in moist ammonia at 500°. The analogous plutonium compound is conveniently prepared[90] by heating the dioxide at 750° in moist hydrogen bromide or, less satisfactorily, by hydrolysis of the tribromide.

Crystal structures. The oxybromides possess the lead fluorochloride-type structure[107,109,110] and there is no evidence for a second crystallographic form as observed for the lanthanide oxychlorides (p. 161). The structure comprises layers, each of which has a central sheet of coplanar oxygen atoms with a sheet of bromine atoms on each side the metal atoms being situated between the Br–O–Br sheets. The metal atom within these layers is surrounded by four oxygen atoms and four bromine atoms. Table 4.8 lists the known metal–bromine distances and unit cell parameters.

New oxybromide phases of the type M_3O_4Br have recently been reported[111] for trivalent samarium, europium and ytterbium. The first two are obtained when the oxide, M_2O_3, and the oxybromide, MOBr, are heated together at 950° and the last by heating YbOBr at 440–450°. These compounds possess orthorhombic symmetry, space group D_{2h}^{17}–$Bbmm$ or C_{2v}^{12}–$Bb2_1m$ (Table 4.9).

TABLE 4.9

Crystallographic Properties of the Oxybromides[111] of the Type M_3O_4Br

Compound	Lattice dimensions (Å)			Density (g cm^{-3})
	a_0	b_0	c_0	
Sm_3O_4Br	12.049	11.928	4.141	6.640
Eu_3O_4Br	11.986	11.854	4.124	6.799
Yb_3O_4Br	11.507	11.406	4.010	8.367

DIVALENT

The dibromides of samarium, europium and ytterbium have been known for many years and more recently evidence for the existence of thulium dibromide has been obtained. An intermediate phase $PrBr_{2.38}$ can be isolated from the Pr–$PrBr_3$ system and there is some evidence for the existence of neodymium dibromide. Lower bromides of the actinide elements are unknown; the available information[64] on $ThBr_2$ is unreliable, particularly since silica apparatus was used.

Dibromides

Samarium, europium and ytterbium dibromide are conveniently prepared by hydrogen reduction of the corresponding tribromide, for example, see references 67, 69, 71, 112, and 113, although undoubtedly reduction with the metal itself or possibly with zinc (cf. the dichlorides, p. 162) would be a useful method.

Phase diagrams characterizing the M–MBr₃ systems have been reported[39,114] for La–LaBr₃, Ce–CeBr₃, Pr–PrBr₃ and U–UBr₃. No

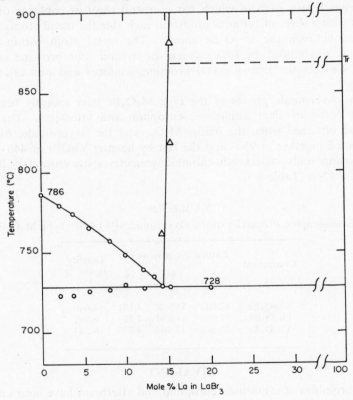

O Thermal analysis, cooling

△ Equilibration

Figure 4.4   The system LaBr₃–La.[114]   (*After* R. A. Sallach and J. D. Corbett, *Inorg. Chem.*, **2**, 457 (1963))

evidence of a lower uranium bromide is observed[39] and the lanthanum and cerium systems[114] are very similar to the corresponding M–MCl₃ systems (p. 163) indicating the absence of stable intermediate bromides. Lanthanum and cerium exhibit simple solution[114] of the metals with 14 and 12 mole % respectively dissolved at the eutectic temperatures of

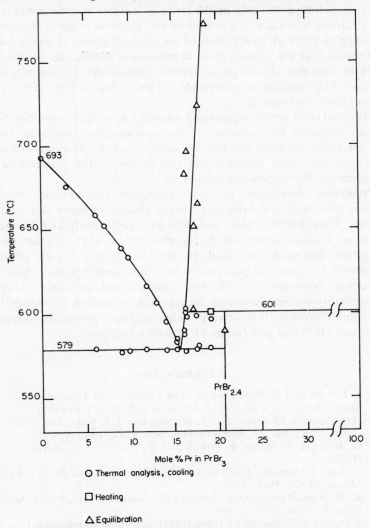

Figure 4.5   The system PrBr₃–Pr.[114] (*After* R. A. Sallach and J. D. Corbett, *Inorg. Chem.*, **2**, 457 (1963))

728 and 687°. These values are higher than those obtained in the analogous chloride systems, approximately 9 mole % in each case. The results obtained for the La–LaBr$_3$ system are illustrated in Figure 4.4. The praseodymium system (Figure 4.5) is also[114] similar to that observed with the corresponding trichloride (p. 165) in that an intermediate phase, PrBr$_{2.38}$, is formed but this, unlike PrCl$_{2.31}$, is stable at room temperature; the electrical resistance of a sample of this material suggests the possible presence of Pr^{2+}. A comparison of the liquidus curves (Figures 4.4 and 4.5) shows that the melting point depression of lanthanum tribromide is markedly less than that of praseodymium tribromide. Cerium tribromide behaves like lanthanum tribromide. These observations are, as yet, incompletely understood.

Electrical conductivity studies and solubility behaviour in the Nd–NdBr$_3$ system suggest[79] the existence of the non-metallic dibromide, NdBr$_2$, but the phase diagram has not yet been reported. Thulium dibromide has recently[115] been prepared but full details of the preparation and properties of this compound are not yet available.

Properties. Samarium (brown), europium (white) and ytterbium (green) dibromide are stable in air in the absence of water and oxidizing agents. They melt[94] at 669°, 683° and 613° respectively and the melting point of thulium dibromide is reported to be 619°. Samarium and europium dibromide are said to be[113] isostructural with strontium dibromide but unit cell parameters have not been reported; ytterbium dibromide possesses[113] a different, undetermined structure. Magnetic susceptibility studies show that, as expected, samarium dibromide[71] and europium dibromide[112] have very similar magnetic properties to trivalent europium ($4f^6$) and gadolinium ($4f^7$) salts respectively.

REFERENCES

1. J. J. Katz and E. Rabinowitch, 'The Chemistry of Uranium', *Nat. Nucl. Energy Ser.*, *Div. VIII*, Vol. 5, McGraw-Hill, New York, 1951.
2. 'The Actinide Elements' (G. T. Seaborg and J. J. Katz, Eds.), *Nat. Nucl. Energy Ser.*, *Div. IV*, McGraw-Hill, New York, **14A**, 1954.
3. O. Johnson, T. Butler, J. Powell and R. Nottorf, U.S. Report CC-1974 (1955).
4. J. Huré, in *Nouveau Traité de Chimie Minérale* (P. Pascal, Ed.), Vol. XV, Masson et Cie, Paris, 1961.
5. K. W. Bagnall in *Halogen Chemistry* (V. Gutmann, Ed.), Vol. 3, Academic Press, London, 1967, p. 303.
6. J. Powell, U.S. Report CC-1504 (1944), according to reference 1, p. 590.
7. J. Powell and R. C. Nottorf, U.S. Reports CC-1496 and 1500 (1944), according to reference 1, p. 590.

8. F. H. Spedding, A. S. Newton, R. Nottorf, J. Powell and V. Calkins, U.S. Report TID-5290, p. 91 (1958).
9. S. A. Shchukarev, I. V. Vasil'kova and V. M. Drozdova, *Zh. Neorgan. Khim.*, **3**, 2651 (1958).
10. J. Prigent, *Ann. Chim. (Paris)*, **5**, 65 (1960); *Compt. Rend.*, **238**, 102 (1954).
11. S. Peterson, *J. Inorg. Nucl. Chem.*, **17**, 135 (1961).
12. R. Sendtner, *Ann.*, **195**, 325 (1879).
13. J. P. Day, Thesis, Oxford University (1965).
14. G. V. Ellert, V. V. Tsapkin, Yu. N. Mikhailov and V. G. Kuznetsov, *Russ. J. Inorg. Chem.*, **10**, 858 (1965).
15. H. Loebel, Dissertation, Berlin University (1907).
16. J. Lucas, *Rev. Chim. Minérale*, **1**, 479 (1964).
17. J. Prigent and J. Lucas, *Compt. Rend.*, **253**, 474 (1961).
18. J. G. Allpress and A. D. Wadsley, *Acta Cryst.*, **17**, 41 (1964).
19. J. Prigent and J. Lucas, *Compt. Rend.*, **251**, 388 (1960).
20. Yu. N. Mikhailov, V. G. Kuznetsov and E. S. Kovaleva, *J. Struct. Chem. (USSR)*, **6**, 752 (1965).
21. A. G. Maddock and D. J. Toms, unpublished observations.
22. D. Brown and P. J. Jones, *J. Chem. Soc. (A)*, **1966**, 262.
23. D. Brown and P. J. Jones, *J. Chem. Soc. (A)*, **1967**, 247.
24. J. Prigent, *Compt. Rend.*, **239**, 424 (1954).
25. G. Kaufmann and R. Rohmer, *Bull. Soc. chim. France*, **1961**, 1969.
26. D. Brown, unpublished observations.
27. J. Prigent, *Compt. Rend.*, **236**, 710 (1953).
28. J. C. Levet, *Compt. Rend.*, **260**, 4775 (1965).
29. D. Brown, A. J. Petcher and A. J. Smith, *Acta Cryst.*, (1968) in press; *Nature*, **217**, 737 (1968).
30. L. Brewer, L. Bromley, P. W. Giles and N. Logfren, in 'The Transuranium Elements' (G. T. Seaborg, J. J. Katz and W. H. Manning, Eds.), *Nat. Nucl. Energy Ser., Div. IV*, McGraw-Hill, New York, **14B**, 861 (1949).
31. J. L. Ryan and C. K. Jørgenson, *Mol. Phys.*, **7**, 17 (1963).
32. Reference 2, p. 84.
33. L. F. Nilsen, *Ber.*, **9**, 1142 (1876); **15**, 2537 (1882); **16**, 153 (1883); *Compt. Rend.*, **95**, 727 (1882).
34. H. Moissan and A. Étard, *Compt. Rend.*, **122**, 573 (1896); *Ann. Chim. Phys.*, **12**, 427 (1897).
35. K. W. Bagnall, D. Brown and P. J. Jones, *J. Chem. Soc. (A)*, **1966**, 1743.
36. H. Lipkind and A. S. Newton, U.S. Report TID-5223, p. 398 (1952).
37. C. Zimmerman, *Ann.*, **216**, 2 (1882).
38. N. W. Gregory, U.S. Report TID-5290, p. 498 (1958).
39. J. D. Corbett, R. J. Clark and T. F. Munday, *J. Inorg. Nucl. Chem.*, **25**, 1287 (1963).
40. D. Brown and P. J. Jones, *Chem. Commun.*, **1966**, 280; *J. Chem. Soc. (A)*, **1967**, 719.
41. S. Fried and N. Davidson, reference 2, p. 472; *J. Am. Chem. Soc.*, **70**, 3539 (1948).
42. L. Brewer, L. Bromley, P. W. Gilles and N. Logfren, Met. Prog. Report CN-3306 (1945); and in 'The Transuranium Elements' (G. T. Seaborg,

J. J. Katz and W. H. Manning, Eds.), *Nat. Nucl. Energy Ser.*, *Div. IV*, McGraw-Hill, New York, **14B**, 1117 (1949).
43. R. W. M. D'Eye, *J. Chem. Soc.*, **1950**, 2764.
44. R. M. Douglas and E. Staritzky, *Anal. Chem.*, **29**, 459 (1957).
45. D. E. Scaife, *Inorg. Chem.*, **5**, 162 (1965).
46. G. W. Watt, W. A. Jenkins and J. M. McGuiston, *J. Am. Chem. Soc.*, **72**, 2260 (1950).
47. J. K. Dawson, *J. Chem. Soc.*, **1951**, 429.
48. N. G. Rambidi, P. A. Akashin and E. Z. Zasorin, *Russ. J. Phys. Chem.*, **35**, 375 (1961).
49. K. W. Bagnall, D. Brown, P. J. Jones and J. G. H. du Preez, *J. Chem. Soc. (A)*, **1966**, 737.
50. J. P. Day and L. M. Venanzi, *J. Chem. Soc. (A)*, **1966**, 197.
51. D. Brown, *J. Chem. Soc. (A)*, **1966**, 766.
52. J. Aloy, *Ann. Chim. Phys.*, **24**, 412 (1901).
53. R. A. Satten, C. L. Schrieber and E. Y. Wong, *J. Chem. Phys.*, **42**, 162 (1965).
54. A. Rosenheim, V. Samter and J. Davidsohn, *Z. Anorg. Chem.*, **35**, 324 (1903).
55. A. Rosenheim and J. Schilling, *Ber.*, **33**, 977 (1900).
56. C. K. Jørgensen, *Acta Chem. Scand.*, **17**, 251 (1963).
57. D. Brown and P. J. Jones, *J. Chem. Soc. (A)*, **1967**, 243.
58. H. Moissan and A. Martinsen, *Compt. Rend.*, **140**, 1510 (1905).
59. E. Chauvenet, *Ann. Chim. Phys.*, **23**, 425 (1911).
60. F. Bourion, *Compt. Rend.*, **145**, 243 (1906).
61. J. Powell and R. W. Nottorf, U.S. Report CC-1778 (1944).
62. K. W. Bagnall, D. Brown and J. F. Easey, *J. Chem. Soc. (A)*, **1968**, 288.
63. W. H. Zachariasen, in 'The Transuranium Elements' (G. T. Seaborg, J. J. Katz and W. H. Manning, Eds.), *Nat. Nucl. Energy Ser.*, *Div. IV*, McGraw-Hill, New York, **14B**, 1473 (1949).
64. E. Hayek, Th. Rehner and A. Frank, *Monatsh.*, **82**, 575 (1951).
65. G. I. Novikov and A. V. Suvorov, *Zh. Neorgan. Khim.*, **1**, 1948 (1956), according to *Chem. Abs.*, **51**, 5610c.
66. C. Matignon, *Compt. Rend.*, **140**, 1637 (1905).
67. W. Klemm and J. Rockstroh, *Z. Anorg. Chem.*, **176**, 181 (1928).
68. F. Ephraim and R. Block, *Ber.*, **61B**, 65 (1928).
69. W. Prandtl and H. Kögel, *Z. Anorg. Chem.*, **172**, 265 (1928).
70. G. P. Baxter and E. E. Behrens, *J. Am. Chem. Soc.*, **54**, 591 (1932).
71. P. W. Selwood, *J. Am. Chem. Soc.*, **56**, 2392 (1934).
72. E. R. Harrison, *J. Appl. Chem. (London)*, **2**, 601 (1952).
73. F. Ephraim and P. Ray, *Ber.*, **62B**, 1520 (1929).
74. G. Jantsch, H. Jawarek, N. Skalla and H. Gawalowski, *Z. Anorg. Chem.*, **207**, 353 (1932).
75. M. D. Taylor and C. P. Carter, *J. Inorg. Nucl. Chem.*, **24**, 387 (1962).
76. C. V. Banks, O. N. Carlson, A. H. Daane, V. A. Fassel, R. W. Fischer, E. H. Olsen, J. E. Powell and F. H. Spedding, U.S. Report IS-1 (1959).
77. N. H. Kiess, *J. Res. Nat. Bur. Std.*, **67A**, 343 (1963).
78. G. Jantsch, N. Skalla and H. Jawarek, *Z. Anorg. Chem.*, **201**, 219 (1931).
79. A. S. Dworkin, H. R. Bronstein and M. A. Bredig, *J. Phys. Chem.*, **67**, 2715 (1963).

80. F. E. Block and T. T. Campbell, in *The Rare Earths* (F. H. Spedding and A. H. Daane, Eds.), Wiley, New York, 1961, Chap. 7.
81. M. F. Bourion, *Compt. Rend.*, **145**, 243 (1907).
82. M. Barre, *Bull. Soc. Chim. France*, **11**, 433 (1912).
83. T. Moeller and G. Griffin, P.B. Report 136.162 (1959).
84. G. W. Cullen, Thesis, University of Illinois (1956).
85. S. Fried, F. Hagemann and W. H. Zachariasen, *J. Am. Chem. Soc.*, **72**, 771 (1950).
86. S. Fried, *J. Am. Chem. Soc.*, **73**, 416 (1951).
87. L. B. Asprey, T. K. Keenan and F. H. Kruse, *Inorg. Chem.*, **4**, 985 (1965).
88. E. D. Eastman, B. J. Fontana and R. A. Webster, U.S. Report TID-5290 190 (1958).
89. Reference 2, p. 388.
90. N. R. Davidson, F. Hagemann, E. K. Hyde, J. J. Katz and I. Sheft, in 'The Transuranium Elements' (G. T. Seaborg, J. J. Katz and W. H. Mannings, Eds.) *Nat. Nucl. Energy Ser.*, *Div. IV*, McGraw-Hill, New York, **14B**, 759 (1949).
91. V. V. Fomin, V. E. Reznikova and L. L. Zaïtseva, *Zh. Neorgan. Khim.*, **3**, 2231 (1958).
92. B. A. Bluestein and G. S. Garner, U.S. Report LA-116 (1944), according to reference 90.
93. J. G. Reavis, K. W. R. Johnson, J. A. Leary, A. N. Morgan, A. E. Ogard and K. A. Walsh, *Extract Phys. Met. Plutonium Alloys Symp.*, San Fransisco, Calif., p. 89 (1959).
94. F. H. Spedding and A. H. Daane, *Metallurgical Rev.*, **5**, 297 (1960).
95. J. J. Katz and I. Sheft, in *Advances in Inorganic and Radiochemistry* (H. J. Emeléus and A. G. Sharpe, Eds.), Vol. 2, Academic Press, New York, 1951, p. 195.
96. A. A. Men'kov and L. N. Kommissarova, *Zh. Neorgan. Khim.*, **9**, 1759 (1964).
97. W. H. Zachariasen, *Acta Cryst.*, **1**, 265 (1948); U.S. Report TID-5212, p. 157 (1955).
98. R. Müller, *Z. Elektrochem.*, **38**, 227, 232 (1932).
99. S. Herzog, K. Gustav, E. Kruger, H. Oberender and R. Schuster, *Z. Chem.* **3**, 428 (1963).
100. Reference 1, p. 521.
101. I. V. Tananaev and V. S. Orlovskiï, *Zh. Neorgan. Khim.*, **7**, 2022, 2299 (1962); **8**, 1104 (1963).
102. E. Shimazaki and K. Niwa, *Z. Anorg. Chem.*, **314**, 21 (1962).
103. W. Fischer, R. Gewehr and H. Wingchen, *Z. Anorg. Chem.*, **242**, 161 (1939).
104. N. N. Sakharova, *Uch. Zap. Saratovsk. Gos. Univ.*, **75**, 12 (1962), according to *Chem. Abs.*, **60**, 5055b.
105. J. L. Ryan and C. K. Jørgensen, *J. Phys. Chem.*, **70**, 2845 (1966).
106. F. Petru and F. Kůtek, *Collection Czech. Chem. Commun.*, **25**, 1143 (1960), according to *Chem. Abs.*, **54**, 12856i.
107. I. Mayer, S. Zolotov and F. Kassierer, *Inorg. Chem.*, **4**, 1637 (1965).
108. I. Mayer and S. Zolotov, *J. Inorg. Nucl. Chem.*, **27**, 1905 (1965).
109. L. G. Sillén and A. L. Nylander, *Svensk. Kem. Tidskr.*, **53**, 367 (1941).
110. W. H. Zachariasen, *Acta Cryst.*, **2**, 388 (1949).

111. V. H. Bärnighausen, G. Brauer and N. Schultz, *Z. Anorg. Chem.*, **338**, 250 (1965).
112. W. Klemm and W. Döll, *Z. Anorg. Chem.*, **241**, 233 (1939).
113. W. Döll and W. Klemm, *Z. Anorg. Chem.*, **241**, 239 (1939).
114. R. A. Sallach and J. D. Corbett, *Inorg. Chem.*, **2**, 457 (1963).
115. A. H. Daane, personal communication (1966).
116. D. Brown, T. S. Petcher and A. J. Smith, *Acta Cryst.*, (1968) in press; Nature **217**, 738 (1968).
117. D. Brown and J. F. Easey, unpublished observations (1967).
118. F. Weigel and V. Scherrer, *Radiochimica Acta*, **7**, 40 (1967).
119. F. H. Spedding and A. H. Daane, U.S. Report IS-350, p. 30 (1961).
120. M. E. Kost, *Zh. Neorgan. Khim.*, **2**, 2689 (1957).
121. J. L. Ryan, unpublished observations (1967).
122. D. Brown, S. Fletcher, and D. G. Holah, *J. Chem. Soc.* (A) (1968) in press.
123. K. Rossmanith, *Monatsh.*, **97**, 1357 (1966).
124. D. Brown, D. G. Holah and C. E. F. Rickard, *Chem. Commun.* **1968**, 651.
125. K. W. Bagnall 'Coordination Chemistry Reviews', **2**, 145 (1967).
126. D. M. Gruen, C. W. DeKock and R. L. McBeth in 'Lanthanide/Actinide Chemistry, *Advances in Chemistry Series* 71' (R. F. Gould, Ed.), *Amer. Chem. Soc. Publications*, Washington 1967, p. 102.

Chapter 5

Iodides and Oxyiodides

HEXAVALENT

Uranyl Iodide

The only actinide (VI) iodide for which there is evidence is uranyl iodide, UO_2I_2, and it is doubtful whether the pure anhydrous compound has ever been obtained. Solutions in aqueous or organic solvents are obtained[1-6] by metathesis of uranyl chloride, nitrate or sulphate with an appropriate soluble iodide. Evaporation of such an ethereal solution yields an unstable, red, deliquescent solid etherate which is soluble in water and various donor solvents, and which reacts[3,4] in ether with N,N-dimethylformamide to form the adduct $UO_2I_2 \cdot 4DMF$. The reported[7] triphenylphosphine adduct, $UO_2I_2 \cdot 2TPP$, has recently been shown[5,8] to be the phosphine oxide complex $UO_2I_2 \cdot 2TPPO$. Amine complexes have been reported[2] but this work has not been substantiated. Although the iodo complexes UO_2BiI_5 and UO_2HgI_4 have been claimed[9] the only reliable work on iodo complexes is that recently reported[5,6] by Day and Venanzi who prepared the complex $(Ph_3BuP)_2UO_2I_4$ from reaction in anhydrous methyl cyanide. The uranium–oxygen stretching vibration in this complex occurs at 925 cm^{-1} and in $UO_2I_2 \cdot 2TPPO$ at 943 cm^{-1}.

PENTAVALENT

Only protactinium is known to form a pentaiodide and pentavalent oxyiodides. In view of the instability of uranium tetraiodide (p. 213) it is unlikely that further actinide pentaiodides will be characterized.

Protactinium Pentaiodide

PaI_5, a black, crystalline solid, was first prepared[10] in milligram amounts by reacting the pentoxide with aluminium triiodide at 400° in a vacuum. This reaction is not satisfactory on a larger scale and the compound is best prepared[11] by direct union of the elements at 450° (cf. UI_4 by this

211

method, p. 213) or by metathesis of the pentachloride or pentabromide using excess silicon tetraiodide. The last reactions are conveniently carried out at about 180° *in vacuo* and the product is then purified by vacuum sublimation at 450°.

Protactinium pentaiodide is reported[12] to crystallize with orthorhombic symmetry, $a_0 = 7.22$, $b_0 = 21.20$, $c_0 = 6.85$ Å. It is extremely moisture-sensitive, rapidly hydrolysing, with hydroxide precipitation, in water; it dissolves in anhydrous methyl cyanide but is virtually insoluble in non-polar solvents such as isopentane and carbon tetrachloride. The hexaiodo complex $Ph_3MeAsPaI_6$ has been isolated[11] from methyl cyanide solution as have the analogous[13] hexabromoprotactinates (v).

Protactinium Oxyiodides

Protactinium oxytriiodide is best prepared[11] by reacting the pentaiodide with antimony trioxide in a vacuum at 150°. It is a dark brown solid which, like the oxytribromide $PaOBr_3$ (p. 187), is thermally unstable, disproportionation occurring above 450°,

$$2PaOI_3 \rightarrow PaI_5 + PaO_2I$$

The resulting protactinium dioxymonoiodide, PaO_2I, is obtained as a fluffy, yellow-brown solid from this reaction but when prepared[11] by reacting the pentaiodide with antimony trioxide at 150° it is a relatively dense material of much darker colour.

Both oxyiodides have been observed[11] as by-products during the preparation of the pentaiodide by reacting protactinium pentoxide and silicon tetraiodide at 600° in a vacuum, a reaction which gives a 70% yield of the pentaiodide. The metal–oxygen stretching vibrations are found at 480, 339 and 276 cm⁻¹ for $PaOI_3$ and at 555, 469, 386 and 281 cm⁻¹ for PaO_2I suggesting oxygen-bridged structures but this has not been confirmed by an x-ray structure analysis owing to the microcrystalline nature of the solids prepared in the above manner. It may, however, be possible to grow suitable crystals of $PaOI_3$ by prolonged heating in the presence of excess pentaiodide in a manner analogous to that employed for $PaOBr_3$.

TETRAVALENT

The decreasing stability of the higher oxidation states of the actinides coupled with the decreasing stability in progressing from fluorides to iodides limits the number of actinide tetraiodides which are stable and, as one might expect, there are no known lanthanide tetraiodides. Thorium, protactinium and uranium tetraiodide have been prepared and hexaiodo complexes of these three elements, $M_2^I M^{IV} I_6$ (M^I = variously Ph_3MeAs,

Ph_4As and NBu_4), are known. Thorium (IV) and protactinium (IV) oxydiiodides, MOI_2 (M = Th and Pa), have been reported but the uranium analogue, although probably capable of existence, is presently unknown.

Tetraiodides

Of the numerous methods, summarized recently elsewhere[14-18], used in the preparation of thorium and uranium tetraiodide undoubtedly the best is direct union of the elements. Thorium tetraiodide, which is obtained[19-22] in this manner at 300–500°, can also be prepared by the action of hydrogen iodide on the hydride[23] or by heating the metal in a mixture of hydrogen and iodine. However, more care is necessary in the preparation of pure uranium tetraiodide owing to the ease with which it undergoes decomposition to the triiodide and iodine and it is probable that the product described by Guichard[24] as being non-volatile and melting above 500° was the triiodide. The tetraiodide has, in fact, been observed[25] to evolve iodine when stored for several weeks in an atmosphere of dry argon at room temperature. Gregory[26] has described an apparatus suitable for the preparation of large amounts of pure tetraiodide in which the iodine pressure is maintained between 100 and 200 mm Hg and the uranium tetraiodide, obtained from the metal at 500°, is condensed at about 300°. At lower iodine vapour pressures the major product is uranium triiodide (p. 222). More recently[25] gram amounts of pure uranium tetraiodide have been prepared by heating metal turnings in the presence of excess iodine. The metal, at one end of a sealed evacuated tube, is maintained at 400–500°, whilst the remaining length of reaction vessel is heated to 140°. Under similar conditions improved yields of the thermally unstable neptunium tetrabromide have been achieved[104]. Uranium tetraiodide can also be prepared[17,26] by reaction of the triiodide with excess iodine.

As mentioned earlier (p. 211) iodine oxidizes protactinium metal to the pentaiodide; the tetraiodide has been prepared[29] by reduction of the last either by hydrogen, or better, aluminium at 450° in a vacuum. It is a dark green solid which reacts with silica on sublimation above 500° to yield small amounts of the pink oxydiiodide, $PaOI_2$.

Chaigneau claims[27,28] that both thorium and uranium tetraiodide can be obtained by heating the respective dioxides with aluminium triiodide, thorium dioxide at 230° and uranium dioxide at 400°, but since the uranium compound was sublimed[28] at 550° it would undoubtedly be mainly triiodide. He also claims[28] to have isolated a white, water soluble diiodide, ThI_2, from the residue remaining from the thoria reaction; a most surprising statement in view of other reports on the lower thorium iodides (p. 229). The product was probably the oxydiiodide, $ThOI_2$.

Crystal structures. No structural data are available for protactinium or uranium tetraiodide but single crystal studies show thorium tetraiodide to possess monoclinic symmetry[30] with $a_0 = 13.216$, $b_0 = 8.068$, $c_0 = 7.766$ Å, and $\beta = 98.68°$; the space group is C_{2h}^5–$P2_1/n$ with $n = 4$ and $\rho_{cal} = 6.00$ g cm^{-3}. In the novel structure each thorium atom is surrounded by eight iodine atoms, at an average distance of 3.20 Å, situated at the corners of a deformed square antiprism (Figure 5.1). The polyhedra share edges and triangular faces to form layers which are weakly bonded to each other. The x-ray powder data reported earlier by D'Eye and his associates[31] are consistent with the monoclinic cell described above but there is no simple relationship between it and the orthorhombic one proposed on the basis of the powder data. Thorium tetraiodide is not isostructural with any of the structurally characterized *d*-transition metal tetraiodides and it will be interesting to have information on the uranium and protactinium tetraiodides.

Properties. Thorium tetraiodide is a yellow, moisture-sensitive compound easily soluble in water, probably with associated hydrolysis, which melts at 566° and boils at 837°. The vapour pressure data for the liquid are represented by[22] the equation,

$$\log p_{mm} = \frac{-31,500}{4.57T} + 9.09$$

Vapour pressure data are not available for protactinium tetraiodide and such studies with the uranium compound are complicated by the dissociation to uranium triiodide and iodine. Determinations by the effusion method[17] indicate a pressure of 2.4×10^{-5} mm Hg at 300° rising to 4.25×10^{-2} mm Hg at 410°. Other workers have reported that their results can be represented[17] by the equation,

$$\log p_{mm} = -\frac{11,520}{T} + 15.53$$

Obviously these results can only be used as a rough indication of the volatility in view of the ready thermal decomposition of uranium tetraiodide. Its melting point is reported as 506° in an atmosphere of iodine (or[32] 518° under uncertain conditions) and the boiling point is estimated as 762°.

Uranium tetraiodide is a black solid which dissolves in water to give the characteristic green uranium (IV) solution. It is converted to U_3O_8 on heating in oxygen and although it has been suggested[33] that uranyl iodide is formed at room temperature this compound is itself unstable with respect to evolution of iodine. Carbon tetrachloride converts[34]

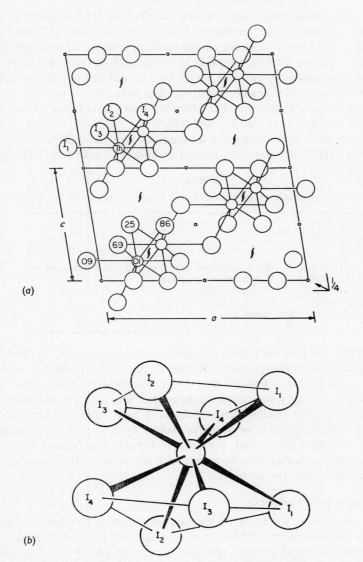

Figure 5.1 The structure of thorium tetraiodide[30]. (*a*) Projection of the ThI₄ structure onto *ac*-plane. The numbers on atoms in the lower part of the figure are *y*-coordinates (×100). (*b*) The approximately square antiprism arrangement of iodine about thorium in ThI₄. (*After* A. Zalkin, J. D. Forrester and D. H. Templeton. *Inorg. Chem.*, **3**, 639 (1964))

thorium tetraiodide to the tetrachloride at 100° but this obviously is not a suitable preparative method for the latter. Chlorine reacts with uranium tetraiodide at room temperature but the reaction products are not specified. Hydrogen reduces it to the triiodide at moderately elevated temperatures and the conversion to uranium metal by thermal decomposition using the hot-wire method[35] has been studied. The reduction of thorium tetraiodide by metallic thorium and the formation of stable lower iodides are discussed in detail later (pp. 222 and 228)

Like the other tetrahalides the tetraiodides form adducts with oxygen donor ligands[3,36,109] and with methyl cyanide[36] (Table 5.1). Thorium

TABLE 5.1

Complexes of Thorium and Uranium Tetraiodide

Complex	Properties
$ThI_4 \cdot 4CH_3CN$	White, insoluble methyl cyanide, moisture-sensitive
$ThI_4 \cdot 6DMA$	Yellow, moisture-sensitive
$UI_4 \cdot 4DMA$	Green, moisture-sensitive
$UI_4 \cdot 4DMF$	Green, stable in air (?)
$UI_4 \cdot 8CO(NH_2)_2$	Green

tetraiodide reacts[37] with potassamide in liquid ammonia to give an amidoiodide, $Th(NH_2)I_2 \cdot 3NH_3$. Rather surprisingly $UI_4 \cdot 4DMF$ is said[3] to be stable in air whereas[36] the DMA complexes, $ThI_4 \cdot 6DMA$ and $UI_4 \cdot 4DMA$ are very susceptible to moisture. The physical and chemical properties of protactinium tetraiodide have been little investigated. It reacts[29] with anhydrous, oxygen-free methyl cyanide to form a reddish-purple, slightly soluble solid, presumably $PaI_4 \cdot 4CH_3CN$, and is hydrolysed by oxygen-free aqueous ammonia with formation of the characteristic black protactinium (IV) hydroxide.

Tetravalent Iodo Complexes

The absorption spectrum of the hexaiodouranate (IV) ion in methyl cyanide has been recorded[38] and more recently solid hexaiodothorates (IV), protactinates (IV) and uranates (IV) have been prepared[25,39,107]. The yellow thorium, blue protactinium and red uranium complexes $M_2^I M^{IV} I_6$ ($M^I =$ variously NBu_4, Ph_3MeAs and Ph_4As) are obtained by reacting the tetraiodides with the appropriate cation iodide in anhydrous methyl cyanide. During the preparation of $(Ph_3MeAs)_2PaI_6$ an interesting series of colour changes occurs[39]. The dark green tetraiodide powder reacts initially with oxygen-free, anhydrous methyl cyanide to form a reddish-purple complex, presumably $PaI_4 \cdot 4CH_3CN$, which is slightly soluble in the solvent to form burgundy-coloured solutions. On the addition of Ph_3MeAsI such solutions turn an intense blue colour and vacuum evaporation of excess solvent leads to the crystallization of the blue hexaiodo complex. Ryan[107] has recently prepared hexaiodo complexes of thorium (IV) and uranium (IV) by condensing anhydrous liquid hydrogen iodide onto hexachloro- or hexabromo complexes. When prepared in this manner the hexaiodothorates (IV) are white and not yellow. This technique of halide replacement involving anhydrous liquid hydrogen halides has also been used for the preparation of numerous lanthanide and actinide chloro- and bromo complexes.

The hydrated complexes $Hg_2ThI_8 \cdot 12H_2O$ and $Hg_5ThI_{14} \cdot 18H_2O$ have been claimed[40] but this work has not been confirmed and iodo complexes are unlikely to be stable towards water.

Tetravalent Oxydiiodides

It is claimed[41] that $ThOI_2 \cdot 3.5H_2O$ and $ThOI_2$ have been isolated from the reaction of thorium hydroxide with alcoholic hydrogen iodide solution but the anhydrous oxydiiodide, $ThOI_2$, is more easily obtained in a pure state[42,43] by heating together thoria and thorium tetraiodide at 600° or[44] by reacting the latter with antimony trioxide. It is a white, hygroscopic solid readily soluble in water and hydrolytic solvents and it possesses a remarkably low bulk density. X-ray powder diffraction data[43] have not been interpreted but from the position[44] of the Th–O stretching vibrations 500 (*sh*), 444 (*m*) and 325 (*s*) cm^{-1} in the infrared spectrum it is likely that the structure comprises an infinite chain of thorium atoms linked by oxygen bridges. The heat of formation[43] of $ThOI_2$, calculated from the heat of solution, is -237.4 kcal/mole, being, as expected, lower than those measured for $ThOF_2$ (-398.8 kcal/mole) and $ThOCl_2$ (-295.4 kcal/mole) and that calculated for $ThOBr_2$ (-268.4 kcal/mole).

Protactinium oxydiiodide has been observed to form[29] during the sublimation of the tetraiodide in silica. It is a pink solid, isostructural with $ThOI_2$ and the Pa–O stretching vibrations are found at 517 (*m*), 444 (*m*) and 315 (*s*), positions similar to those observed for the thorium compound (Table B.2).

TRIVALENT

Pure triiodides have been characterized for scandium, yttrium and the lanthanides (Table 5.2) apart from promethium and europium; the latter owing to the instability of the triiodide with respect to the diiodide and

TABLE 5.2

Crystallographic Properties and Some Physical Data
for the Lanthanide Triiodides

Compound	Colour	m.p.[a 103] (°C)	Structure type	Lattice parameters (Å)[64,82]			Mol. vol. (Å³)
				a_0	b_0	c_0	
LaI$_3$	Grey	772	PuBr$_3$	4.37	14.01	10.04	153.6
CeI$_3$	Yellow	766	PuBr$_3$	4.341	14.00	10.015	152.2
PrI$_3$	—	737	PuBr$_3$	4.309	13.98	9.958	150.0
NdI$_3$	Light green	784	PuBr$_3$	4.284	13.979	9.948	148.9
(PmI$_3$)	—	(797)	—	—	—	—	—
SmI$_3$	Orange-yellow	850 d	BiI$_3$	7.49	—	20.80	168.4
(EuI$_3$)	—	d	—	—	—	—	—
GdI$_3$	Yellow	925	BiI$_3$	7.539	—	20.83	170.8
TbI$_3$	—	957	BiI$_3$	7.526	—	20.838	170.4
DyI$_3$	Dark green	978	BiI$_3$	7.488	—	20.833	168.6
HoI$_3$	Light yellow	994	BiI$_3$	7.474	—	20.817	167.8
ErI$_3$	Violet-red	1015	BiI$_3$	7.451	—	20.78	166.5
TmI$_3$	Bright yellow	1021	BiI$_3$	7.415	—	20.78	164.9
YbI$_3$	White	d	BiI$_3$	7.434	—	20.72	165.3
LuI$_3$	Brown	1050	BiI$_3$	7.395	—	20.71	163.5
YI$_3$	White	965	BiI$_3$	7.505	—	20.88	169.8
ScI$_3$	White	920	BiI$_3$	7.135	—	20.360	149.6

[a] Corbett[65,80] reports melting points of 997, 778, 787 and 931°C respectively for
YI$_3$, LaI$_3$. NdI$_3$, and GdI$_3$.

d = decomposes.

TABLE 5.3

Crystallographic Properties and Some Physical Data
for the Actinide Triiodides

Compound	Colour	m.p. (°C)	Structure type	Lattice parameters (Å)			Refer- ence
				a_0	b_0	c_0	
ThI$_3$	Black	—	—	—	—	—	21
PaI$_3$	Black	—	PuBr$_3$	4.33	14.00	10.02	106
UI$_3$	Black	766	PuBr$_3$	4.32	14.01	10.01	81
NpI$_3$	Brown	770	PuBr$_3$	4.30	14.03	9.95	81
PuI$_3$	Bright green	(777)	PuBr$_3$	4.33	13.95	9.96	64
AmI$_3$	Yellow	—	BiI$_3$	7.42	—	20.55	71
CmI$_3$	White	—	BiI$_3$	7.44	—	20.40	71

iodine and the former presumably owing to lack of investigation. With the increased stability of the lower valence states of the higher actinides we find that the elements thorium to curium inclusive form triiodides (Table 5.3) and with the availability of the actinide elements beyond curium further triiodides will undoubtedly be characterized. In view of the many similarities observed in the two series of triiodides it is convenient to discuss their preparation and properties in general terms making specific comments as necessary. However, since thorium triiodide presents a rather special case this will be dealt with separately. Complex iodides of the trivalent elements are unknown but several oxyiodides of the type MOI have been prepared.

Triiodides

Dehydration of lanthanide triiodide hydrates by heating them in a stream of hydrogen iodide, a procedure analogous to that used for the preparation of the trichlorides (p. 149) is not a useful method of preparation of the anhydrous triiodides. They can be prepared directly from the oxide by reaction with aluminium iodide[27,45] or ammonium iodide (e.g. references 46 and 47) but neither of these reactions is particularly effective owing to the low yields and simultaneous formation of the oxyiodides[48] although the latter reaction has been employed[49] to produce pure single crystals of certain lanthanide triiodides under carefully controlled conditions. Neptunium[50] and americium triiodide[51] have been obtained on the microgram scale by reacting the dioxides with aluminium iodide above 500° and a volatile compound believed to be actinium triiodide has been obtained[52] by reacting either aluminium or ammonium iodide with the oxide or oxalate at 500–700°. A recent modification of the method originally used by Jantsch and Skalla[53] to prepare the lanthanide triiodides involves[54] atmospheric evaporation of a hydriodic acid solution of the iodide containing 12 moles of ammonium iodide followed by careful removal of excess water and ammonium iodide by heating in a vacuum. Under these conditions the less stable samarium and europium triiodides are converted to the diiodides. Americium triiodide[64] has been prepared by heating the trichloride with ammonium iodide at 400° in hydrogen, a method also employed[71] for the curium compound.

A more general method for the preparation of the lanthanide triiodides, one of the oldest[55] and most satisfactory procedures[56–62], is the conversion of the trichloride by heating it in a mixture of hydrogen iodide and hydrogen or in hydrogen iodide alone at high temperature. Not unnaturally europium diiodide is obtained[57,60,61] under such conditions and extreme care is necessary to obtain pure samarium and ytterbium triiodides.

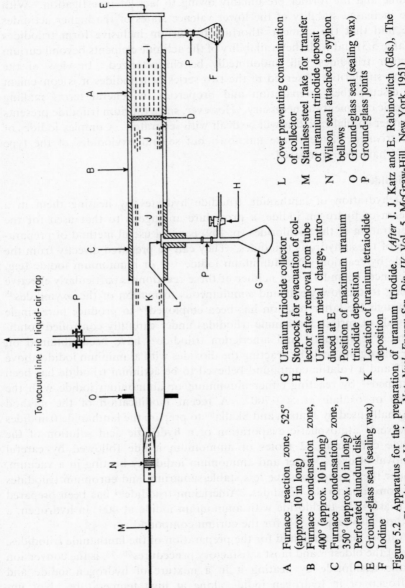

A Furnace reaction zone, 525°
   (approx. 10 in long)
B Furnace condensation zone,
   400° (approx. 10 in long)
C Furnace condensation zone,
   350° (approx. 10 in long)
D Perforated alundum disk
E Ground-glass seal (sealing wax)
F Iodine

G Uranium triiodide collector
H Stopcock for evacuation of collector after removal from tube
I Uranium metal charge, intro-
   duced at E
J Position of maximum uranium
   triiodide deposition
K Location of uranium tetraiodide
   deposition

L Cover preventing contamination
   of collector
M Stainless-steel rake for transfer
   of uranium triiodide deposit
N Wilson seal attached to syphon
   bellows
O Ground-glass seal (sealing wax)
P Ground-glass joints

Figure 5.2   Apparatus for the preparation of uranium triiodide. *(After J. J. Katz and E. Rabinowitch (Eds.), The
Chemistry of Uranium, Nat. Nucl. Energy Ser. Div. IV, Vol. 5, McGraw-Hill, New York, 1951)*

It is probable that the method recently employed to convert protactinium pentachloride to the pentaiodide (p. 212) could also be used for the preparation of the lanthanide and the more stable actinide triiodides.

With the availability of the pure metals, however, the most attractive preparations of the triiodides are those involving direct combination of

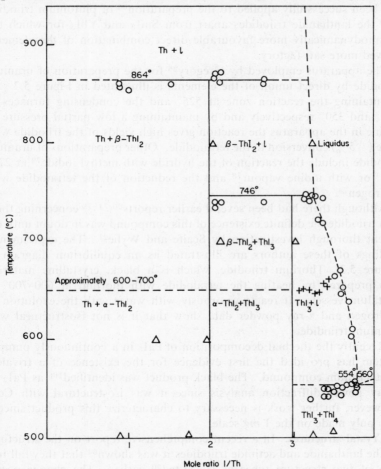

O Results for D.T.A; horizontals represent the average for many results, not all of which are shown.

△ Equilibrated samples quenched from these points.

Figure 5.3 The system thorium–iodine.[21] (*After* D. E. Scaife and A. W. Wylie, *J. Chem. Soc.*, **1964**, 5450)

the elements (e.g. lanthanides[63-66], uranium triiodide[17,26,32]), and reaction of the metal with ammonium iodide (e.g. YI_3[67]), hydrogen iodide (e.g. PuI_3[68,69]) or with mercuric iodide[64]. This last reaction, first used by Asprey and Kruse[70] to prepare TmI_2 (p. 227), is particularly attractive since the only by-product, mercury, is easily removed by vacuum distillation. It has been successfully applied to the preparation[64] of plutonium triiodide and the lanthanide triiodides apart from SmI_3 and YbI_3 for which the thermodynamically more favourable direct combination of the elements proved more satisfactory.

The apparatus employed by Gregory[26] for the preparation of uranium triiodide by direct union of the elements is illustrated in Figure 5.2. By maintaining the reaction zone at 525° and the condensing furnaces at 400° and 350° respectively and by maintaining a low partial pressure of iodine in the apparatus the reaction gives high yields of the triiodide with little ($<5\%$) conversion to the tetraiodide. Other preparations of uranium triiodide include the reaction of the hydride with methyl iodide[72] at 275–300° or with iodine vapour[73] and the reduction of the tetraiodide with hydrogen[74].

Although there had been several earlier reports[20,75-78] concerning thorium triiodide the definite existence of this compound was in doubt until the recent thorough investigation by Scaife and Wylie[21]. The experimental findings of these authors are illustrated as an equilibrium diagram in Figure 5.3. Thorium triiodide, which is a black, crystalline material, was prepared by heating the tetraiodide with thorium at 600–700° in tantalum vessels. It reacts vigorously with water with the evolution of hydrogen and x-ray powder data show that it is not isostructural with uranium triiodide.

Recently the thermal decomposition of PaI_5 in a continuously pumped system has provided the first evidence for the existence of a trivalent protactinium compound. The black product was identified[106] as PaI_3 by x-ray power diffraction analysis since it was isostructural with CeI_3. However, further work is necessary to characterize this product since it was only made on the 1 *mg* scale.

Crystal structures. In a recent comprehensive report on the structures of the lanthanide and actinide triiodides it was shown[64] that they fall into one of two structure types as suggested[60] earlier. The change occurs between NdI_3 and SmI_3 and between PuI_3 and AmI_3 in the two series (Tables 5.2 and 5.3). Others have reported[79] that neodymium triiodide undergoes a phase change at 574° and there is some evidence[80] that the gadolinium compound may behave similarly above 740°. The triiodides of the lighter elements in each series are orthorhombic, possessing the plutonium tribromide-type structure (p. 198), whilst the heavier members possess the close-packed hexagonal bismuth triiodide-type structure (Figure 5.4) in which each metal atom is at the centre of an almost perfect octahedron of halogen atoms. This structure change is, of course, consistent with the contraction in ionic radius occurring in each series. The unit cell data listed in Tables 5.2 and 5.3 (apart from those for PaI_3[106] UI_3[81], NpI_3[81] and ScI_3[82]) are taken from a single recent report[64] to permit

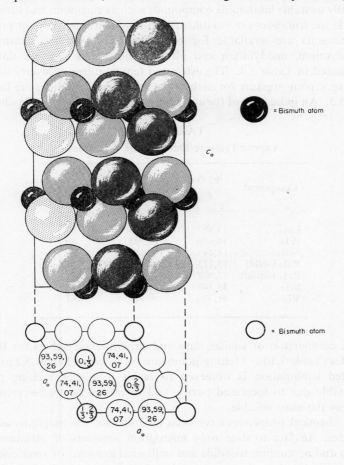

Figure 5.4 Two projections of the hexagonal cell of BiI_3. (*After* R. W. G. Wyckoff, *Crystal Structures*, Vol. 2, Wiley, New York, 1963)

a convenient comparison of molecular volumes. Slightly different parameters have been reported by others for YI_3[83], LaI_3[84], and PuI_3[84]. An
orthorhombic form of americium triiodide ($a_0 = 4.30$, $b_0 = 14.00$ and
$c_0 = 9.90$ Å) has been reported by Zachariasen[84] but was not observed by
Asprey and co-workers[64] during their recent study. Unfortunately
preparative details are lacking in the former case.

Properties. Apart from thorium and protactinium triiodide and the
thermally unstable lanthanide compounds such as europium and samarium
triiodide the triiodides are volatile at high temperatures. Vapour pressure
measurements are available for only scandium, yttrium, lanthanum,
praseodymium, neodymium and plutonium triiodide and the data are
summarized in Table 5.4. The estimated temperatures necessary to yield
2mm Hg vapour pressure for each of the lanthanide triiodides are listed in
Table 5.5. An indication of the accuracy of these estimates can be obtained

TABLE 5.4

Vapour Pressure Data for the Triiodides

Compound	$\log p_{mm} = -A/T + B$		Reference
	A	*B*	
LaI_3	15,397	12.845	89
PrI_3	14,640	12.703	89
NdI_3	14,495	12.475	89
PuI_3 (solid)	15,173	$29.18 - 5.035 \log T$	81
PuI_3 (liquid)	13,008	$30.16 - 6.042 \log T$	81
ScI_3	13,348	14.170	22
YI_3	11,706	9.540	67

from a comparison of similar data and experimental values for the trichlorides (Table 3.10). Melting points are recorded in Tables 5.2 and 5.3.
Repeated sublimation is observed to result in higher melting points,
presumably due to increased purity, and therefore the higher values are
probably the most reliable.

Few chemical properties have been recorded for the moisture-sensitive
triiodides. In fact to date only microgram amounts of actinium, plutonium and neptunium triiodide and milligram amounts of americium and
curium triiodide have been prepared and there is scope for investigation
into many aspects of the chemistry of the triiodides. For example, iodo
complexes analogous to the hexachloro and hexabromo salts (pp. 158
and 201) are still unknown.

TABLE 5.5
Estimated Temperatures for 2 mm Hg Vapour Pressure of the Lanthanide Triiodides[48]

Halide	Estimated temperature (°C)	Boiling point (°C)	Halide	Estimated temperature (°C)	Boiling point (°C)
LaI$_3$	910	1405	TbI$_3$	890	1330
CeI$_3$	940	1400	DyI$_3$	880	1320
PrI$_3$	870	1380	HoI$_3$	880	1300
NdI$_3$	860	1370	ErI$_3$	880	1280
PmI$_3$	860	1370	TmI$_3$	840	1260
SmI$_3$	d	d	YbI$_3$	d	d
EuI$_3$	d	d	LuI$_3$	800	1210
GdI$_3$	870	1340	YI$_3$	800	1310

d = decomposes.

The lanthanide triiodides are soluble in aqueous acid media and they react with dimethylformamide[85] (DMF) to form complexes of the type MI$_3 \cdot$8DMF (M = La, Pr, Nd, Sm and Gd). Uranium and neptunium triiodides are soluble in aqueous solutions and unless oxygen is excluded undergo oxidation to the tetravalent state; the former shows the transient red colour characteristic of trivalent uranium halide solutions.

Hydrogen reduction of europium, samarium and ytterbium triiodide yields their respective diiodides but americium triiodide, the most likely actinide compound to undergo reduction to a salt-like dihalide, is unaffected[64] by hydrogen at high temperatures. Similarly lanthanum, thulium and lutetium triiodide are not reduced in hydrogen. Metal–metal triiodide phase studies are discussed later (p. 227).

The magnetic susceptibility of uranium triiodide[86] exhibits Curie–Weiss dependence between 200° and 400°K but deviations are observed below 200°; $\theta = +5°$ and the effective magnetic moment is 3.31 B.M. At lower temperatures an antiferromagnetic transition has been[87] observed. Neodymium triiodide[88] has an effective magnetic moment of 3.72 B.M. ($\theta = 9°$) which is close to the theoretical value of 3.62 B.M. for the $^4I_{9/2}$ ground state of Nd^{3+}.

Trivalent Oxyiodides

Although few oxyiodides have been characterized it is likely that further lanthanide oxyiodides could be prepared and certain of the higher actinides may form such compounds as the stability of the M (III) state increases. Lanthanum oxyiodide, LaOI, has been obtained[90] in an impure state by heating the oxide with lithium, sodium or calcium iodide and evaporation of solutions of the oxide in hydriodic acid yields a poorly crystalline product which reverts to the oxide on being heated above 740°. Samarium, thulium and ytterbium oxyiodides[91] are obtained in a pure state by heating an 'evaporated' triiodide sample in moist air or alternatively by evaporating dry an aqueous solution of the triiodide, saturated with ammonium iodide, and heating the product in a vacuum at 550°. Oxidation of europium diiodide with moist oxygen yields[92] EuOI and interaction of Er_2O_3 and ErI_3 at 1050° leads to the formation[94] of ErOI. Hydrolysis of actinium triiodide at 700° with moist ammonia gas yields[52] what is believed to be an oxyiodide but only PuOI has been identified with certainty for the actinide elements. This is a bright green, volatile material, which was first observed during[68,69] various attempts to prepare the triiodide on the microgram scale. Yttrium oxyiodide has been observed in an impure state during[67] the dehydration of the triiodide hydrate in the presence of ammonium iodide.

It is obvious from the above preparations that no single method has been applied to the preparation of the trivalent oxyiodides. Possibly the triiodide–oxide interaction, e.g. Er_2O_3/ErI_3, may be useful in this respect but a more attractive procedure would be to heat the triiodide with antimony (or bismuth) trioxide, Sb_2O_3, as for the preparation of protactinium (v) oxyiodides (p. 212) and various tetravalent actinide oxyhalides (pp. 194 and 147).

The oxyiodides all crystallize with the tetragonal PbFCl-type of structure as do many of the other oxyhalides of the trivalent lanthanides and

TABLE 5.6

Crystallographic Data[a] for Lanthanide and Actinide (III) Oxyiodides

Compound	Colour	Symmetry	Lattice parameters (Å)		Reference
			a_0	c_0	
LaOI	White	Tetragonal	4.144	9.126	90
PuOI	Green	Tetragonal	4.034	9.151	93
PmOI	—	Tetragonal	4.000	9.180	105
SmOI	—	Tetragonal	4.008	9.192	91
EuOI	—	Tetragonal	3.993	9.186	92
TmOI	—	Tetragonal	3.887	9.166	91
YbOI	—	Tetragonal	3.870	9.161	92

[a] All possess the PbFCl-type of structure.

actinides. Unit cell dimensions are listed in Table 5.6; metal–iodine distances in LaOI, TmOI and PuOI are reported as 3.48, 3.28 and 3.44 Å respectively. Thulium oxyiodide again exhibits the relatively short halogen–halogen separation (3.89 Å) observed in the other isostructural oxyhalides.

Recently a few thiohalides have been reported[95,96] for certain of the lanthanide elements. These compounds are conveniently prepared by direct union of the appropriate elements at 500° in a sealed vessel. Cerium (III) thioiodide, CeSI, disproportionates above 800° to form CeI_3 and Ce_2S_3 and reacts at 500° with lead dichloride to form the thiochloride CeSCl and lead diiodide. The lanthanum and cerium compounds crystallize with orthorhombic symmetry, space group *Pcab* (Table 3.17). CeSI is reported to be dimorphic. The second form, which is monoclinic, is isostructural with PrSI, NdSI and SmSI.

DIVALENT

The diiodides of samarium, europium and ytterbium have been known for several years but thulium diiodide, which is isostructural with the last, neodymium diiodide and thorium diiodide were prepared only recently. Gadolinium diiodide and the metallic diiodides of lanthanum, cerium, praseodymium and thorium have also been characterized by $M-MI_3$ phase studies. Other divalent iodides of the lanthanides and actinides are unknown; attempted[64] hydrogen reduction of americium triiodide has so far met with no success.

Diiodides

The diiodides of samarium, europium and ytterbium are readily obtained[53,55,57,60-62,92,97] by hydrogen reduction or thermal decomposition of their respective triiodides and, in fact, attempts to prepare europium triiodide invariably yield the diiodide. Thulium triiodide is not reduced by hydrogen but it reacts with metallic thulium at 550° to form[70] the diiodide, which is a black, reactive solid evolving gas when dissolved in water in which a transient red colour is observed. Alternatively, thulium metal reacts with HgI_2 at 300–400° to form the diiodide. A new samarium diiodide phase which is deficient in iodine, $SmI_{1.90}$, has been observed on heating the diiodide with the metal; it is not, however, isostructural with neodymium diiodide, $NdI_{1.95}$, described below.

In addition to their work on the metal–metal trichloride and the analogous tribromide systems (pp. 162 and 204 respectively) Corbett and his collaborators have reported some interesting investigations[65,66,79,80,88,94,99,100,108] on the corresponding iodide systems. Their studies show that stoicheiometric diiodides of lanthanum, cerium and praseodymium exist[65,99,100] together with a phase intermediate between MI_2 and MI_3 in each case (e.g. Figure 5.5). Neodymium[66,79,88] forms a phase of composition $NdI_{1.95}$ (Figure 5.6), termed the diiodide, and an impure gadolinium phase, $GdI_{2.04}$ (Figure 5.7), has been observed[80]; the impurity of the latter is probably a consequence of the unfortunate proximity of the melting point of the phase, 831°, and the eutectic temperature, 825°.

Compounds designated ThI_2 have been reported by Hayek and colleagues[76], D'Eye and Anderson[20] and Watt and co-workers[78]. The first authors claimed that the triiodide disproportionated at 550° to yield the diiodide and the volatile tetraiodide whilst the others obtained their black product by reduction of the tetraiodide with metallic thorium. In view of the recent work on thorium diiodide, discussed below, and the

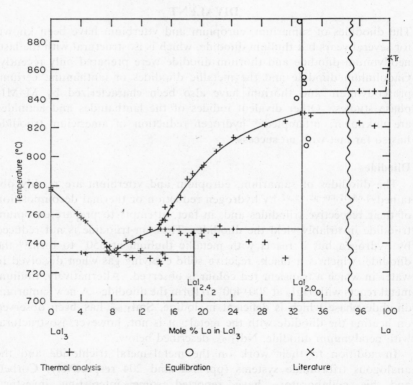

Figure 5.5   The system $LaI_3 + La$[99].   (*After* J. D. Corbett, L. F. Druding, W. J. Burkhard and C. B. Lindahl, *Discussions Faraday Soc.*, **32**, 79 (1962))

lack of physical properties, particularly x-ray powder diffraction data, in the report of Hayek and colleagues[76] it is not certain that they actually prepared the diiodide. Similarly the observations on which Watt and co-workers[78] based their conclusions, in particular the statements that thorium triiodide but not the diiodide reduces liquid ammonia and that the diiodide is soluble in dimethylformamide, indicate that these data should be treated with caution.

Recent publications by Clark and Corbett[101] and by Scaife and Wylie[21] have done much to clarify the confusion surrounding the existence of thorium diiodide. The former authors[101] reacted thorium tetraiodide

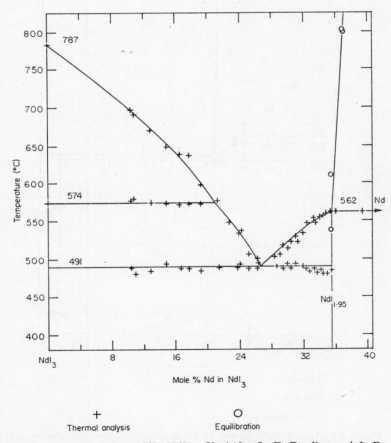

Figure 5.6 The system NdI_3–$NdI_{1.95}$.[79] (*After* L. F. Druding and J. D. Corbett, *J. Am. Chem. Soc.*, **83**, 2462 (1961))

with high purity thorium metal in sealed tantalum containers at 700–850° and obtained lustrous golden crystals of ThI₂ on the metal surfaces. In the presence of oxygen the oxydiiodide, ThOI₂, was formed. Scaife and Wylie[21] also carried out their investigations using sealed tantalum, or less satisfactorily platinum, vessels and examined the products after rapid quenching of the reaction. A diagrammatic representation of their results is shown in Figure 5.3. Two forms of the diiodide have been observed, the black, low-temperature, α-form undergoes a phase change to give a golden β-form between 600° and 700°. A peritectic decomposition,

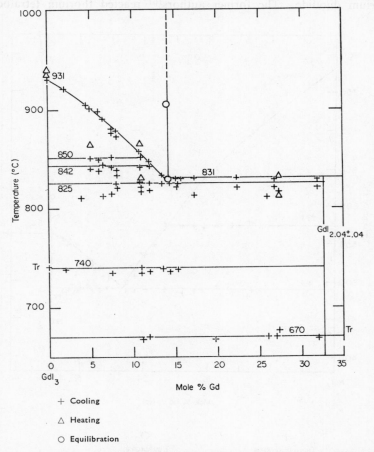

Figure 5.7 The salt-rich portion of the equilibrium phase diagram for the system GdI₃–Gd. The 831° and 670° horizontals extend to the right to pure metal. (*After* J. E. Mee and J. D. Corbett, *Inorg. Chem.*, **4**, 88 (1965))

$2ThI_2 \rightarrow ThI_4 + Th$, occurs at 864°. β-ThI_2 is identical with the phase reported by Clark and Corbett[101] and the α-phase, which is best prepared at 600°, has an x-ray powder diffraction pattern in close agreement with that previously reported for the diiodide by D'Eye and co-workers[31]. This pattern cannot, as suggested by Clark and Corbett[101], be attributed to the oxydiiodide $ThOI_2$.

Properties. Samarium and europium diiodide were reported[97] to be isostructural several years ago and the latter was recently shown[92] to possess monoclinic symmetry with the probable space group C_{2h}^5–$P2_1/c$. Thulium and ytterbium diiodide, however, possess[70] the hexagonal cadmium (II) hydroxide-type of structure (Figure 5.8). Neodymium

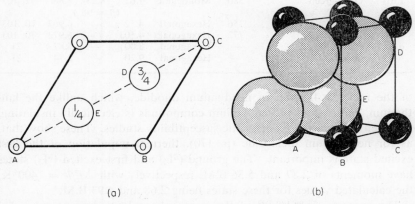

(a)   (b)

Figure 5.8 The structure of $Cd(OH)_2$. (*a*) A basal projection of the atomic positions within the hexagonal unit prism of the $Cd(OH)_2$ arrangement. Letters refer to the correspondingly marked atoms of (*b*). (*b*) A perspective packing drawing of the atomic arrangement in $Cd(OH)_2$. The large and small spheres have been given the relative sizes of the I′ and Cd^{2+} ions. (*After* R. W. G. Wyckoff, *Crystal Structures*, Vol. 1, Wiley, New York, 1963)

diiodide is isostructural[79] with the samarium and europium dibromides and GdI_2[80], α-ThI_2 and[21] β-ThI_2 have all been provisionally assigned hexagonal unit cells (Table 5.7) but no structural data are available.

Ytterbium diiodide, which is stable in the absence of water or oxidizing agents, is volatile above 750° in a vacuum. Samarium diiodide initially forms a red solution in water, with which it reacts with the evolution of hydrogen, and europium diiodide forms stable solutions in water (in the absence of oxygen). Above 600° samarium diiodide disproportionates to yield the metal and triiodide whereas europium diiodide is volatile above 800°. The magnetic moment[98] of the latter, 7.9 B.M. is the same as that of the dichloride and dibromide and of gadolinium (III) salts. The presence

TABLE 5.7

The Lanthanide Diiodides

Com-pound	Colour	m.p. (°C)	Symmetry	Lattice parameters (Å)			Refer-ence
				a_0	b_0	c_0	
LaI_2	Black	820	—	—	—	—	65
CeI_2	Dark bronze	808	—	—	—	—	65
PrI_2	Golden-bronze	758	—	—	—	—	65
NdI_2	Red-violet	562	—	—	—	—	79
GdI_2	Brass coloured	831	Hexagonal	8.67	—	5.75	80
SmI_2	Dark green or yellow	520	Monoclinic	—	—	—	103
EuI_2	Olive-green	580	Monoclinic	7.62	8.23	7.88	92, 103
					($\beta = 98°$)		
TmI_2	Black	756	Hexagonal	4.52	—	6.967	70, 103
YbI_2	Black	772	Hexagonal	4.503	—	6.972	70, 103
α-ThI_2	Black	—	Hexagonal	8.00	—	7.87	21
β-ThI_2	Golden	—	Hexagonal	9.20	—	3.74	21

of the reduced $4f^4$ ion in neodymium diiodide, which unlike the lanthanum, cerium and praseodymium compounds is electrically insulating, has been confirmed[88] by magnetic susceptibility studies. These show that, as for neodymium dichloride (p. 170), thermal population of the first excited state is important. The ground ($^5I_4$) and first excited ($^5I_5$) states have moments of 2.87 and 5.36 B.M. respectively with $\Delta E/k = 1400°K$, the calculated values for these states being 2.68 and 4.93 B.M.

The properties[65,99,100,108] of the isomorphous lanthanum, cerium and praseodymium diiodides, which are respectively black, dark bronze and golden-bronze in colour, indicate that they probably do not contain the M^{2+} ion but should be formulated as $M^{3+}(e^-)I_2^-$. Thus, they are all good electrical conductors (cf. GdI_2) with low specific resistances (6–8 × 10⁻⁴ ohms) and lanthanum diiodide has been shown to be only weakly paramagnetic ($\chi_m = 220 \pm 50 \times 10^{-6}$ e.m.u.). It is also noteworthy that they have melting points, 830°, 808° and 758° respectively, which are higher than those of the corresponding triiodides (772°, 766° and 737° respectively) whereas the salt-like neodymium, samarium and thulium diiodides melt at lower temperatures (562°, 520° and 756° respectively) than their triiodides (784°, 850° (d) and 1021° respectively). The variation in melting point and the degree of decomposition observed on fusion indicate that these diiodides decrease in stability from lanthanum to praseodymium. Their behaviour is similar to that observed with thorium diiodide (below) but the sharp change found at neodymium is not clearly understood at present.

The brass-coloured gadolinium diiodide[80] melts at 831° (cf. GdI_3, m.p. 931°) and possesses hexagonal symmetry (Table 5.7). Like europium diiodide[102] it exhibits ferromagnetism at low temperature and whilst it might be expected by analogy with cerium diiodide to contain the Gd^{3+} ion the reported resistance measurements appear to be inconclusive. Rather surprisingly[94] yttrium and erbium react with their molten triiodides without the formation of intermediate phases. Information on the $Sc–ScI_3$ system is not available at present.

It has been suggested[80] that one factor determining whether a salt-like $(M^{2+}(I^-)_2)$ or a 'metallic' $(M^{3+}(I^-)_2(e^-))$ diiodide is to be formed may be the electronic state of the reduced metal. Thus for the formation of a metal-like solid the divalent ion derived from a M^{3+} ion with a $4f^n$ configuration should probably be $(Xe)4f^n 5d^1$ in order for the extra electron to be delocalized into a conduction band which probably has a substantial amount of $5d$ character. In this light NdI_2 and the diiodides of Sm, Eu, Tm and Yb are salt-like simply because the added electron is buried in the $4f$ shell.

Both forms of thorium diiodide evolve hydrogen on dissolution in water, are hydrolysed in moist air and require careful hydrolysis in an atmosphere of ammonia in order to prevent loss of iodine during analysis. They disproportionate to metallic thorium and the tetraiodide above 550° in a vacuum and β-ThI_2 has been found[21] to be insoluble in dimethylformamide (cf. reference 78). The low specific resistance and diamagnetism $(\chi_m = (-80 \pm 20) \times 10^{-6}$ e.m.u.) exhibited by β-ThI_2 suggest[101] that the compound does not contain Th^{2+} ions but is best represented as $Th^{4+}(e^-)_2 I_2$ in a similar manner to the diiodides of lanthanum, cerium and praseodymium.

REFERENCES

1. J. Aloy, *Ann. Chim. Phys.*, **24,** 417 (1901).
2. A. von Unruh, Dissertation, Rostock (1909), according to Katz and Rabinowitch, reference 17, p. 596.
3. M. Lamisse, R. Heimburger and R. Rohmer, *Compt. Rend.*, 258, 2078 (1964).
4. M. Lamisse and R. Rohmer, *Bull. Soc. Chim. France*, **1963,** 24.
5. P. J. Day, Thesis, Oxford University (1965).
6. P. J. Day and R. Venanzi, *J. Chem. Soc. (A)*, **1966,** 1363.
7. A. K. Majumdar, A. K. Mukherjee and R. G. Bhattacharya, *J. Inorg. Nucl. Chem.*, **26,** 386 (1964).
8. B. W. Fitzsimmons, P. Gans, B. Hayton and B. C. Smith, *J. Inorg. Nucl. Chem.*, **28,** 915 (1966).
9. H. Truttwin, German Patent D.R.P. No. 420,391, according to Katz and Rabinowitch, reference 17, p. 596.

10. A. G. Maddock and D. J. Toms, unpublished observations.
11. D. Brown, J. F. Easey and P. J. Jones, *J. Chem. Soc. (A)*, **1967**, 1698.
12. A. G. Maddock, D. J. Toms and A. C. Fox, unpublished observations.
13. D. Brown and P. J. Jones, *J. Chem. Soc. (A)*, **1967,** 247.
14. J. Flahaut in *Nouveau Traité de Chimie Minérale* (P. Pascal, Ed.), Vol. IX, Masson et Cie, Paris, 1963, p. 1072.
15. J. Elston in *Nouveau Traité de Chimie Minérale* (P. Pascal, Ed.), Vol. XV, Masson et Cie, Paris, 1961, p. 209.
16. 'The Actinide Elements' (G. T. Seaborg and J. J. Katz, Eds.) *Nat. Nucl. Energy Ser. Div. IV*, McGraw-Hill, New York, **14A,** 85 (1954).
17. J. J. Katz and E. Rabinowitch, 'The Chemistry of Uranium', *Nat. Nucl. Energy Ser. Div. VIII*, McGraw-Hill, New York, **5,** 532 (1951).
18. K. W. Bagnall in *Halogen Chemistry* (V. Gutmann, Ed.), Vol. 3, Academic Press, London, 1967, p. 303.
19. H. Moissan and A. Étard, *Ann. Chim. Phys.*, **12,** 427 (1897).
20. J. S. Anderson and R. W. M. D'Eye, *J. Chem. Soc.*, **1949,** S.244.
21. D. E. Scaife and A. W. Wylie, *J. Chem. Soc.*, **1964,** 5450.
22. W. Fischer, R. Gewehr and H. Wingchen, *Z. Anorg. Chem.*, **242,** 161 (1939).
23. H. Lipkind and A. S. Newton, U.S. Report TID-5223, p. 398 (1952).
24. W. Guichard, *Compt. Rend.*, **145,** 921 (1907); *Bull. Soc. Chim. France*, (4), 3 (1908).
25. K. W. Bagnall, D. Brown, P. J. Jones and J. G. H. du Preez, *J. Chem. Soc.*, **1965,** 350.
26. N. W. Gregory, U.S. Report TID-5290, p. 465 (1958).
27. M. Chaigneau, *Compt. Rend.*, **242,** 263 (1956).
28. M. Chaigneau, *Bull. Soc. Chim. France*, **1957,** 886.
29. D. Brown and P. J. Jones, *J. Chem. Soc. (A)*, **1967,** 719.
30. A. Zalkin, J. D. Forrester and D. H. Templeton, *Inorg. Chem.*, **3,** 639 (1964).
31. R. W. M. D'Eye, I. F. Ferguson and E. J. McIver, *Proc. Intern. Congr. Pure Appl. Chem.*, *16th Paris* (1958), p. 341.
32. M. M. Popov and M. D. Senin, *Zh. Neorgan. Khim.*, **2,** 1479 (1957).
33. Reference 17, p. 538.
34. G. W. Watt and S. C. Malhotra, *J. Inorg. Nucl. Chem.*, **14,** 184 (1960).
35. C. H. Prescott, F. L. Reynolds and J. A. Holmes, U.S. Report TID-5290, p. 511 (1958).
36. K. W. Bagnall, D. Brown, P. J. Jones and J. G. H. du Preez, *J. Chem. Soc.*, **1965,** 3594.
37. G. W. Watt and S. C. Malhotra, *J. Inorg. Nucl. Chem.*, **11,** 255 (1959).
38. J. L. Ryan and C. K. Jørgensen, *Mol. Phys.*, **7,** 17 (1963).
39. D. Brown and P. J. Jones, *J. Chem. Soc. (A)*, **1967,** 243.
40. A. Duboin, *Ann. Chim. Phys.*, **16,** 258 (1909); *Compt. Rend.*, **146,** 1027 (1908).
41. E. Chauvenet, *Ann. Chim. Phys.*, **23,** 425 (1911).
42. E. Hayek, Th. Rehner and A. Frank, *Monatsh.*, **82,** 475 (1951).
43. D. E. Scaife, A. G. Turnbull and A. W. Wylie, *J. Chem. Soc.*, **1965,** 1432.
44. K. W. Bagnall, D. Brown, J. F. Easey, and P. J. Jones, *J. Chem. Soc. (A)*, **1968,** 188.
45. K. W. Foster, G. Pish, H. W. Schamp, J. M. Goode and T. E. Eyler, Report MLM 686 (1952).

46. B. S. Hopkins, J. B. Reed and L. F. Audrieth, *J. Am. Chem. Soc.*, **57**, 1159 (1935).
47. V. K. Val'tsev and L. K. Solov'er, *Izvest. Sib. Otdel Akad. Nauk SSSR*, **1960**, 80, according to *Chem. Abs.*, **54**, 19256i.
48. F. E. Block and T. T. Campbell, in *The Rare Earths* (F. H. Spedding and A. H. Daane, Eds.), Wiley, New York, 1961, Chap. 7.
49. N. H. Kiess, *J. Res. Nat. Bur. Std.*, **67A**, 343 (1963).
50. S. Fried and N. Davidson, *J. Am. Chem. Soc.*, **70**, 3539 (1948).
51. S. Fried, *J. Am. Chem. Soc.*, **73**, 416 (1951).
52. S. Fried, F. Hagemann and W. H. Zachariasen, *J. Am. Chem. Soc.*, **72**, 771 (1950).
53. G. Jantsch and N. Skalla, *Z. Anorg. Chem.*, **193**, 391 (1930).
54. M. D. Taylor and C. P. Carter, *J. Inorg. Nucl. Chem.*, **24**, 387 (1962).
55. C. Matignon, *Ann. Chim. Phys.*, **8**, 243 (1906); *Compt. Rend.*, **40**, 1637 (1905).
56. G. Jantsch, H. Grübitsch, F. Hoffmann and H. Alber, *Z. Anorg. Chem.*, **185**, 49 (1929).
57. G. Jantsch, *Naturwiss.*, **18**, 155 (1930).
58. G. Jantsch, H. Jawarek, N. Skalla and H. Gawalowski, *Z. Anorg. Chem.*, **207**, 353 (1932).
59. G. Jantsch, N. Skalla and H. Grübitsch, *Z. Anorg. Chem.*, **212**, 65 (1933).
60. E. Hohmann and H. Bommer, *Z. Anorg. Chem.*, **248**, 383 (1941).
61. G. Jantsch, H. Alber and H. Grübitsch, *Monatsh.*, **53/54**, 305 (1929).
62. G. Jantsch, N. Skalla and H. Jawarek, *Z. Anorg. Chem.*, **201**, 207 (1931).
63. A. Zaïdel and L. Lipis, *J. Expt. Theoret. Phys.* (*USSR*), **13**, 101 (1943).
64. L. B. Asprey, T. K. Keenan and F. H. Kruse, *Inorg. Chem.*, **3**, 1137 (1964).
65. J. D. Corbett, L. F. Druding and C. B. Lindahl, *J. Inorg. Nucl. Chem.*, **17**, 176 (1961).
66. L. F. Druding and J. D. Corbett, *J. Am. Chem. Soc.*, **81**, 5512 (1959).
67. C. V. Banks, O. N. Carlsen, A. H. Daane, V. A. Fassel, R. W. Fischer, E. H. Olsen, J. E. Powell and F. H. Spedding, U.S. Report IS-1 (1959).
68. F. Hagemann, B. M. Abraham, N. R. Davidson, J. J. Katz and I. Sheft, in 'The Transuranium Elements' (G. T. Seaborg, J. J. Katz and W. H. Manning, Eds.), *Nat. Nucl. Energy Ser. Div. IV*, McGraw-Hill, New York, **14B**, 957 (1949).
69. Reference 16, p. 391.
70. L. B. Asprey and F. H. Kruse, *J. Inorg. Nucl. Chem.*, **13**, 32 (1960).
71. L. B. Asprey, T. K. Keenan and F. H. Kruse, *Inorg. Chem.*, **4**, 985 (1965).
72. Reference 17, p. 536.
73. Reference 17, p. 535.
74. Reference 17, p. 538.
75. E. Hayek and Th. Rehner, *Experienta*, **5**, 114 (1949).
76. E. Hayek, Th. Rehner and A. Frank, *Monatsh.*, **82**, 575 (1951).
77. G. Jantsch, J. Homayr and F. Zemek, *Monatsh.*, **85**, 526 (1954).
78. G. W. Watt, D. M. Sowards and S. C. Malhotra, *J. Am. Chem. Soc.*, **79**, 4908 (1957).
79. L. F. Druding and J. D. Corbett, *J. Am. Chem. Soc.*, **83**, 2462 (1961).
80. J. E. Mee and J. D. Corbett, *Inorg. Chem.*, **4**, 88 (1965).
81. J. J. Katz and I. Shaft, in *Advances in Inorganic and Radiochemistry*,

(H. J. Emeleus and A. G. Sharpe, Eds.) Vol. 2, Academic Press, New York, 1960, p. 195.

82. A. A. Men'kov and L. N. Kommissarova, *Zh. Neorgan. Khim.*, **9**, 766 (1964).
83. B. H. Krause, A. B. Hook and F. Wawner, *Acta Cryst.*, **16**, 848 (1963).
84. W. H. Zachariasen, *Acta Cryst.*, **1**, 265 (1948).
85. T. Moeller and V. Galosyn, *J. Inorg. Nucl. Chem.*, **12**, 259 (1959).
86. J. K. Dawson, *J. Chem. Soc.*, **1951**, 429.
87. L. B. Roberts and R. B. Murray, *Phys. Rev.*, **100**, 650 (1955).
88. R. A. Sallach and J. D. Corbett, *Inorg. Chem.*, **3**, 993 (1964).
89. E. Shimazaki and K. Niwa, *Z. Anorg. Chem.*, **314**, 21 (1962).
90. L. G. Sillen and A. L. Nylander, *Svensk Kem. Tidskr.*, **53**, 367 (1941).
91. F. H. Kruse, L. B. Asprey and B. Morosin, *Acta Cryst.*, **14**, 541 (1961).
92. H. Baernighausen, *J. Prakt. Chem.*, **14**, 313 (1961); *Angew. Chem.*, **75**, 1109 (1963).
93. W. H. Zachariasen, *Acta Cryst.*, **2**, 388 (1949).
94. J. D. Corbett, D. L. Pollard and J. E. Mee, *Inorg. Chem.*, **5**, 761 (1966).
95. M. C. Dagron, *Compt. Rend.*, **260**, 1422 (1965).
96. M. C. Dagron, *Compt. Rend.*, **262**, 1575 (1966).
97. W. Klemm and W. Döll, *Z. Anorg. Chem.*, **241**, 239 (1939).
98. W. Döll and W. Klemm, *Z. Anorg. Chem.*, **241**, 233 (1939).
99. J. D. Corbett, L. F. Druding, W. J. Burkhard and C. B. Lindahl, *Discussions Faraday Soc.*, **32**, 79 (1962).
100. A. S. Dworkin, R. A. Sallach, H. R. Bronstein, M. A. Bredig and J. D. Corbett, *J. Phys. Chem.*, **67**, 1145 (1963).
101. R. J. Clark and J. D. Corbett, *Inorg. Chem.*, **2**, 460 (1963).
102. T. R. McGuire and M. V. Shafer, *J. Appl. Phys.*, **35**, 984 (1964).
103. F. H. Spedding and A. H. Daane, *Metallurgical Rev.*, **5**, 297 (1960).
104. D. Brown and J. F. Easey, unpublished observations.
105. F. Weigel and V. Scherrer, *Radiochim. Acta*, **7**, 40 (1967).
106. V. Scherrer, F. Weigel and M. Van Ghemen, *Inorg. Nucl. Chem. Letters*, **3**, 589 (1967).
107. J. L. Ryan, personal communication (1968).
108. J. D. Corbett, R. A. Sallach and D. A. Lokken in 'Lanthanide/Actinide Chemistry', *Advances in Chemistry Series* 71 (R. F. Gould, Ed.), *Amer. Chem. Soc. Publ.* Washington 1967, p. 56.
109. V. A. Golovnya and G. T. Bolotova, *Russ. J. Inorg. Chem.* **11**, 1419 (1966).

Appendix A

Thermochemical Properties

SCANDIUM, YTTRIUM AND THE LANTHANIDE HALIDES

Only limited experimental thermochemical data are available for these compounds. Selected values are listed in tabular form (Tables A.1–A.4) and these will be discussed only briefly.

Bommer and Hohmann first reported[1,2] the heats of formation for the trichlorides and triiodides but their values have recently been shown[3,4] to be in error. However, certain of their results are included in Table A.1 because no other experimentally determined values are yet available. Such values are probably too high by 5–10 kcal/mole. Estimated heats of formation for the trifluorides and tribromides are given in Table A.2 and similar data for the lanthanide dihalides are listed in Table A.3 together with the few experimental values available for certain of the dichlorides. Corbett and co-workers[5] have pointed out that the estimated values for the dichlorides[6,7] are based on incorrect values of the successive ionization constants and consequently are in error. Recalculation with present data[5] should lead to more reliable values.

Dworkin and Bredig[8] have reported the heat capacities and heats and entropies of fusion for certain of the lanthanide trihalides (Table A.4). Although there have been several reports[9-12] concerning the heats of vaporization of the trichlorides the results are not in agreement and further work is necessary to clarify the situation. Estimated values for other thermodynamic functions of the trihalides are available. These include estimates of the heats and entropies of fusion, heats and entropies of vaporization and the data necessary to estimate free energies of formation. This information is tabulated[13-16] in other books or reviews and will not be repeated here.

As discussed earlier (p. 161) the hydrolysis,

$$MCl_3(s) + H_2O(g) \rightleftharpoons MOCl(s) + 2HCl(g)$$

of several lanthanide trichlorides has been studied in detail by Koch and

Cunningham[17]. However, in view of the unreliable heats of formation of the lanthanide trichlorides available at that time no heats of formation were calculated for the oxychlorides. Such information could now be obtained by using the more accurate heats of formation reported[3,4] for certain of the lanthanide trichlorides.

ACTINIDE HALIDES

More experimental data are available for the actinide halides than for those of the lanthanide elements. The information on thorium and plutonium halides has recently been critically assessed by Rand[18,19] and that on the uranium compounds by Rand and Kubaschewski[20]. The values listed for the halides of these elements (Tables A.5–A.9) are those recommended by these authors who have provided complete literature coverage. No experimental data are available for the halides of the elements actinium, protactinium and neptunium although the heats of formation for certain of the tetrahalides of the last two elements are

TABLE A.1

Experimentally Determined Values for the Heats and Free Energies of Formation for Lanthanide Trichlorides and Triiodides

| Element | Trichlorides[a] | | | | Triiodides | |
	$-\Delta H_{f\,298°}$ kcal/mole	$-\Delta G_{f\,298°}$ kcal/mole	$-S°$ e.u.	Reference	$-\Delta H_{f\,298°}$ kcal/mole	Reference
La	255.9	238.3	34.5	3	167.4	2
Ce	252.8	235.2	34.5	4	164.4	2
Pr	252.1	234.5	34.5	3	162.0	2
Nd	245.6	227.9	34.6	4	158.9	2
Pm	—	—	—	—	—	—
Sm	244.8	232	—	1	153.4	2
Eu	247.1	—	—	1	—	—
Gd	240.1	222.5	34.9	3	147.6	2
Tb	241.6	—	—	1	—	—
Dy	β 237.8	—	—	1	144.5	2
	γ 234.8	—	—	1		
Ho	232.8	—	—	1	141.7	2
Er	229.1	211.4	35.1	3	140.0	2
Tm	235.8	—	—	28	137.8	2
Yb	229.4	—	—	28	—	—
Lu	227.9	—	—	1	133.2	2
Sc	225.4	—	—	29	127.0	22
Y	232.7	215	32.7	3	143.4	2

[a] See comment in text concerning the accuracy of Bommer and Hohmann's results.

currently being measured[21]. Estimated heats of formation for numerous tri- and tetravalent actinide mixed halides have been listed by Maslov and Maslov[31].

In the following tables the solid state is implied for the halides and oxyhalides unless contrary indication is made (viz. { }, liquid; ( ) gas) and in Table A.8 the halogens are assumed to be in the gaseous state.

TABLE A.2

Estimated Heats of Formation for Lanthanide Trifluorides
and Tribromides (kcal/mole)[a]

Element	Trifluorides		Tribromides	
	$-\Delta H_{f\,298°}$[13]	$-\Delta H_{f\,298°}$[7]	$-\Delta H_{f\,298°}$[13]	$-\Delta H_{f\,298°}$[22]
La	421	435	223	214
Ce	416	435	228	212
Pr	413	436	225	218
Nd	410	431	223	208
Pm	408	432	219	—
Sm	405	431	216	209
Eu	391	429	202	—
Gd	404	430	214	200
Tb	400	428	211	—
Dy	398	418	209	197
Ho	395	427	207	193
Er	392	426	205	190
Tm	391	423	203	—
Yb	376	425	185	—
Lu	392	425	200	—
Sc	367	429	190	—
Y	397	445	208	196

[a] Recent experimentally determined[32] values for LaF_3, PrF_3, NdF_3 and ErF_3 are 405, 401, 395 and 378 kcal/mole respectively

TABLE A.3

Heats of Formation of the Lanthanide Dihalides[a] (kcal/mole)

Element	Difluorides[b] $-\Delta H_{f\,298°}$[6,7]	Dichlorides[b] $-\Delta H_{f\,298°}$[6,7]	$-\Delta H_{f\,298°}$	Dibromides $-\Delta H_{f\,298°}$[13]	Diiodides $-\Delta H_{f\,298°}$[13]
La	(213)	(128)	—	—	—
Ce	(217)	(131)	—	—	—
Pr	(244)	(156)	—	—	—
Nd	(250)	(161)	163.2[6]	—	—
Pm	(268)	(178)	—	—	—
Sm	(282)	(192)	195.6[23]	(182)	(155)
Eu	(294)	(200)	195.8[24]	(187)	(160)
Gd	(209)	(117)	—	—	—
Tb	(212)	(118)	—	—	—
Dy	(243)	(148)	—	—	—
Ho	(241)	(145)	—	—	—
Er	(247)	(150)	—	—	—
Tm	(257)	(159)	—	—	—
Yb	(286)	(185)	184.5[23]	(157)	(135)
Lu	(235)	(133)	—	—	—
Sc	(216)	(119)	—	—	—
Y	(272)	(151)	—	—	—

[a] Estimated values in parentheses.
[b] See comments in the text concerning the accuracy of the estimated values.

TABLE A.4

Experimentally Determined Heat Capacities and Heats and Entropies of Fusion for Certain Lanthanide Trihalides[8]

Compound	C_p solid cal/mole	C_p liquid cal/mole	ΔH_m kcal/mole	ΔS_m kcal/mole
LaCl$_3$	34.7	37.7	13.0	11.5
PrCl$_3$	32.3	32.0	12.1	11.4
NdCl$_3$	35.4	35.0	12.0	11.6
GdCl$_3$	29.1	33.7	9.6	11.0
HoCl$_3$	29.0	35.3	7.0	7.1
ErCl$_3$	32.0	33.7	7.8	7.4
LaBr$_3$	33.0	34.5	13.0	12.3
PrBr$_3$	31.5	37.0	11.3	11.7
NdBr$_3$	31.0	35.5	10.8	11.3
GdBr$_3$	32.1	32.3	8.7	8.2
CeI$_3$	36.5	36.5	12.4	12.0
PrI$_3$	31.3	34.2	12.7	12.6
α-NdI$_3$	27.6	36.3	9.7	9.2
β-NdI$_3$	30.4	—	—	—

TABLE A.5

Heats of Formation and Standard Entropies
for the Actinide Halides[a] [18-20,25,26,30]

Compound	$-\Delta H_{f\,298°}$ (kcal/mole)	$S_{298°}$ (e.u.)	Compound	$-\Delta H_{f\,298°}$ (kcal/mole)	$S_{298°}$ (e.u.)
Fluorides			**Chlorides—***contd.*		
AcF_3	(420)	—	$PaCl_4$	—	—
UF_3	345 ± 10	(28 ± 2)	UCl_4	251.3 ± 1.0	47.4 ± 0.3
NpF_3	(360)	—	$NpCl_4$	(237)	—
PuF_3	371 ± 3.0	(30.7 ± 3.0)	$PaCl_5$	—	—
AmF_3	(394)	—	UCl_5	261.5 ± 2.0	(58.0 ± 1.5)
CmF_3	—	(28.1)	UCl_6	270.7 ± 3.0	68.3 ± 0.4
ThF_4	504.6	33.95			
PaF_4	(477)	—	**Bromides**		
UF_4	450 ± 5	36.3 ± 0.1	$AcBr_3$	(220)	—
NpF_4	(428)	—	UBr_3	172.3 ± 2.0	(45.0 ± 2.0)
PuF_4	425 ± 8.0	(38.7 ± 0.5)	$NpBr_3$	(174)	—
AmF_4	(400)	—	$PuBr_3$	187.7 ± 1.0	(45.7 ± 5.0)
CmF_4	—	—	$AmBr_3$	—	—
U_4F_{17}	461.5 ± 6	(37.7 ± 1.0)	$CmBr_3$	—	—
U_2F_9	472.5 ± 6	(39.4 ± 2.0)	$ThBr_4$	231.1	54.5
PaF_5	—	—	$PaBr_4$	—	—
UF_5	491.5 ± 6	45.0 ± 3.0	UBr_4	197.5 ± 1.0	(56.0 ± 2.0)
UF_6	523 ± 6	54.4 ± 0.5	$NpBr_4$	(183)	—
(UF_6)	511 ± 6	90.4 ± 0.2	$PaBr_5$	—	—
(NpF_6)	(463)	—	UBr_5	—	—
PuF_6	430.0 ± 8.0	53.0 ± 0.5			
(PuF_6)	418.0 ± 8.0	88.4 ± 0.2	**Iodides**		
			AcI_3	(169)	—
			ThI_3	~123	—
Chlorides			UI_3	114.2 ± 2.0	(57.0 ± 3.0)
$AcCl_3$	(260)	—	NpI_3	(120)	—
UCl_3	213.5 ± 2.0	38.0 ± 0.2	PuI_3	(130)	(56)
$NpCl_3$	(216)	—	AmI_3	—	—
$PuCl_3$	229.8 ± 0.8	(39.2 ± 3.0)	CmI_3	—	—
$AmCl_3$	249.2	—	ThI_4	160.3	63.5
$CmCl_3$	226.4	—	PaI_4	—	—
$BkCl_3$	—	—	UI_4	126.5 ± 1.0	(67.0 ± 3.0)
$CfCl_3$	—	—	PaI_5	—	—
$ThCl_4$	284.9	45.5			

[a] Estimated values are in parentheses and are taken from reference 25.

TABLE A.6

Heats of Formation and Standard Entropies
for the Actinide Oxyhalides[a] [18-20,25,26]

Compound	$-\Delta H_{f\,298°}$ (kcal/mole)	$S°$ (e.u.)	Compound	$-\Delta H_{f\,298°}$ (kcal/mole)	$S°$ (e.u.)
Oxyfluorides			Oxybromides		
AcOF	(265)	—	AcOBr	—	—
PuOF	—	—	PuOBr	206.4 ± 3.0	(27.0 ± 3.0)
ThOF$_2$	401 ± 2.0	24.2 ± 2.0	ThOBr$_2$	270 ± 3.0	33.0
Pa$_2$OF$_8$	—	—	PaOBr$_2$	—	—
U$_2$OF$_8$	—	—	UOBr$_2$	240.2 ± 2.0	37.7 ± 0.1
UO$_2$F$_2$	399.0 ± 4.0	32.4 ± 0.2	NpOBr$_2$	—	—
NpO$_2$F$_2$	—	—	PaOBr$_3$	—	—
PuO$_2$F$_2$	—	—	PaO$_2$Br	—	—
			UOBr$_3$	236.0 ± 2.5	(49.0 ± 3.0)
Oxychlorides			UO$_2$Br	—	—
AcOCl	—	—	UO$_2$Br$_2$	276.6 ± 3.0	(40.5 ± 2.5)
UOCl	—	—			
PuOCl	222.7 ± 1.0	(25.2 ± 3.0)	Oxyiodides		
AmOCl	227.6 ± 2.7	—	AcOI	—	—
CfOCl	—	—	PuOI	(183)	—
ThOCl$_2$	296.1 ± 1.0	27.7	ThOI$_2$	237.6	40.0
PaOCl$_2$	—	—	PaOI$_2$	—	—
UOCl$_2$	260.0 ± 2.0	33.1 ± 0.1	PaOI$_3$	—	—
NpOCl$_2$	—	—	PaO$_2$I	—	—
Pa$_2$OCl$_8$	—	—	UO$_2$I$_2$	—	—
Pa$_2$O$_3$Cl$_4$	—	—			
PaO$_2$Cl	—	—			
UOCl$_3$	284.2 ± 3.0	(42.0 ± 3.0)			
UO$_2$Cl$_2$	302.9 ± 3.0	36.0 ± 0.1			

[a] Estimated values are in parentheses and are taken from reference 25.

TABLE A.7
Heat Capacities for Certain Actinide Halides and Oxyhalides[18-20]
$$Cp = a + bT + cT^{-2} \text{ (cal/mole/deg)}$$

Compound	Cp at 298°K	a	$b \times 10^{-3}$	$c \times 10^{-5}$	Temperature range (°K)
ThF_4	—	26.75	5.854	−1.805	298
UF_4	27.7	25.7	7.00	−0.06	298–1309
UF_5	31.6				
UF_6	40.0	12.6	92.0		273–337
(UF_6)	31.0	22.3	28.5		273–400
(PuF_6)	31.0	37.23	0.274	−5.65	298–1500
UCl_3	24.4	20.8	7.75	1.05	298–900
$ThCl_4$	—	28.75	5.561	−1.470	298–1043
UCl_4	28.9	27.2	8.75	−0.79	298–800
(UCl_4)	—	25.8	14.40	—	890–920
UCl_6	42.0				
$ThBr_4$	—				
UBr_4	30.0	31.4	4.92	−3.15	350–750
ThI_4	—				
UI_4	31.0	34.8	2.38	−4.72	380–720
$\{UI_4\}$	—	39.6			820–870
UO_2F_2	24.7				
$UOCl_2$	22.7				
UO_2Cl_2	25.8				
$UOBr_2$	23.4				

{ }, liquid.   ( ), gas.

Halides of the Lanthanides and Actinides

TABLE A.8
Free Energies of Reaction[18-20]
$$\Delta G = A + BT + CT \log T \text{ (cal)}$$

Reaction	A	B	C	Temperature range (°K)
$U + \frac{3}{2}F_2 \rightarrow UF_3$	$-343{,}000$	52.7		$298-1405$
$Pu + \frac{3}{2}F_2 \rightarrow PuF_3$	$-371{,}000$	56.0		$298-913$
$\{Pu\} + \frac{3}{2}F_2 \rightarrow \{PuF_3\}$	$-370{,}000$	54.6		$913-1700$
$\{Pu\} + \frac{3}{2}F_2 \rightarrow \{PuF_3\}$	$-350{,}200$	43.1		$1700-2230$
$Th + 2F_2 \rightarrow ThF_4$	$-502{,}590$	71.04		$298-1383$
$U + 2F_2 \rightarrow UF_4$	$-448{,}500$	67.4		$298-1309$
$U + 2F_2 \rightarrow \{UF_4\}$	$-433{,}100$	55.6		$1309-1405$
$\{U\} + 2F_2 \rightarrow \{UF_4\}$	$-433{,}100$	55.6		$1405-1730$
$Pu + 2F_2 \rightarrow PuF_4$	$-424{,}000$	70.0		$298-1500$
$UF_4 + \frac{1}{2}F_2 \rightarrow UF_5$	$-41{,}200$	14.8		$298-565$
$UF_4 + F_2 \rightarrow UF_6$	$-73{,}000$	30.4		$298-337$
$UF_4 + F_2 \rightarrow (UF_6)$	$-62{,}300$	-2.8		$298-1309$
$\{UF_4\} + F_2 \rightarrow (UF_6)$	$-74{,}900$	7.1		$1309-1730$
$PuF_4 + F_2 \rightarrow (PuF_6)$	$6{,}800$	—		$298-1500$
$PuF_4 + F_2 \rightarrow (PuF_6)$	$5{,}200$	2.6		$620-1500$
$U + \frac{3}{2}Cl_2 \rightarrow UCl_3$	$-213{,}000$	51.0		$298-1110$
$U + \frac{3}{2}Cl_2 \rightarrow \{UCl_3\}$	$-200{,}700$	40.0		$1110-1405$
$Pu + \frac{3}{2}Cl_2 \rightarrow PuCl_3$	$-229{,}400$	52.8		$298-1040$
$\{Pu\} + \frac{3}{2}Cl_2 \rightarrow \{PuCl_3\}$	$-215{,}400$	39.4		$1040-2000$
$Th + 2Cl_2 \rightarrow ThCl_4$	$-282{,}310$	67.59		$298-1043$
$Th + 2Cl_2 \rightarrow \{ThCl_4\}$	$-256{,}910$	42.90		$1043-1190$
$U + 2Cl_2 \rightarrow UCl_4$	$-253{,}100$	112.8	-14.3	$298-861$
$U + 2Cl_2 \rightarrow \{UCl_4\}$	$-236{,}700$	52.0		$861-1060$
$PuCl_3 + \frac{1}{2}Cl_2 \rightarrow (PuCl_4)$	$39{,}300$	-30.0		$298-1040$
$\{PuCl_3\} + \frac{1}{2}Cl_2 \rightarrow (PuCl_4)$	$26{,}000$	-17.0		$1040-1500$
$UCl_4 + \frac{1}{2}Cl_2 \rightarrow UCl_5$	$-10{,}100$	15.3		$298-600$
$UCl_4 + Cl_2 \rightarrow UCl_6$	$-19{,}200$	31.3		$298-450$
$UCl_4 + Cl_2 \rightarrow (UCl_6)$	$2{,}600$	-1.5		$450-900$
$U + \frac{3}{2}Br_2 \rightarrow UBr_3$	$-182{,}050$	53.1		$298-1000$
$U + \frac{3}{2}Br_2 \rightarrow \{UBr_3\}$	$-190{,}500$	60.9		$1000-1405$
$Pu + \frac{3}{2}Br_2 \rightarrow PuBr_3$	$-197{,}900$	53.4		$298-954$
$Pu + \frac{3}{2}Br_2 \rightarrow \{PuBr_3\}$	$-182{,}500$	37.2		$954-1736$
$Th + 2Br_2 \rightarrow ThBr_4$	$-243{,}780$	70.42		$298-997$
$Th + 2Br_2 \rightarrow \{ThBr_4\}$	$-222{,}670$	47.65		$997-1132$
$U + 2Br_2 \rightarrow UBr_4$	$-211{,}000$	70.9		$298-792$
$U + 2Br_2 \rightarrow \{UBr_4\}$	$-194{,}000$	49.2		$792-1050$
$Th + 2I_2 \rightarrow ThI_4$	$-188{,}600$	69.68		$298-864$
$Th + 2I_2 \rightarrow \{ThI_4\}$	$-169{,}380$	46.55		$864-1113$
$U + 2I_2 \rightarrow UI_4$	$-153{,}500$	62		$298-779$
$U + 2I_2 \rightarrow \{UI_4\}$	$-134{,}000$	37		$779-1030$
$UI_3 + \frac{1}{2}I_2 \rightarrow UI_4$	$-20{,}000$	28.3	-2.5	$298-779$

{ }, liquid.   ( ), gas.

TABLE A.9

Free Energies, Heats and Entropies of Vaporisation and Sublimation of Certain Actinide Halides[18-20,27]

Compound	$\Delta G = -RT \ln p_{atm}$ $= A + BT + CT \log T$ (cal)			Temperature range (°K)	$\Delta H_{298°}$ (kcal/mole)	$\Delta S_{298°}$ (e.u./mole)
	A	B	C			
$\langle PuF_3 \rangle$	101,400	− 90.6	13.8	298–1700	99.6	50.5
$\{PuF_3\}$	91,200	−114.4	23.0	1700–2500		
$\langle AmF_3 \rangle$	112,650	−155.5	32.2			
$\langle ThF_4 \rangle$	85,950	− 91.55	13.8		82.4	49.9
$\{ThF_4\}$	80,970	−116.88	23.03			
$\langle UF_4 \rangle$	75,100	− 90.3	13.8	298–1309	73.2	50.2
$\{UF_4\}$	70,100	−115.2	23.0	1309–1720		
$\langle PuF_4 \rangle$	73,400	− 93.0	13.8	298–1300	71.6	52.8
$\{PuF_4\}$	50,000	− 32.0	—	1300–1550		
$\langle UF_6 \rangle$	15,150	−109.6	25.3	273– 337		
$\{UF_6\}$	6,870	− 21.13	23.0	33– 400		
$\langle PuF_6 \rangle$	15,000	−109.0	25.3	260– 325	11.7	35.4
$\{PuF_6\}$	7,200	− 21.5	—	325– 350		
$\langle UCl_3 \rangle$	72,000	− 81	13.8	298–1110	70.2	47.3
$\{UCl_3\}$	65,000	−109	23.0	1110–1930		
$\langle PuCl_3 \rangle$	75,500	− 92.2	13.8	298–1040	73.7	48.7
$\{PuCl_3\}$	69,100	−110.5	23.0	1040–2000		
$\langle ThCl_4 \rangle$	63,730	− 98.33	13.82	298–1043	63.14	59.9
$\{ThCl_4\}$	49,090	−119.04	25.33	1043–1183		
$\langle UCl_4 \rangle$	51,900	− 93.0	13.8	298– 863	50.1	52.9
$\{UCl_4\}$	45,500	−119.3	25.3	863–1062		
$\langle UCl_6 \rangle$	18,300	− 33.5	—	298– 400	18.8	35.1
$\langle UBr_3 \rangle$	75,100	− 91.8	13.8	298–1000	73.4	51.6
$\{UBr_3\}$	68,600	−112.8	23.0	1000–1810		
$\langle PuBr_3 \rangle$	77,600	−117.6	20.7	298– 954	74.9	57.4
$\{PuBr_3\}$	66,500	−112.8	23.0	954–1736		
$\langle ThBr_4 \rangle$	57,190	− 95.46	13.82	298– 952	56.4	56.72
$\{ThBr_4\}$	45,950	−117.94	25.33	952–1124		
$\langle UBr_4 \rangle$	49,400	− 92.7	13.8	298– 792	47.6	52.6
$\{UBr_4\}$	40,100	−114.6	25.3	792–1050		
$\langle ThI_4 \rangle$	52,960	− 94.21	13.82	298– 839	51.18	54.02
$\{ThI_4\}$	43,320	−123.12	27.64	839–1105		
$\langle UI_4 \rangle$	56,400	−108.6	16.1	298– 779	54.3	61.7
$\{UI_4\}$	42,600	−117.5	25.3	779–1030		

$\langle \ \rangle$, solid. $\{ \ \}$, liquid.

TABLE A.10

The Heat and Free Energy of Formation and the
Entropy of Certain Uranium Mixed Halides[33]

Compound	$-\Delta H_{298}°$ (kcal/mole)	$-\Delta G_{298}°$ (kcal/mole)	$S_{298}°$ (e.u.)
$UBrCl_3$	241.6	220.3	50
UBr_2Cl_2	230.6	210.5	57.7
UBr_3Cl	220.1	200.5	60.7
$UICl_3$	227.3	206.4	53.1
$UIBr_3$	195.4	177.5	70.9
$UBrCl_2$	202.1	187.2	44.8
UBr_2Cl	191.9	178.0	50.6

REFERENCES

1. H. Bommer and E. Hohmann, *Z. Anorg. Chem.*, **248**, 357 (1941).
2. H. Bommer and E. Hohmann, *Z. Anorg. Chem.*, **248**, 383 (1941).
3. F. H. Spedding and J. P. Flynn, *J. Am. Chem. Soc.*, **76**, 1474 (1954).
4. F. H. Spedding and C. F. Millar, *J. Am. Chem. Soc.*, **74**, 4195 (1952).
5. J. D. Corbett, D. L. Pollard and J. E. Mee, *Inorg. Chem.*, **5**, 761 (1966).
6. O. G. Polyachenok and G. I. Novikov, *Russ. J. Inorg. Chem.*, **8**, 816 (1963).
7. G. I. Novikov and O. G. Polyachenok, *Russ. Chem. Rev.*, **33**, 342 (1964).
8. A. S. Dworkin and M. A. Bredig, *J. Phys. Chem.*, **67**, 697 (1963).
9. E. R. Harrison, *J. Appl. Chem.*, **2**, 601 (1952).
10. V. E. Shimazaki and S. Niwa, *Z. Anorg. Chem.*, **314**, 21 (1962).
11. J. L. Moriaty, *J. Chem. Eng. Data*, **8**, 422 (1963).
12. O. G. Polyachenok and G. I. Novikov, *Zh. Neorgan. Khim.*, **8**, 2818 (1963).
13. L. Brewer, L. A. Bromley, P. W. Gilles and N. L. Logfren, in 'The Chemistry and Metallurgy of Miscellaneous Materials' (L. L. Quill, Ed.), *Nat. Nucl. Energy Ser. Div. IV*, McGraw-Hill, New York, **19B**, 76 (1950).
14. L. Brewer, reference 13, p. 193.
15. *The Properties of the Rare Earth Metals and Compounds*, compiled for the Rare Earth Research Group, Ohio (1959).
16. A. Glasner, U.S. Report ANL-5750 (1958).
17. C. W. Koch and B. B. Cunningham, *J. Am. Chem. Soc.*, **75**, 796 (1953); **76**, 1471 (1954); U.S. Report UCRL-2286 (1953).
18. M. Rand, to be published.
19. M. Rand, *At. Energy Rev.*, **4**, Special Issue No. I, 7 (1966).
20. M. Rand and O. Kubaschewski, *The Thermochemical Properties of Uranium Compounds*, Oliver and Boyd, London, 1963.
21. J. Fuger, personal communication (1967).
22. D. E. Wilcox and L. A. Bromley, *Ind. Eng. Chem.*, **55** (7), 32 (1963).
23. G. R. Machlan, C. T. Stubblefield and L. Eyring, *J. Am. Chem. Soc.*, **77**, 2975 (1955).

24. C. T. Stubblefield, J. L. Rutledge and R. Phillips, *J. Phys. Chem.*, **69**, 991 (1965).
25. J. J. Katz and G. T. Seaborg, *The Chemistry of the Actinide Elements*, Methuen, London, 1957.
26. J. Fuger and B. B. Cunningham, *J. Inorg. Nucl. Chem.*, **25**, 1423 (1963).
27. S. C. Carniglia and B. B. Cunningham, *J. Am. Chem. Soc.*, **77**, 1451 (1955).
28. J. M. Stuve, Report BM-RI-6705 (1965).
29. J. M. Stuve, *U.S. Bur. Mines, Rep. Invest.*, No. 6902 (1966).
30. J. C. Wallmann, J. Fuger, H. Haug, S. A. Marei and B. M. Bansal, *J. Inorg. Nucl. Chem.*, **29**, 2097 (1967).
31. P. G. Maslov and Yu. P. Maslov, *Zh. Obshch. Khim.* **35**, 2112 (1965).
32. O. G. Polyachenok, *Russ. J. Inorg. Chem.* **12**, 449 (1967).
33. J. J. Katz and E. Rabinowitch, 'The Chemistry of Uranium', *Nat. Nucl. Energy Ser. Div. VIII*, McGraw-Hill, New York, **5**, 540 (1951).

Appendix B

Metal–Halogen and Metal–Oxygen Vibrational Frequencies

Available information on the infrared spectra of the halides, oxyhalides and halide complexes is summarized in Tables B.1 to B.4 inclusive. Raman and infrared data on the actinide hexafluorides, lanthanum trifluoride and lanthanum trichloride have been presented and discussed at the appropriate places in the text (pp. 25, 81 and 157 respectively) and will not be repeated here. In those specific cases the spectra have been interpreted in terms of the vibrational modes giving rise to the bands observed but in general this is not the case and the observed bands are merely presented under the headings ν_{M-O} and ν_{M-X}. Some pertinent comments concerning the position of the metal–oxygen vibrations in relation to the nature of the bonding present in the oxyhalides are given at the foot of Table B.2. It must be remembered, however, that structural information is lacking in almost every instance and that such comments are the results of comparisons of the spectra with those of compounds of known structure.

The tabulated data will not be discussed in detail but it is appropriate to make one or two general observations. Thus, a unit change of valence state of the central metal ion in the hexahalo complexes (e.g. $PaX_6^- \rightarrow PaX_6^{2-}$, Table B.4) results in a change of 30–50 cm^{-1} in the position of the M–X stretching vibration and increased coordination within a given valence state (e.g. PaX_6^-, PaX_7^{2-} and PaX_8^{3-}) results in a lowering of the M–X vibrational frequency. One other point of general interest is that the position of the metal–oxygen stretching vibration in the actinide penta-valent and hexavalent oxychloro complexes is not necessarily indicative of the valence state of the complex. For example ν_{M-O} for $Cs_2NpO_2Cl_4$, Cs_2NpOCl_5 and $Cs_3NpO_2Cl_4$ occurs at 921, 919 and 800 cm^{-1} respectively.

For detailed discussions of metal–halogen vibrational frequencies the reader is referred to recent review articles[1,2].

TABLE B.1

Infrared Stretching Vibrations of Certain
Lanthanide and Actinide Binary Halides

Compound	ν_{M-X} (cm^{-1})	Reference
MF$_3$	400–500s, b	3
ThF$_4$ (g)	520s, b	4
PaF$_4$	400s, b	5
ThCl$_4$	245s, b	6
UCl$_4$	254s, b	6
PaCl$_5$	323s; 362m	7

M = lanthanide element
(g) = gaseous state
s = strong
m = medium
b = broad

TABLE B.2
Infrared Stretching Vibrations of Certain Lanthanide and Actinide Oxyhalides

Compound	ν_{M-O} (cm^{-1})^b	ν_{M-X} (cm^{-1})^b	Comments^a	Reference
MOF	400–500 v.b. (? assignment)		–	8, 9
ThOCl$_2$	571–246 (several peaks)	298w; 282w	a	10
PaOCl$_2$	555–243 (several peaks)	290w; 278w	a	10
UOCl$_2$	555–242 (several peaks)	—	a	10
NpOCl$_2$	551–242 (several peaks)	—	a	10
ThOBr$_2$	543–243 (several peaks)	—	a	10
PaOBr$_2$	546–240 (several peaks)	—	a	10
UOBr$_2$	538–253 (several peaks)	—	a	10
NpOBr$_2$	526–250 (several peaks)	—	a	10
ThOI$_2$	500sh; 444m; 325s	—	a	10
PaOI$_2$	515m; 444m; 315s	—	a	10
Pa$_2$OF$_8$	790m; 740m; 690m	450s, b	b	5
NpOF$_3$	985s	852w., 350s.b., 300s.b.	—	11
NpO$_2$F	800s, b	277s	c	11
α-Pa$_2$OCl$_8$	460s; 500m	326m; 370s	a	7
β-Pa$_2$OCl$_8$	458s; 506m	324m; 370s	a	7
Pa$_2$O$_3$Cl$_4$	426sh; 540s, b	342w; 378w	a	7
PaO$_2$Cl	520s, b; 624sh	396m	a	7
PaOBr$_3$	515m; 476m; 364s; 303w	—	a	12, 14
PaO$_2$Br	575m; 386s; 286w	—	a	12, 14
UO$_2$Br	940; 890; 850	—	e	13
PaOI$_3$	480m; 339s; 276w	—	a	14
PaO$_2$I	555m; 469w; 386s; 281w	—	a	14
UO$_2$F$_2$	990s	—	d	11, 15
NpO$_2$F$_2$	980s	446w; 277s; 250s	d	11
PuO$_2$F$_2$	975s	—	d	16
UO$_2$Cl$_2$	948w; 905s	—	d	17
UO$_2$Br$_2$	948w; 930m; 905s; 825w	—	d	17

^{a} The positions of the metal–oxygen stretching vibrations indicate:

    (a) a polymeric oxygen-bridged structure
    (b) a dimeric oxygen-bridged structure
    (c) the presence of discrete MO$_2^+$ groups
    (d) the presence of discrete MO$_2^{2+}$ groups
    (e) see p. 187
^{b} s = strong, m = medium, w = weak, b = broad, sh = shoulder

TABLE B.3

Infrared Stretching Vibrations of Certain Lanthanide
and Actinide Fluoro and Oxyfluoro Complexes

Complex	$\nu_{M-O}{}^a$ (cm^{-1})	$\nu_{M-F}{}^a$ (cm^{-1})	Reference
Na_2CeF_6	—	430; 405	36
K_2CeF_6	—	430; 403	36
Rb_2CeF_6	—	430; 408	36
Cs_2CeF_6	—	430; 408	36
Rb_2PrF_6	—	424; —	36
Cs_3CeF_7	—	425; 408	36
Rb_3PrF_7	—	435; 408	36
Cs_3PrF_7	—	435; 406	36
Cs_3TbF_7	—	430; 405	36
Cs_3DyF_7	—	425; 405	36
NH_4PaF_6	—	513s; 444w	18
$KPaF_6$	—	523s; 454w	19
$CsUF_6$	—	503	20
$NOUF_6$	—	550s; 509sh	21
$(NH_3OH)UF_6$	—	526s, b	37
$N_2H_6(UF_6)_2$	—	526s, b	37
K_2PaF_7	—	430s; 356w	19
Rb_2PaF_7	—	438s; 356w	19
Cs_2PaF_7	—	438s; 356w	19
$(NH_4)_2PaF_7$	—	434s; 357w	19
$NOUF_7$	—	550sh; 509s	22
$N_2H_6UF_7$	—	435s, b	37
Li_3PaF_8	—	404s	19
Na_3PaF_8	—	468sh; 422s	19
K_3PaF_8	—	401s	19
Cs_3PaF_8	—	395s	19
Rb_3NpF_8	—	401s, b	11
$CsUO_2F_3$	935sh; 895s	—	11
$CsNpO_2F_3$	935s	—	11
$K_3UO_2F_5$	863; 789	—	23
$Cs_3NpO_2F_5$	870sh; 840s	357s; 294w	11

^{a} s = strong, w = weak, b = broad, sh = shoulder

TABLE B.4

Infrared Stretching Vibrations of Actinide
Chloro, Bromo and Oxyhalogeno Complexes

Complex type	ν_{M-O} (cm^{-1})	$\nu_{3M-X}{}^{a}$ (cm^{-1})	Reference
$AmCl_6^{3-}$	—	241–242	35
$ThCl_6^{2-}$	—	255–263	6, 24
$PaCl_6^{2-}$	—	255–266	25
UCl_6^{2-}	—	253–267	6, 24–27
$NpCl_6^{2-}$	—	265–267	6, 28
$ThBr_6^{2-}$	—	177–179	6
$PaBr_6^{2-}$	—	180–182	25
UBr_6^{2-}	—	178–181	6
$PaCl_6^{-}$	—	305–310	29
UCl_6^{-}	—	305–310	30
$^{b}PaCl_8^{3-}$	—	290	29
$PaBr_6^{-}$	—	215–216	31
$NpOCl_5^{2-}$	907–921	271–275	28
$^{c}NpO_2Cl_4^{3-}$	810, 794	264	28
$AmO_2Cl_4^{3-}$	800	290	35
$UO_2Cl_4^{2-}$	904–930	270	27, 32–34
$^{c}NpO_2Cl_4^{2-}$	919	271	28
$^{c,e}AmO_2Cl_4^{2-}$	902	313, 244	35
$^{c,f}AmO_2Cl_4^{2-}$	902	303, 230	35
$UO_2Br_4^{2-}$	921–934	—	34
$^{d}UO_2I_4^{2-}$	925	—	34

[a] The assignment ν_3 refers only to the hexa- and octahalogeno complexes.
[b] Tetramethylammonium complex only.
[c] Caesium complex only.
[d] Triphenylbutylphosphonium complex only.
[e] Cubic modification.
[f] Monoclinic modification.

REFERENCES

1. D. M. Adams, 'Metal-Ligand and Related Vibrations', Arnold, London, 1967.
2. R. J. H. Clark, *Intern. Rev. Halogen Chem.* (V. Gutmann, Ed.), **3**, 85 (1967).
3. L. P. Batsanova, G. H. Grigor'eva and S. S. Batsanova, *Russ. J. Struct. Chem.*, **1**, 33 (1963).
4. A. Büchler, J. B. Berkowitz-Muttuck and D. H. Dugre, *J. Chem. Phys.*, **34**, 2202 (1961).
5. L. Stein, *Proc. Intern. Conf. Phys. Chim. Protactinium, Centre Natl. Rech. Sci. Paris*, p. 101 (1966).
6. D. Brown, *J. Chem. Soc.* (*A*), **1966**, 766.
7. D. Brown and P. J. Jones, *J. Chem. Soc.* (*A*), **1966**, 874.
8. L. R. Batsanova and G. N. Kustova, *Zh. Neorgan. Khim.*, **9**, 330 (1964).
9. F. Kutek, *Zh. Neorgan. Khim.*, **9**, 2784 (1964).
10. K. W. Bagnall, D. Brown and J. F. Easey, *J. Chem. Soc.* (*A*), **1968**, 288.
11. K. W. Bagnall, D. Brown and J. F. Easey, *J. Chem. Soc.* (*A*), **1968**, in press.
12. D. Brown and P. J. Jones, *J. Chem. Soc.* (*A*), **1966**, 262.
13. J. C. Levet, *Compt. Rend.*, **260**, 4775 (1965).
14. D. Brown, J. F. Easey and P. J. Jones, *J. Chem. Soc.* (*A*), **1967**, 1698.
15. H. R. Hoekstra, *Inorg. Chem.*, **2**, 492 (1963).
16. I. F. Alenchikova, L. L. Zaitseva, L. V. Lipis, N. S. Nikolaev, V. V. Fomin and N. T. Chebotarev, *Zh. Neorgan. Khim.*, **3**, 951 (1958).
17. J. Prigent, *Compt. Rend.*, **247**, 1737 (1958).
18. D. Brown, unpublished observations.
19. D. Brown and J. F. Easey, *J. Chem. Soc.* (*A*), **1966**, 254.
20. M. J. Reisfeld and G. A. Crosby, *Inorg. Chem.*, **4**, 65 (1965).
21. J. R. Geichmann, E. A. Smith, S. S. Trond and P. R. Ogle, *Inorg. Chem.*, **1**, 661 (1962).
22. J. R. Geichmann, E. A. Smith and P. R. Ogle, *Inorg. Chem.*, **2**, 1012 (1963).
23. S. P. McGlynn, J. K. Smith and W. C. Neely, *J. Chem. Phys.*, **35**, 105 (1961).
24. D. M. Adams, J. Chatt, J. M. Davidson and J. Gerratt, *J. Chem. Soc.*, **1963**, 2189.
25. D. Brown and P. J. Jones, *J. Chem. Soc.* (*A*), **1967**, 243.
26. J. P. Day and L. M. Venanzi, *J. Chem. Soc.* (*A*), **1966**, 197.
27. J. P. Day, Thesis, Oxford University (1965).
28. K. W. Bagnall and J. B. Laidler, *J. Chem. Soc.* (*A*), **1966**, 516.
29. K. W. Bagnall and D. Brown, *J. Chem. Soc.*, **1964**, 3021.
30. K. W. Bagnall, D. Brown and J. G. H. du Preez, *J. Chem. Soc.*, **1964**, 2603.
31. D. Brown and P. J. Jones, *J. Chem. Soc.* (*A*), **1967**, 247.
32. P. Gans, Thesis, London University (1964).
33. P. Gans and B. C. Smith, *J. Chem. Soc.*, **1964**, 4172.
34. J. P. Day and L. M. Venanzi, *J. Chem. Soc.* (*A*), **1966**, 1363.
35. K. W. Bagnall, J. B. Laidler and M. A. A. Stewart, *J. Chem. Soc.* (*A*), **1968**, 133.
36. K. Rödder, Thesis, Westfälischen Wilhelms-Universität, Münster (1963).
37. B. Frlec and H. H. Hyman, *Inorg. Chem.* **6**, 2233 (1967).

Appendix C

Mixed Halides of Uranium and Protactinium

Several mixed halides of trivalent and tetravalent uranium have been known for many years. The uranium (III) compounds can be prepared by thermal decomposition or hydrogen reduction of a mixed uranium (IV) halide or by fusion of two trivalent halides. The tetravalent compounds are conveniently prepared either by heating a trivalent uranium halide with a halogen of higher atomic number or by heating together stoicheiometric amounts of the appropriate tetrahalides. No new work has been reported since the compounds were first characterized and the reader is therefore referred to the original articles[1,2] and other reviews[3,4] for complete discussions of their preparation and properties. Table C.1 lists these uranium compounds together with recommended preparative methods. A few thermodynamic properties have been listed in Table A.10 (p. 246) and vapour pressure data are given in Table C.2. Maslov[5] has discussed the calculation of the thermodynamic properties of the uranium (IV) chlorofluorides $UClF_3$, UCl_2F_2 and UCl_3F.

More recently a single protactinium (v) mixed halide, $PaBr_3I_2$, has been prepared[6] by melting together PaI_5 and $PaBr_5$. Other mixed halides of the pentavalent actinides are unknown but many protactinium and uranium compounds could undoubtedly be prepared by similar methods as indeed could mixed tetravalent halides of thorium, protactinium and neptunium.

Only a few mixed halogeno complexes are presently known and these are listed in Table C.3, together with brief details of their preparation. The N,N-dimethylacetamide (DMA) complexes $UI_2Cl_2 \cdot 5DMA$ and $UI_3Cl \cdot 5DMA$ have also been prepared[12] recently and the urea complex $UICl_3 \cdot 8 \, CO(NH_2)_2$ is known[13].

TABLE C.1

Uranium Mixed Halides[1-4]

	Compound	Colour	m.p. ($^\circ$C)[a]	Preparation
Uranium (III)	UCl_2Br	Black	(800)	$2UCl_3 + UBr_3$ (fusion)
	$UClBr_2$	Black	(775)	$UClBr_3 + H_2$
	UCl_2I	Black	(750)	$UCl_2I_2 \rightarrow UCl_2I + \frac{1}{2}I_2$
	$UClI_2$	Black	(725)	$UCl_3 + 2UI_3$ (fusion)
	UBr_2I	Black	(700)	$UBr_2I_2 \rightarrow UBr_2I + \frac{1}{2}I_2$
	$UBrI_2$	Black	(690)	$UBrI_3 \rightarrow UBrI_2 + \frac{1}{2}I_2$
Uranium (IV)	UF_3Cl	Green	—	$UF_3 + \frac{1}{2}Cl_2$ (310°)
	UF_2Cl_2	Green	(460)	$UO_2F_2 + 2CCl_4$ (450°)
	UF_3Br	Dark green	—	$UF_3 + \frac{1}{2}Br_2$ (250°)
	UF_3I	Brownish-black	—	$UF_3 + \frac{1}{2}I_2$ (250°)
	UCl_3Br	Dark green	521	$UCl_3 + \frac{1}{2}Br_2$ (500°)
	UCl_2Br_2	Dark green	510	$UCl_4 + UBr_4$ (fusion)
	$UClBr_3$	Greenish-brown	502	$UCl_4 + 3UBr_4$ (fusion)
	UCl_3I	Black	<490	$UCl_3 + \frac{1}{2}I_2$ (500°)
	UCl_2I_2	Black	<500	$UCl_4 + UI_4$ (fusion)
	$UClI_3$	Black	<500	$UCl_4 + 3UI_4$ (fusion)
	UBr_3I	Dark brown	478	$UBr_3 + \frac{1}{2}I_2$ (500°)
	UBr_2I_2	Dark brown	<500	$UBr_4 + UI_4$ (fusion)
	$UBrI_3$	Black	<500	$UBr_4 + 3UI_4$ (fusion)
	UCl_2BrI	Black	<500	$UCl_2Br + \frac{1}{2}I_2$
	$UClBr_2I$	Black	<500	$UClBr_2 + \frac{1}{2}I_2$

[a] Estimated values are in parentheses.

TABLE C.2

Available Vapour Pressure Data for
Uranium Mixed Halides

Compound	$\log P_{mm} = A - (B/T)$		Temperature range ($^\circ$C)
	A	B	
$UBrCl_3$	13.852	10,526	320–430
UBr_2Cl_2	13.149	9,901	330–404
UBr_3Cl	13.280	10,000	320–420
$UIBr_3$	13.416	9,901	315–382

TABLE C.3

Mixed Halogeno Complexes of Uranium

Compound	Colour	Preparation	Reference
$(Ph_3BuP)_2UCl_2Br_2$	Green	$UCl_4 + 2Ph_3BuPBr$ in CH_3CN	7
$(Ph_3BuP)_2UO_2Cl_2Br_2$	Yellow	$UO_2Cl_2 + 2Ph_3BuPBr$ in CH_3CN	8
$(Ph_3BuP)_2UO_2Br_2I_2$	Red	$UO_2Br_2 + 2Ph_3BuPI$ in CH_3CN	8
$Cs_2UO_2ClBr_3$	Yellow	$UO_2Cl_2 \cdot H_2O + CsBr$ in aqueous HBr	9
$Cs_2UO_2Cl_2Br_2$	Yellow	$UO_2Cl_2 \cdot H_2O + CsBr$ in aqueous HBr	9
$Cs_2UO_2Cl_3Br$	Yellow	$UO_2Cl_2 \cdot H_2O + CsBr$ in aqueous HBr	9
$(NH_4)_2UO_2Cl_2Br_2$	Yellow	$NH_4UO_3Br_2 + HCl$ (g) at 150°	10, 11
$K_2UO_2Cl_2Br_2$	Yellow	$UO_2Cl_2 + 2KBr$ at 270°	10, 11

REFERENCES

1. J. C. Warf and N. Baenziger, U.S. Report TID-5290, p. 120 (1958).
2. N. W. Gregory, U.S. Report TID-5290, p. 465 (1958).
3. J. J. Katz and E. Rabinowitch, 'The Chemistry of Uranium', *Nat. Nucl. Energy Ser. Div. VIII*, Vol. 5, McGraw-Hill, New York, 1951, p. 539.
4. J. Elston in *Nouveau Traite de Chimie Minérale* (P. Pascal, Ed.), Vol. XV, Masson et Cie, Paris, 1961.
5. P. G. Maslov, *Zh. Neorgan. Khim.*, **9**, 2076 (1964).
6. D. Brown, J. F. Easey and P. J. Jones, *J. Chem. Soc. (A)*, **1968**, 1698.
7. J. P. Day and L. M. Venanzi, *J. Chem. Soc. (A)*, **1966**, 197.
8. J. P. Day and L. M. Venanzi, *J. Chem. Soc. (A)*, **1966**, 1363.
9. G. V. Ellert, V. V. Tsapkin, Yu. N. Mikhailov and V. G. Kuznetsov, *Zh. Neorgan. Khim.*, **10**, 1572 (1965).
10. J. Prigent and J. Lucas, *Compt. Rend.*, **251**, 388 (1960); **253**, 474 (1961).
11. J. Lucas, *Rev. Chim. Minerale*, **1**, 479 (1964).
12. K. W. Bagnall, D. Brown, P. J. Jones and J. G. H. du Preeze, *J. Chem. Soc.*, **1965**, 3594.
13. V. A. Golovnya and G. T. Bolotova, *Russ. J. Inorg. Chem.* **11**, 1419 (1966).

Index

In the subject index specific compounds are listed for each element under the two subheadings Halides and Halogeno Complexes. For convenience, Halogeno Complexes are listed in order of increasing valence state. Classes of compounds are cross referenced under Actinide and Lanthanide as appropriate. Reference to phase diagrams is made under each compound and also under the general heading Phase Studies. A small supplementary index lists all those compounds for which structural information is available.

Subject Index

Index of Structures